Problem Books in Mathematics

Series Editor

Peter Winkler
Department of Mathematics
Dartmouth College
Hanover, NH
USA

More information about this series at http://www.springer.com/series/714

Ravi P. Agarwal · Simona Hodis ·
Donal O'Regan

500 Examples and Problems of Applied Differential Equations

 Springer

Ravi P. Agarwal
Department of Mathematics
Texas A&M University–Kingsville
Kingsville, TX, USA

Simona Hodis
Department of Mathematics
Texas A&M University–Kingsville
Kingsville, TX, USA

Donal O'Regan
Department of Mathematics
National University of Ireland
Galway, Ireland

ISSN 0941-3502   ISSN 2197-8506   (electronic)
Problem Books in Mathematics
ISBN 978-3-030-26386-7   ISBN 978-3-030-26384-3   (eBook)
https://doi.org/10.1007/978-3-030-26384-3

Mathematics Subject Classification (2010): 34-XX, 34-01, 35-XX, 35-01, 65-XX, 65L06

This Springer imprint is published by the registered company Springer Nature Switzerland AG
The registered company address is: Gewerbestrasse 11, 6330 Cham, Switzerland

Dedicated to our Parents

Preface

This book supplements some of our (Agarwal and O'Regan) previous books published by Springer-Verlag. These books have been downloaded by almost 400,000 mathematicians and engineers. Over the last number of years, we realized that in order to make these books more complete, we should include applications in a variety of areas. Initially, we thought this would be a short project but as you can see in this book, it amounted to a huge project of over 385 pages. In addition, some of the applications we consider hopefully will motivate future research for engineers and mathematicians, especially in the area of nonlinear boundary value problems. In this present book, we include only the minimum required theory (without proofs) and illustrate the theory from real-world problems with 500 or so examples and problems.

In Chap. 1, we discuss first-order linear differential equations and a few of the many applications in this chapter include the growth of bacteria colonies, sound waves traveling through air, commodity prices, suspension bridges, and the motion of oil in a cylinder. Nonlinear first-order differential equations are considered in Chap. 2 and some of the applications here include the study of rumor spread, the flow of water in an open tank through a small hole in the bottom, and the model of the shape of a tsunami. Chapters 3 and 4 discuss second (and higher)-order differential equations via Wronskian's and power series. Among the many applications in these chapters, the authors discuss problems that arise in simple mechanical spring–mass systems, wave motion, investment models, electrical circuits, planetary motion, models of the circulation of blood in blood vessels, quantum mechanics, the response of the accelerogram in recording earthquakes, chemical reaction models, and statistical mechanics. In Chap. 5, we discuss first-order differential systems via Wronskian's and the fundamental matrix. Linear and nonlinear systems are both discussed. Applications in this chapter to name but a few include models on love, price–demand–supply models, radioactive decay, predator–prey examples, interacting SIR disease models, and probability examples. Unfortunately, one of the major problems in differential equations is that it is usually impossible to obtain solutions to problems in closed (exact) form. As a result, one is forced to consider numerical methods in differential equations and this is discussed in Chap. 6. Here, numerical solutions are considered in a variety of problems, for example, the effect of malaria and other diseases, fox populations in which rabies is present and the heat flow in an electron tube filled with an inert gas. In Chap. 7, we discuss how one could obtain qualitative information (in particular stability information) for solutions to differential equations. This arises since usually one cannot obtain solutions in closed form and numerical solutions might give no information on stability (that is, on what happens to the solution when one makes small changes to the initial data). Again, a variety of applications are presented, for example, in the study of the variation of host and parasite populations with time and the model of a simple food chain in a chemostat (a system in which the chemical composition is kept at a controlled level). Linear boundary value problems are discussed in Chap. 8 and among the many examples we include a variety of elastic string (or beam) models and heat conduction problems. Our final chapter, Chap. 9 is the most interesting for potential future research in differential equations. We consider nonlinear boundary value problems and we pay particular attention to the theory of upper and lower solutions. This chapter is a compendium of applications, including examples from electricity, problems involving non-Newtonian fluid flow, examples involving shape deformation of a membrane cap, or problems on the analysis of the performance of solid-propellant rockets.

The field of differential equations focuses mainly on its theory regarding properties of the equations or the existence and uniqueness of their solutions, but lacks in real-life applications that can be solved at the undergraduate level. Students would be more motivated to study the material of differential equations and retain it better if they are exposed to plenty of world applications. This book presents an unprecedented number of real-life applications of differential equations together with the underlying theory and techniques.

In writing a book of this nature no originality is claimed, only a humble attempt is made to present the applications as simply, clearly, and accurately as possible. It is hoped that it will motivate an inquisitive reader to generate future research (motivated from the applications included in this book) in this vast and ever-expanding field.

Kingsville, USA   Ravi P. Agarwal
Kingsville, USA   Simona Hodis
Galway, Ireland   Donal O'Regan

Contents

1 First-Order Linear Differential Equations   1
   References   19

2 Some First-Order Nonlinear Differential Equations   21
   References   45

3 Second- and Higher Order Differential Equations   47
   References   83

4 Power Series Solutions   85
   References   114

5 Systems of First-Order Differential Equations   115
   References   161

6 Runge–Kutta Method   163
   References   182

7 Stability Theory   183
   References   220

8 Linear Boundary Value Problems   221
   References   291

9 Nonlinear Boundary Value Problems   293
   References   375

Correction to: Systems of First-Order Differential Equations C1

Author Index   379

Subject Index   383

Chapter 1
First-Order Linear Differential Equations

Consider the first-order linear differential equation

$$y' + p(t)y = q(t), \quad ' = \frac{d}{dt} \tag{1.1}$$

where the functions p and q are continuous in an open interval $I = (\alpha, \beta)$ [1]. We can find the general solution of (1.1) in terms of the known functions p and q by multiplying both sides of (1.1) by an integrating factor $e^{P(t)}$, where $P(t)$ is a function such that $P'(t) = p(t)$. Indeed, we have

$$e^{P(t)}y' + e^{P(t)}p(t)y = e^{P(t)}q(t),$$

or

$$\left(e^{P(t)}y\right)' = e^{P(t)}q(t)$$

and hence an integration gives

$$e^{P(t)}y(t) = \int^{t} e^{P(s)}q(s)ds + c,$$

or

$$y(t) = e^{-P(t)}\left[\int^{t} e^{P(s)}q(s)ds + c\right], \tag{1.2}$$

where c is an arbitrary constant.

If along with the differential equation (1.1) an initial condition

$$y(t_0) = y_0, \quad t_0 \in I \tag{1.3}$$

© Springer Nature Switzerland AG 2019
R. P. Agarwal et al., *500 Examples and Problems of Applied Differential Equations*, Problem Books in Mathematics,
https://doi.org/10.1007/978-3-030-26384-3_1

is given, then it is convenient to choose $P(t)$ such that $P'(t) = p(t)$, $P(t_0) = 0$, i.e., $P(t) = \int_{t_0}^t p(s)ds$. Then, the unique solution of the initial value problem (1.1), (1.3) can be written as

$$y(t) \; = \; e^{-P(t)}y_0 + e^{-P(t)} \int_{t_0}^t e^{P(s)}q(s)ds. \tag{1.4}$$

If $p(t) = p$, $q(t) = q$ are constants, then $P(t) = p(t - t_0)$ and (1.4) reduces to

$$y(t) \; = \; e^{-p(t-t_0)}y_0 + \frac{q}{p}\left(1 - e^{-p(t-t_0)}\right). \tag{1.5}$$

In many natural processes, the rate of change of a physical quantity is proportional to the current amount of the quantity. This leads to solving a differential equation of the form $y' = ky$, where k is a constant.

Example 1.1 A pesticide sprayed onto tomatoes decomposes into a harmless substance at a rate proportional to the amount $M(t)$ still unchanged at time t. If an initial amount of 10 pounds sprayed onto an acre reduce to 5 pounds in 6 days, when will 80% of the pesticide be decomposed?

Clearly, $M(t)$ satisfies the differential equation

$$\frac{dM}{dt} = -kM,$$

where $k > 0$ is a constant. The solution of this differential equation is $M(t) = M_0 e^{-kt}$, where $M_0 = M(0) = 10$ and t is the time in days. Since $M(6) = 5$ is given, we have $5 = 10e^{-6k}$, or $k = -\ln(0.5)/6$. Let T be the time when $M(T) = 2$, i.e., 80% decomposed, then we have $2 = 10e^{-kT}$, or $T = -\ln(0.2)/k = 6(-\ln(0.2))/(-\ln(0.5)) \simeq 13.9$ days.

Example 1.2 Under certain conditions such as when enough food and space is available, bacteria colonies grow at a rate that is directly proportional to the population of the colony. If $P(t)$ denotes the population of the colony at a time t, this means that $P' = kP$, where k is a constant. If the population of a colony of bacteria was 120 initially ($t = 0$) and after 3 h ($t = 3$), the population was 200 and what was the population when $t = 2$, i.e., 2 h after the initial time?

Since $P(0) = 120$ it follows that $P(t) = 120e^{kt}$. We also know $P(3) = 200$, and hence $k = (1/3)\ln(5/3) \simeq 0.1703$. Thus, $P(t) = 120e^{0.1703t}$, and from this we find $P(2) = 120e^{0.3406} \simeq 169$.

Example 1.3 When sound waves travel through air (or any other medium), their intensity I is governed by the differential equation

$$\frac{dI}{dx} = -aI,$$

where x is the distance traveled, and $a > 0$ is a constant. The solution, $I = I_0 e^{-ax}$ show that the intensity decreases exponentially with distance. A similar law, known as Lambert's law (Johann Heinrich Lambert (1728–1777); however, earlier discovered by Pierre Bouguer (1698–1758) in the year 1729), holds for the absorption of light in a transparent medium.

Example 1.4 Consider the absorption of X-rays through a homogeneous partially opaque body (see Fig. 1.1). The decay of X-ray intensity denoted as I is a function of penetration distance r. In a thin slice of the object perpendicular to the direction of incidence of the x-rays, the absorption, i.e., $I(r + \Delta r) - I(r)$ is proportional to the intensity, the density of the medium, and the thickness. Thus, we can write

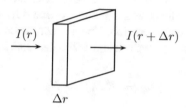

$$I(r) \qquad\qquad I(r + \Delta r)$$

$$\Delta r$$

Fig. 1.1

$$\Delta I = I(r + \Delta r) - I(r) = -DI(r)\Delta r,$$

where D is a constant. This relation immediately leads to the differential equation

$$\frac{dI}{dr} = -DI.$$

Example 1.5 Suppose that a sum of money P_0 is deposited in a bank that pays interest r at an annual rate. The value $P(t)$ of the investment at any time t depends on the frequency with which interest is compounded. Banks have different schemes for compounding such as annually, biannually, quarterly, monthly, weekly, or even daily. If we assume that compounding is done continuously, then we can describe the growth of the investment with the help of the following initial value problem

$$\frac{dP}{dt} = rP, \quad P(0) = P_0$$

where dP/dt denotes the rate of change of the value of the investment. Clearly, its solution $P(t) = P_0 e^{rt}$ indicates that a bank account with continuously compounding interest grows exponentially. We also note that if the interest is compounded m times per year, then

$$P(t) = P_0 \left(1 + \frac{r}{m}\right)^{mt},$$

and from calculus, we have

$$\lim_{m\to\infty} P_0\left(1+\frac{r}{m}\right)^{mt} = P_0 e^{rt}.$$

Example 1.6 Let $N(t)$ denote the number of atoms of a radioactive substance present at time t. The physicist Lord Ernest Rutherford (1871–1937) showed that the number of atoms that disintegrate per unit time dN/dt is proportional to N, i.e.,

$$\frac{dN}{dt} = kN,$$

and hence $N(t) = N_0 e^{kt}$, where N_0 is the number of atoms present at time $t = 0$. The disintegration of a radioactive substance governed by the differential equation $N' = kN$ is called a first-order reaction. The most common measure as to how fast a radioactive substance decomposes is its half-life, i.e., time $t = t_1$ when $N(t)$ reduces to $N_0/2$. This means

$$\frac{1}{2}N_0 = N(t_1) = N_0 e^{kt_1},$$

which gives

$$k = -t_1^{-1}\ln 2. \tag{1.6}$$

Thus, $N(t)$ can be rewritten as

$$N(t) = N_0 e^{-\ln(2)t/t_1}. \tag{1.7}$$

For Uranium U-238 the half-life is known to be 4.5 billion years, whereas for Carbon C-14 it is 5730 years, and for white lead it is only 22 years (see Table 1.1).

This table explains why some of the unstable elements in the periodic table are not found in natural minerals: whatever quantity may have been present when the Earth was born has long since been transformed into more stable elements.

Now we shall show that if $N'(T)$ and $N'(0)$ are known then T can be determined. For this, we have $N'(T) = kN(T)$ and $N'(0) = kN(0)$ and hence

$$\frac{N'(T)}{N'(0)} = \frac{N(T)}{N(0)} = \frac{N_0 e^{kT}}{N_0} = e^{kT},$$

which gives

$$T = \frac{1}{k}\ln\left(\frac{N'(T)}{N'(0)}\right) = -\frac{t_1}{\ln 2}\ln\left(\frac{N'(T)}{N'(0)}\right). \tag{1.8}$$

Table 1.1

Element	Nuclide	Half-life
Aluminum	^{26}Al	7.4×10^5 years
Beryllium	^{10}Be	1.51×10^6 years
Carbon	^{14}C	5730 years
Chlorine	^{36}Cl	3.01×10^5 years
Iodine	^{131}I	8.05 days
Potassium	^{40}K	1.2×10^9 days
Polonium	^{209}Po	100 years
Polonium	^{210}Po	138 days
Radon	^{222}Rn	3.82 days
Radium	^{226}Ra	1700 years
Radium	^{220}Ra	23 milliseconds
Thorium	^{230}Th	75000 years
Uranium	^{238}U	4.51×10^9 years

An interesting application of radioactive decomposition in archeological research is radiocarbon dating. The time of death of an ancient living thing can be determined by measuring the proportion of radiocarbon remaining in a sample. The technique was developed by Willard Frank Libby(1908–1980) around 1950 and this led to him a Nobel Prize. Radiocarbon C-14 is a radioactive form of carbon that is present in the Earth's atmosphere. All living plants and animals absorb radiocarbon as a part of their life processes. Thus, the radiocarbon that decays in a living thing is replaced, so the level remains essentially constant. After the living thing dies, the C-14 is not replaced and decays with half-life 5730 years. Based on this theory the age of the cave paintings at Lascaux, France has been estimated as 15,500 years. Further, after generations of controversy, in the year 1988, the Shroud of Turin was shown to have been made later than 1200 A.D.

If a sample of charred wood from an excavated composite is found to have 16% of the original amount of C-14, how old is the site? From (1.7) it follows that

$$\frac{16}{100}N_0 = N_0 e^{-\ln(2)t/t_1}$$

and hence

$$t = -\frac{5730}{\ln(2)}\ln(0.16) \simeq 15150 \text{ years}.$$

A piece of charcoal excavated in Nippur in 1950 registered an average of 4.029 counts per minute, (i.e., $N'(T) = 4.029$), while a piece of living wood registered an average of 6.68 counts per minute, (i.e., $N'(0) = 6.68$) (for details, see [7]). Thus, from (1.8), we have

$$T \; = \; -\frac{5730}{\ln(2)} \ln \left(\frac{4.029}{6.68} \right) \; \simeq \; 4180 \; \text{years}.$$

Hence the tree was alive in 2230 B.C.

Example 1.7 Let $S(t)$ denote the number of scientists at time t, $bS(t)\Delta t$ and $dS(t)\Delta t$ be the number of new scientists trained and retire from science in the time period $(t, t + \Delta t)$. Then, we have

$$S(t + \Delta t) - S(t) \; = \; bS(t)\Delta t - dS(t)\Delta t \; = \; (b - d)S(t)\Delta t$$

and this leads to the differential equation $S' = (b - d)S$. Hence, the number of scientists change at an exponential rate. The same model applies to science and technology, and in particular mathematics, and double themselves in a certain period of time. If the doubling time of scientists is less than the doubling period of the human population (which is happening now), and this trend continues, then eventually every human being will become a scientist.

Example 1.8 Around 1800, Rev Thomas Robert Malthus (1766–1834), a political economist, studied the population of Europe and noted that it doubled at regular intervals. Continued study indicated that the rate of increase of the population was proportional to the present population. He formulated the law known as Malthusian law of population growth as $P'(t) = kP(t)$. If initially the number of people $P(0)$ is known, then from this differential equation, the population $P(t)$ at a later time t can be obtained as $P(t) = P(0)e^{kt}$. The population of the USA between 1790 and 1890 roughly corresponded to the formula $P(t) = 3.9e^{0.012t}$, where $P(t)$ is the population in millions t years after 1790.

To accommodate birth and death rates in Malthus model, we can take $k = b - d$, where b is the number of individuals born and d is the number of individuals die per unit time. This gives $P(t) = P(0)e^{(b-d)t}$. Thus, if $b > d$ the population grows exponentially and will become double of its present $(t = 0)$ size at time $t_1 = \ln 2/(b - d)$. If $b < d$ the population will decay exponentially and will become half of its present size at time $t_2 = \ln 2/(d - b)$. Clearly, t_1 and t_2 are independent of $P(0) = 0$. If $b = d$, then the population will remain constant.

If there is immigration from outside at a rate proportional to the size of the population, the effect will be the same as increasing the birth rate. Similarly, if there is emigration at a rate proportional to the population size, the effect will be the same as increasing the death rate. However, if immigration and emigration take place at constant rate i and e, respectively, then the differential equation we need to solve is

$$\frac{dP}{dt} \; = \; bP - dP + i - e \; = \; kP + q, \tag{1.9}$$

whose solution in view of (1.5) can be written as

$$P(t) = e^{kt}\left(P(0) + \frac{q}{k}\right) - \frac{q}{k}. \tag{1.10}$$

If $k < 0$, then from (1.10) it follows that $\lim_{t\to\infty} P(t) = -q/k$, i.e., eventually the population will stabilize at $(-q/k)$.

The above model can be applied to a wide variety of applications. For example, consider the timber in a forest. Planting of new plants will correspond to immigration and cutting of trees will correspond to emigration.

Example 1.9 (*Isaac Newton's (1642–1727) law of cooling*) Law of cooling states that the rate of change of the temperature of an object is directly proportional to the difference between its temperature $\theta(t)$ and the temperature of its surroundings T, i.e.,

$$\frac{d\theta}{dt} = -k(\theta - T), \tag{1.11}$$

where $k > 0$ (if $\theta > T$ then the object will become cooler with time, i.e., $\theta' < 0$).

In the investigation of a homicide or accidental death, it is often important to estimate the time of death. We shall show that Eq. (1.11) can be used to estimate the time of death. For this, let $t = 0$ be time when the corpse was discovered and its temperature measured to be θ_0. Let t_d be the time of death and the body temperature θ_d was normal $98.6°F$. Finally, let after time t_1 the temperature of the body be θ_1. From the differential equation (1.11) and from these measurements, we find

$$\theta(t) = T + e^{-kt}(\theta_0 - T),$$
$$\theta_d = T + e^{-kt_d}(\theta_0 - T),$$
$$\theta_1 = T + e^{-kt_1}(\theta_0 - T)$$

and hence

$$t_d = t_1 \ln\left(\frac{\theta_d - T}{\theta_0 - T}\right) \bigg/ \ln\left(\frac{\theta_1 - T}{\theta_0 - T}\right). \tag{1.12}$$

In particular, if $T = 60°F$, $\theta_0 = 85°F$, $t_1 = 4\,h$ and $\theta_1 = 72°F$ then from (1.12), we find $t_d \simeq -2.367\,h$. Thus, the body was discovered approximately $2\,h$, $22\,min$ after death.

Example 1.10 Let $p(t)$ denote the price of a commodity at time t, then its rate of change is proportional to the difference between the demand $d(t)$ and the supply $s(t)$ of the commodity in the market, i.e.,

$$\frac{dp}{dt} = k[d(t) - s(t)]. \tag{1.13}$$

Clearly, if demand is more then the supply, the price increases and hence the constant $k > 0$. Now we assume that $d(t)$ and $s(t)$ are linear functions of $p(t)$, i.e.,

$$d(t) = d_0 + d_1 p(t) \quad \text{and} \quad s(t) = s_0 + s_1 p(t), \quad d_1 < 0, \quad s_1 > 0.$$

Then, the Eq. (1.13) takes the form

$$\frac{dp}{dt} = k[\alpha - \beta p(t)], \quad \alpha = (d_0 - s_0), \quad \beta = (s_1 - d_1) > 0. \tag{1.14}$$

The solution of (1.14) can be written as

$$p(t) = \frac{\alpha}{\beta} - \left(\frac{\alpha}{\beta} - p(0)\right) e^{-k\beta t}.$$

Clearly, $p(t) \to \alpha/\beta$ (equilibrium price) as $t \to \infty$.

Example 1.11 Often, it is convenient to divide a physical or biological process into several distinct stages. The process is then described by the interactions between the individual stages. Each such stage is named a compartment, and the process is called a compartmental system. It is always assumed that the contents of each compartment are homogeneous. The substance transferred from one compartment to another is immediately merged into the latter. Here, we shall study one-compartment systems which consist an amount $y(t)$ of substance, an input rate $I(t)$ at which substance enters the system, and a fractional transfer coefficient k indicating the fraction of the substance removed from the system per unit time (see Fig. 1.2). Thus, the rate at which the amount $y(t)$ changes depends on the difference between the input and output at any time t, i.e.,

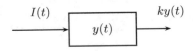

Fig. 1.2

$$y'(t) = I(t) - ky(t). \tag{1.15}$$

This equation is often called an *equation of continuity*, or a conservation equation.

Consider a tank containing 1000 gallons of water in which 50 pounds of salt is dissolved. Suppose 10 gallons of brine, each containing 1 pound of salt, runs into the tank per minute, and the mixture is kept uniform by stirring, runs out of the tank at the rate of 8 gallons per minute. We shall find the amount of salt $y(t)$ (in pounds) in the tank at any time t (after t minutes).

Clearly, here $I(t) = 10$ lb/min, and k is not a constant rather

$$k = k(t) = \frac{8}{1000 + 10t - 8t} = \frac{4}{500 + t} \quad \text{lb/min.}$$

Thus, the differential equation (1.15) takes the following form:

$$\frac{dy}{dt} = 10 - \frac{4}{500 + t} y(t),$$

which can be solved to obtain

$$y(t) = 2(500 + t) + c(500 + t)^{-4}.$$

Now since $y(0) = 50$ it follows that $c = -950(500)^4$ and hence

$$y(t) = 2(500 + t) - 950 \left(1 + \frac{t}{500} \right)^{-4}.$$

Example 1.12 A crude way of estimating the volume V of blood in the human body is as follows: Let the initial concentration of glucose in the bloodstream be $g(0)$, which is measured by taking a sample of blood. We inject a given amount of glucose in the bloodstream and then, say, after 2 min find the increase in concentration of glucose in a sample of blood. For greater accuracy, we introduce glucose in the blood stream at a constant rate I. Glucose is also removed from the bloodstream due to the physiological needs of the human body at a rate proportional to $g(t)$. Thus, the continuity principle gives

$$V \frac{dg}{dt} = I - kg. \tag{1.16}$$

Since I and $g(0)$ are known, to find V and k from (1.16) we need to measure concentration of glucose in the blood at two different times, i.e., $g(t_1)$ and $g(t_2)$ where $t_1 < t_2$.

By all methods, the volume of blood in human body turns out to be 5–6 L which is distributed in heart, aorta, main arteries, arterioles, capillaries, veins, etc.

Example 1.13 Let a dose D of a medicine be given to a patient at regular intervals of duration T each. The medicine also disappears from the body according to the following simple rule $A'(t) = -kA(t)$, where $A(t)$ is the amount of the medicine at time t. Since $A(0) = D$, it follows that

$$A(t) = De^{-kt}, \quad 0 \le t < T.$$

At time T, the residue of the first dose is De^{-kT} and now another dose D is given, so that

$$A(t) = \left(De^{-kT} + D \right) e^{-k(t-T)} = De^{-kt} + De^{-k(t-T)}, \quad T \le t < 2T. \tag{1.17}$$

In (1.17), the first term gives the residual of the first dose and the second term gives the residual of the second dose. Continuing in this way, we get

$$A(t) = De^{-kt} + De^{-k(t-T)} + \cdots + De^{-k(t-(n-1)T)}$$
$$= De^{-kt}\left[1 + e^{kT} + \cdots + e^{k(n-1)T}\right] \qquad (1.18)$$
$$= De^{-kt}\frac{e^{knT} - 1}{e^{kT} - 1}, \quad (n-1)T \le t < nT.$$

From (1.18), it follows that $A(t)$ is discontinuous at $T, \ 2T, \cdots$. Moreover,

$$A(nT - 0) = D\frac{1 - e^{-knT}}{e^{kT} - 1} \quad \text{and} \quad A(nT + 0) = D\frac{e^{kT} - e^{-knT}}{e^{kT} - 1}.$$

Hence, the medicine never exceeds the amount $D/(1 - e^{-kT})$. The minimum in an interval occurs at the end of each interval. This also increases, but always lies below $D/(e^{kT} - 1)$. This gives us at least some information as to what happens to the drugs which we take on long term basis for the illnesses such as hypertension and diabetes. For the short term medicinal courses (say, antibiotics), it indicates the length of the period of residue remaining in the body.

Example 1.14 Newton's second law states that the rate of change of momentum of a body equals the force applied. This law can be applied to bodies with variable mass, e.g., rockets. We shall demonstrate this for motion along a straight line. Let Mv be the momentum of the body at time t, and let $\Delta M(v + u)$ be the momentum which is added to the body at time t because the body is joined by another body of mass ΔM which before joining had the velocity $v + u$ (u is the velocity of ΔM relative to M). Further, assume that at the instant of joining, the velocity of M changes from v to $v + \Delta v$ and ΔM takes on the velocity of M. Thus, the momentum at time t is $Mv + \Delta M(u + v)$, whereas at time $t + \Delta t$ it is $(M + \Delta M)(v + \Delta v)$. Hence, the change of momentum $\Delta(Mv)$ during the time Δt is

$$\Delta(Mv) = (M + \Delta M)(v + \Delta v) - Mv - \Delta M(u + v) = M\Delta v - u\Delta M + \Delta M\Delta v.$$

Therefore, from Newton's second law, we have

$$M\Delta v - u\Delta M + \Delta M\Delta v = \int_t^{t+\Delta t} F(s)ds.$$

Dividing this relation by Δt and then letting $\Delta t \to 0$, and assuming that ΔM, Δv and $\int_t^{t+\Delta t} F(s)ds$ tend to zero in such a way that the resulting limits exist, we obtain

$$M\frac{dv}{dt} - u\frac{dM}{dt} = F(t),$$

which is the same as

$$\frac{d(Mv)}{dt} = (v + u)\frac{dM}{dt} + F(t). \qquad (1.19)$$

Now consider a rocket traveling vertically upward in such a way that its rate of change of mass dM/dt is constant $(-r)$. The lost mass consists of burning the fuel, and maintains the constant exhaust speed c of the rocket. The rocket is acted upon by a gravitational force Mg, where g is gravitational constant, and starts with initial velocity v_0 and initial mass M_0. We need to find the velocity v and distance traveled x as functions of time t. For this, substituting $dM/dt = -r$, $M = M_0 - rt$, $u = -c$, $F = -Mg$ in (1.19), we find the differential equation

$$\frac{dv}{dt} = \frac{cr}{M_0 - rt} - g. \tag{1.20}$$

Since $v(0) = v_0$, an integration of (1.20) gives

$$v(t) = v_0 - c \ln\left(1 - \frac{rt}{M_0}\right) - gt. \tag{1.21}$$

Now since $dx/dt = v$ and $x(0) = 0$, it follows that

$$x(t) = v_0 t + \frac{cM_0}{r}\left\{1 - \left(1 - \frac{rt}{M_0}\right)\left[1 - \ln\left(1 - \frac{rt}{M_0}\right)\right]\right\} - \frac{1}{2}gt^2. \tag{1.22}$$

We shall now find the *burnout velocity* v_1 of the rocket, i.e., the velocity with which the rocket is traveling at time t_1, when the entire fuel supply is exhausted and the remaining mass M_1 of the rocket is that of its structure and payload. Since, $t_1 = (M_0 - M_1)/r$, from (1.21), we have

$$v_1 = v_0 + c \ln \frac{M_0}{M_1} - \frac{(M_0 - M_1)g}{r}. \tag{1.23}$$

If M_f, M_p, and M_s, respectively, denote the mass of the fuel, mass of the payload, and the mass of the rocket structure, then clearly $M_0 = M_f + M_p + M_s$ and $M_1 = M_p + M_s$, and the relation (1.23) can be written as

$$v_1 = v_0 + c \ln\left(1 + \frac{M_f}{M_s + M_p}\right) - \frac{M_f g}{r}. \tag{1.24}$$

Thus, for given fuel and payload, the higher the exhaust velocity c of the fuel and the smaller the structural mass M_s, gives the higher burnout velocity v_1 of the rocket.

Problems

1.1 A cable of a suspension bridge supporting a uniform load of W pounds per horizontal foot and with horizontal tension H in the cable at the origin satisfies the differential equation (see Fig. 1.3)

$$\frac{dy}{dx} = \frac{W}{H}x.$$

Show that the cable hangs in a parabola by solving this differential equation with the condition $y(0) = 0$.

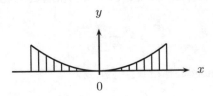

$$y$$

$$x$$

0

Fig. 1.3

1.2 If a piston has orifices, its motion through oil in a cylinder is governed by the differential equation (see Fig. 1.4)

$$\frac{dv}{dt} = -kv,$$

where v is the velocity and $k > 0$ is a constant. If x is the distance that piston has moved in time t, (i.e., $x' = v$), show that $x(t) = v_0(1 - e^{-kt})/k$ where v_0 is the initial velocity. Find the range of distances over which the piston can move.

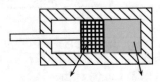

Piston with orifices   oil

Fig. 1.4

1.3 The form of the body of a snake during locomotion gives rise to the following differential equation:

$$\frac{1}{L}\frac{dL}{dt} = \frac{1}{3\eta}a,$$

where η is the density of the body of the snake and a is a constant. Show that the general solution of this equation is $L(t) = ce^{at/3\eta}$. It shows the way a section of the snake's body changes as it propels itself forward by muscle action.

1.4 In cancer research, a vitally important phenomenon is the relation between the time required for an irradiated cell to reach metaphase and the amount and rate of the radiation dose. The principal mechanism producing metaphase delay is the failure of chromosome coiling in prophase that is due to radiation-induced cross-linking of chromosomal structural protein fibers. If ξ represents the amount of coiling agent per chain length, L the length of the chromosome, σ the number of contractile sites per chain length, the constant g depends on the elastic property of the protein fibers, and t is the time after cell birth (cleavage), then it follows that

$$\frac{d}{dt}\left(\frac{L}{g}\right) = -k\xi\sigma\frac{L}{g},$$

where k is a constant. If $L(0) = L_0$, show that $L(t) = L_0 e^{-k\xi\sigma t}$, i.e., each contractile fiber acts like a stretched spring.

1.5 Charcoal from an excavation at Jarmo (Iraq) was dated in 1970 at 5150 B.C. How many counts per minute did the charcoal register? (see [7]).

1.6 Deep ocean water has measurably less radioactivity due to C-14 than surface ocean water. This indicates that the rate of turnover of the ocean takes place in thousands of years. Samples taken on the surface, and at depths of 1829 m and 2743 m had average readings of 6.72, 5.45, and 5.34 counts per minute, respectively (see [4]). Show that the respective apparent ages of the latter two samples are 1732 and 1900 years.

1.7 Suppose a wet sheet in a dryer loses its moisture at a rate proportional to its moisture content, and it loses half of its moisture in 15 min. Find the time of losing 98% of its moisture. (84.66 min)

1.8 The economist Vilfredo Pareto (1848–1923) discovered that the rate of decrease of the number of people N in a stable economy having an income of at least x dollars is directly proportional to the number of such people and inversely proportional to their income. Find a relation (Pareto's law) between N and x. ($N = cx^{-k}$)

1.9 A population is decaying exponentially. Can this decay be stopped or reversed by an immigration at a large constant rate into the population?

1.10 The differential equation $P' = k\cos(t)P$, where k is a constant, is often used as a model of a population that undergoes seasonal fluctuations. Find the solution of this equation satisfying the initial condition $P(0) = P_0$. ($P(t) = P_0 e^{k\sin(t)}$.)

1.11 The solutions of the differential equation

$$\frac{dy}{dx} = \frac{\alpha - x}{\beta + \gamma x + \delta x^2}y$$

give most of the important distributions of statistics for appropriate choices of the constants α, β, γ and δ. Solve this differential equation when

(1) $\alpha = \beta = \delta = 0$, $\gamma > 0$ (exponential distribution),
(2) $\gamma = \delta = 0$, $\beta > 0$ (normal distribution),
(3) $\beta = \delta = 0$, $\gamma > 0$, $\alpha > -\gamma$ (gamma distribution),
(4) $\beta = 0$, $\gamma = -\delta$, $(\alpha - 1)/\gamma < 1$, $\alpha/\gamma > -1$ (beta distribution).

1.12 After dinner, a couple orders coffee. The husband adds a teaspoon of cool cream to the coffee at once but does not drink it immediately. The wife waits for 4 min and then adds the cream (same amount and at the same temperature). They then drink their coffee. Who drinks the hotter coffee?

1.13 Let y denote the equity capital of a company, N dividend payout ratio, r rate of return on equity, and s rate of new stock financing. Then, under certain assumptions y satisfies the differential equation (see [6])

$$\frac{dy}{dt} = (1 - N)ry + s.$$

Suppose that the units have been chosen so that $y(0) = 1$. Find y as a function of t.

1.14 A tank contains 100 gallons of brine whose concentration is 2.5 lb of salt per gallon. Brine containing 2 lb of salt per gallon runs into the tank at a rate of 5 gallons per minute and the mixture (kept uniform by stirring) runs out at the same rate. Show that the amount of the salt y in the tank at any instant is $y(t) = 200 + 50e^{-t/20}$. Note that $y(t) > 200$ for all t, and $y(t) \to 200$ as $t \to \infty$.

1.15 There are 100 million liters of fluoridated water in the reservoir containing a city's water supply, and the water contains 500 kg of fluoride. To decrease the fluoride content, freshwater runs into the reservoir at the rate of 3 million liters per day, and the mixture of water and fluoride kept uniform, runs out of the reservoir at the same rate. Show that the number of kg of fluoride y in the reservoir after t days is $y(t) = 500e^{-0.03t}$.

1.16 In a study of biophysical limitations associated with deep diving, the following differential equation occurs (see [2]):

$$\frac{dy}{dx} - \alpha y = \beta + be^{-ax},$$

where α, β, b and a are constants. Show that the general solution of this differential equation can be written as

$$y(x) = -\frac{\beta}{\alpha} - \frac{b}{a + \alpha}e^{-ax} + ce^{\alpha x},$$

where c is an arbitrary constant.

1.17 An optimal control theory method to the education investment decision leads to the differential equation (see [9])

$$\frac{dy}{dt} = 1 - ky,$$

where y denotes the education of an individual at time t and the constant k is the rate at which education becomes obsolete or being forgotten. Show that the general solution of this differential equation can be written as $y(t) = ce^{-kt} + (1/k)$, where c is an arbitrary constant.

1.18 A simple mathematical model of the memorization process is given by the differential equation

$$\frac{dy}{dt} = a(M - y) - by,$$

where M is the total amount of material to be memorized and a and b are positive constants, $y(t)$ represents the amount of material memorized at time t, $(M - y)$ is the amount remaining to be memorized, and the term $-by$ accounts for the fact that more we memorize, the more rapidly we forget. Solve this equation subject to the initial condition $y(0) = 0$ and find the limiting value of y as $t \to \infty$ also interpret the result. ($y(t) = \frac{aM}{a+b}(1 - e^{-(a+b)t})$. As $t \to \infty$, $y \to \frac{aM}{a+b}$ and hence if $b > 0$ the material will never be completely memorized.)

1.19 A body of mass M is dropped from rest from a great height in the Earth's atmosphere. Assume that it falls in a straight line and that the only forces acting on it are the Earth's gravitational attraction Mg, and air resisting force which is proportional to its velocity. Thus, if $s = s(t)$ denotes the distance the body has fallen at time t, and $v = ds/dt$ represents its velocity, then by Newton's second law the motion of the body is governed by the initial value problem

$$Mv' = Mg - kv, \quad v(0) = 0 \quad (k \text{ is a constant}).$$

Show that

$$v = \frac{Mg}{k}\left(1 - e^{-kt/M}\right) \quad \text{and} \quad s = \frac{Mg}{k}t + \frac{gM^2}{k^2}\left(e^{-kt/M} - 1\right).$$

Further, if the initial velocity is v_0, i.e., $v(0) = v_0$, then

$$v = \frac{Mg}{k}\left(1 - e^{-kt/M}\right) + v_0 e^{-kt/M}.$$

Thus, the limiting velocity, as t increases without bound, is Mg/k, which is independent of v_0. (If the body is a parachutist, then this fact makes a safe landing possible.)

1.20 A constant torque T is applied to a rotating shaft. The shaft has moment of inertia J. There is viscous damping Bw, where w is the shaft speed. Hence by Newton's second law of motion

$$J\frac{dw}{dt} + Bw = T, \quad J > 0, \quad B > 0, \quad T \neq 0.$$

(1) Given $w(0) = w_0$, solve this equation. $(w_0 e^{-Bt/J} + (1 - e^{-Bt/J})T/B)$
(2) Show that $w(t)$ has a limit as $t \to \infty$ and compute this limit. (T/B)

1.21 A drug is infused into a patient's bloodstream at a constant rate of c grams per second. Simultaneously, the drug is removed at a rate proportional to the amount $y(t)$ of the drug present at any time. Show that the differential equation $y' = c - ky$ governs $y(t)$ at any time. Find its general solution.

1.22 Suppose a cell is suspended in a solution containing a solute of constant concentration C_s. Further suppose that the cell has constant volume V and the area of its permeable membrane is the constant A. According to Adolf Eugen Fick's law (1829–1901), the rate of change of its mass M is directly proportional to the area A and the difference $C_s - C(t)$, i.e.,

$$\frac{dM}{dt} = kA(C_s - C(t)),$$

where $C(t)$ is the concentration of the solute inside the cell at any time t, and k is some constant. If $M(t) = VC(t)$ and $C(0) = C_0$, show that $C(t) = C_s + (C_0 - C_s)e^{-kAt/V}$.

1.23 The kinetics of cell growth produces a differential equation governing cell volume V as

$$V' = W + aM + \left(\frac{2c}{r_0}\right)(M - bV),$$

where b is a concentration of effective compounds outside the cell, c is a permeation constant, r_0 is radius of the cylindrical cell, M is quantity of macromolecules per cell, W is the constant pool of precursor, and a is a proportionality constant. If $V(0) = V_0$, show that

$$V(t) = A + (V_0 - A)e^{-kt},$$

where

$$k = \frac{2bc}{r_0}, \quad A = \frac{W}{k} + \frac{M}{k}\left(a + \frac{2c}{r_0}\right).$$

1.24 Active oxygen and free radicals are believed to be exacerbating factors in causing cell injury and aging in living tissue (see [5]). These molecules also accelerate the deterioration of foods. Researchers are therefore interested in understanding the protective role of natural antioxidants. In the study of one such antioxidant (hsian-tsao leaf gum), the antioxidation activity of the substance has been found to depend upon concentration in the following way:

$$\frac{dA}{dc} = k[A^* - A(c)], \quad A(0) = 0,$$

where A is a quantitative measure of antioxidant activity at concentration c, the constant A^* represents a limiting or equilibrium value of this activity and k is a positive rate constant. Show that the activity $A(c)$ never exceeds the value A^*. $(A^*(1 - e^{-kc}))$

1.25 In some microorganisms like algae the following problem occurs (see [3]):

$$\frac{dN}{dt} = \mu[N + A(t)], \quad N(0) = 0,$$

where N represents the cell density, μ the growth rate, and A is the attachment rate. Show that when $A(t) \equiv A$ is a constant, its solution can be written as $N(t) = A(e^{\mu t} - 1)$.

1.26 Assume that a nerve contains two substances (or two groups of substances) whose concentrations are x and y. The ratio x/y determines the excitation of the nerve. Whenever $x/y \geq c$ (c is known as the excitation constant), excitation occurs. Let $c = 1$, so that the excitation occurs whenever $x \geq y$. If x_0 and y_0 be the concentrations of x and y in a resting nerve, then clearly $x_0 < y_0$. Rashevsky [8] has shown that

$$\frac{dx}{dt} = A - a(x - x_0)$$

$$\frac{dy}{dt} = B - b(y - y_0),$$

where A, a, B and b are constants. Find x and y as functions of t, and y as a function of x. $(x = x_0 + (1 + e^{-at})A/a, \ y = y_0 + (1 + e^{-bt})B/b, \ y = y_0 + \left[1 - (1 + ax_0/A - ax/A)^{b/a}\right]B/b)$.

1.27 In order to comply with the existing law on drunken driving, an objective measure of blood alcohol level in drivers has been established. BAC (blood alcohol count) is a number that represents the percentage of bloodstream alcohol level. For example, BAC $= 0.08$ means 0.08% of the bloodstream is occupied by alcohol. An estimated BAC value of 0.05 is high enough to produce the usual impairment of driving abilities. It is illegal to drive with a BAC of 0.05–0.10, depending on a state. BAC values are dependent on several factors such as weight, gender, usual drinking habits, and amount of food in the stomach. In this problem, only weight and amount of alcohol in the system will be considered. We denote by

$$t = \text{time } (h) \text{ since commencement of drinking,}$$
$$t_0 = \text{time at which drinking is stopped,}$$
$$A(t) = \text{number of ounces of alcohol in bloodstream at time } t,$$
$$A_0 = \text{number of ounces of alcohol in bloodstream at the time}$$
$$\text{drinking is stopped,}$$
$$B(t) = BAC \text{ at time } t,$$
$$B_0 = \text{BAC at the time drinking is stopped,}$$
$$C(t) = \text{rate of alcohol intake (ounces/h), we shall assume that}$$
$$C(t) = C \text{ is a constant,}$$
$$W = \text{weight (lbs).}$$

It is known that $B(t) \simeq 7.2A(t)/W$.

Now assume that the body eliminates alcohol from the bloodstream at a rate which is proportional to the amount present at any time. Then, $A(t)$ and $B(t)$ satisfy the following differential equations:

$$A'(t) = \begin{cases} C - kA(t), & 0 < t < t_0, \quad A(0) = 0 \\ -kA(t), & t > t_0, \end{cases}$$

$$B'(t) = \begin{cases} \dfrac{7.2C}{W} - kB(t), & 0 < t < t_0, \quad B(0) = 0 \\ -kB(t), & t > t_0, \end{cases}$$

where k is a constant.

Show that continuous for $t \geq 0$ and differentiable except at t_0 solutions of the above differential equations are

$$A(t) = \begin{cases} \dfrac{C}{k}(1 - e^{-kt}), & 0 \leq t < t_0 \\ \dfrac{C}{k}(1 - e^{-kt_0})e^{-k(t-t_0)}, & t \geq t_0 \end{cases}$$

$$B(t) = \begin{cases} \dfrac{7.2C}{Wk}(1 - e^{-kt}), & 0 \leq t < t_0 \\ \dfrac{7.2C}{Wk}(1 - e^{-kt_0})e^{-k(t-t_0)}, & t \geq t_0. \end{cases}$$

To determine the rate of elimination of alcohol from the bloodstream, i.e., k several methods are known. For example, according to the Traffic Institute at Northwestern University, the rate of elimination of alcohol is 0.5 oz/h. Further, Heublein Inc. has estimated that the rate of elimination reduces the BAC by 0.01 oz/h.

1.28 The secretion of hormones into the blood is often a periodic activity. If a hormone is secreted on a 24h cycle, then the rate of change of the level of the hormone in the blood may be represented by the initial value problem

$$\frac{dy}{dt} = a - b\cos\left(\frac{\pi t}{12}\right) - ky, \quad y(0) = y_0$$

where $y(t)$ is the amount of the hormone in the blood at time t, a is the average secretion rate, b is the amount of variation in the secretion, and k is a positive constant reflecting the rate at which the body removes the hormone from the blood. Find the solution of the above initial value problem. ($\frac{a}{k} - \frac{144bk}{144k^2+\pi^2}\left[\cos\left(\frac{\pi t}{12}\right) + \frac{\pi}{12k}\sin\left(\frac{\pi t}{12}\right)\right] + \left(y_0 - \frac{a}{k} + \frac{144bk}{144k^2+\pi^2}\right)e^{-kt}$)

1.29 For a simple electric circuit with a given value of inductance L, resistance R, and applied electromotive force E, one of the Gustav Robert Kirchhoff's (1824–1887) laws implies that the current $i(t)$ is a solution of the differential equation

$$L\frac{di}{dt} + Ri = E\sin(kt) \quad (k \text{ is a constant}).$$

Show that

$$i(t) = \frac{RE\sin(kt) - kLE\cos(kt)}{R^2 + k^2 L^2} + Ce^{-(R/L)t}.$$

1.30 A chain of weight w per unit length is lying on the ground, piled up in one spot. Beginning at time $t = 0$, one end of the chain is hoisted up with constant velocity v_0. Determine the force F, as a function of time t, which is necessary for the hoisting of the chain against the effect of gravitational acceleration g. $(F = wv_0 t + wv_0^2/g)$

References

1. R.P. Agarwal, D. O'Regan, *An Introduction to Ordinary Differential Equations* (Springer, New York, 2008)
2. H. Bradner, R.S. Mackay, Bull. Math. Biophys. **25**, 251–272 (1963)
3. D.E. Caldwell, Biochemical Engineering V, Annals of the New York Academy of Sciences, vol. 56 (1987), pp. 274–280
4. J.L. Kulp, L.E. Tryin, W.R. Eckclman, W.A. Snell, Science **116**, 409–414 (1952)
5. L.S. Lai, S.T. Chou, W.W. Chao, J. Agric. Food Chem. **49**, 963–968 (2001)
6. J.L. Lebowitz, C.O. Lee, P.B. Linhart, Bell J. Econ. **7**, 463–477 (1976)
7. W.F. Libby, *Radiocarbon Dating*, 2nd edn. (University of Chicago Press, Chicago, 1955)
8. N. Rashevsky, *Mathematical Biophysics*, vol. 1 (Dover Publications, New York, 1960)
9. L. Southwick, S. Zionts, Oper. Res. **22**, 1156–1174 (1974)

Chapter 2
Some First-Order Nonlinear Differential Equations

Certain nonlinear first-order differential equations can be reduced to linear equations by an appropriate change of variables [1, 2]. For example, it is always possible for the James (Jacob) Bernoulli's (1654–1705) equation

$$y' + p(t)y = q(t)y^n, \quad n \neq 0, \ 1. \tag{2.1}$$

In (2.1), $n = 0$ and 1 are excluded because in these cases the Eq. (2.1) is obviously linear. Equation (2.1) is equivalent to the differential equation

$$y^{-n}y' + p(t)y^{1-n} = q(t). \tag{2.2}$$

In (2.2) we introduce a new dependent variable $z = y^{1-n}$, to get

$$\frac{1}{1-n}z' + p(t)z = q(t), \tag{2.3}$$

which is a first-order linear differential equation, and can be solved rather easily.

Another class of nonlinear differential equations which can be solved are of the form

$$Y(y)y' + T(t) = 0.$$

Such equations are called separable equations. The general solutions (usually implicit) of this type of equations can be obtained by direct integration and appear as

$$\int T(t)dt + \int Y(y)dy = c,$$

where c is a constant.

© Springer Nature Switzerland AG 2019
R. P. Agarwal et al., *500 Examples and Problems of Applied Differential Equations*, Problem Books in Mathematics,
https://doi.org/10.1007/978-3-030-26384-3_2

Example 2.1 Malthus law does not give satisfactory results particularly when the population becomes large. This is due to the fact that as population increases, due to overcrowding and limitations of resources, the birth rate b decreases and the death rate increases with the population size P. Thus in Malthus law k is not a constant, but a function of P. The simplest assumption is to take $b = b_1 - b_2 P$, $d = d_1 + d_2 P$ where $b_1, b_2, d_1, d_2 > 0$. Then, the Malthus model becomes

$$\frac{dP}{dt} = (b_1 - b_2 P)P - (d_1 + d_2 P)P = (b_1 - d_1)P - (b_2 + d_2)P^2 = P(a - bP), \quad (2.4)$$

where $a, b > 0$. This model is due to the Belgian mathematician Pierre François Verhulst (1804–1849) and in the literature the term $P(a - bP)$ is referred to as *logistic growth* and the Eq. (2.4) is called the *logistic equation*. Verhulst could not test the accuracy of his model because of inadequate census data. This model was rediscovered independently by two American scientists Raymond Pearl (1879–1940) and Lowell Jacob Reed (1886–1966). In 1920, Pearl and Reed examined how closely the U.S. population growth curve followed a logistic curve, i.e., the solution of the differential equation (2.4). Reasonable agreements between logistic curve and experimental data was demonstrated by Pearl in 1930 for fruit fly populations, and by Georgii Frantsevich Gause (1910–1986) in 1935 for flour beetle populations.

Equation (2.4) is a Bernoulli equation with $p(t) = -a$, $q(t) = -b$ and $n = 2$. In (2.4) the substitution $z = P^{-1}$ leads to the linear differential equation

$$z' + az = b, \quad (2.5)$$

whose solution is

$$z(t) = \alpha e^{-at} + \frac{b}{a},$$

where α is a constant, and hence

$$\frac{1}{P(t)} = \alpha e^{-at} + \frac{b}{a}. \quad (2.6)$$

In (2.6) using the fact that the initial population $P(0)$ is known, we find

$$P(t) = \frac{aP(0)}{bP(0) + [a - bP(0)]e^{-at}}. \quad (2.7)$$

From (2.7), it is clear that $\lim_{t \to \infty} P(t) = a/b$ regardless of the initial population $P(0)$. If $P(0) < a/b$, $P(t)$ increases monotonically and approaches to a/b; and if $P(0) > a/b$, $P(t)$ decreases monotonically and approaches to a/b. Thus, the logistic curve remains bounded (see Figs. 2.1 and 2.2).

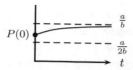

Fig. 2.1

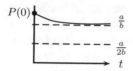

Fig. 2.2

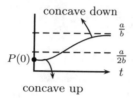

Fig. 2.3

Now a simple differentiation of (2.4) gives

$$\frac{d^2 P}{dt^2} = 2b^2 P \left(P - \frac{a}{b}\right)\left(P - \frac{a}{2b}\right). \tag{2.8}$$

From calculus we know that the points where $P'' = 0$ are possible points of inflection, however, $P = 0$ and $P = a/b$ can obviously be ruled out. For $0 < P < a/(2b)$ we note from (2.8) that $P'' > 0$, and $a/(2b) < P < a/b$ implies $P'' < 0$. Thus, the graph of P changes from concave up to concave down at the point $P = a/(2b)$. Hence, $P = a/(2b)$ is a point of inflection. If $0 < P(0) < a/(2b)$ the graph of $P(t)$ assumes the shape of an S (Fig. 2.3).

Example 2.2 Social scientists study how rumor spread. Consider a population of N people, and let $P(t)$ be the number who have heard the rumor at time t. The more people who have heard the rumor, the more opportunities they will have to pass it on. However, these opportunities also depend on the number of people who have not heard it yet, i.e., $[N - P(t)]$. Thus the rate at which new people hear the rumor is proportional to the product $P(N - P)$, i.e., $P' = kP(N - P)$, where k is some constant. This equation is exactly the same as (2.4) with $a = kN$ and $b = k$, thus $\lim_{t \to \infty} P(t) = N$, i.e., eventually all the people will hear the rumor. (A similar differential equation can be formulated in the spread of cultural changes.)

Example 2.3 Let $N(t)$ be the number of companies, which have adopted a techno-logical innovation till time t, then the rate of change of the number of these companies depends both on the number of companies which have adopted this innovation and on the number of those which have not yet adopted it. Thus, if C is the total number of companies in the region then $N' = kN(C - N)$, where k is some constant. Hence, ultimately all the companies will adopt this innovation.

Example 2.4 Let $y(t)$ be the amount of undissolved solute in a solvent at time t and let c_0 be the maximum amount of the solute that can be dissolved in a unit volume of the solvent. Let V be the volume of the solvent. It is found that the rate at which the solute is dissolved is proportional to the amount of undissolved solute and the difference between the concentration of the solute at time t and the maximum possible concentration, so that

$$\frac{dy}{dt} = ky(t)\left(\frac{y(0) - y(t)}{V} - c_0\right) = \frac{ky(t)}{V}[(y_0 - c_0 V) - y(t)].$$

This equation is also similar to that of (2.4).

Example 2.5 Diseases such as measles, chickenpox, smallpox, and mumps are gen-erally considered to impart immunity for life. To asses the effect of such a disease, we consider a group of individuals born in one specific year. Let $N(t)$ denote the number who have survived to age t, and let $S(t)$ be the number who have not had the disease and are still susceptible to it at age t. Let $0 < p < 1$ be the probability of a susceptible getting the disease, and let $1/m$ be the proportion of those who die due to the disease. To study the effects of smallpox Daniel Bernoulli (1700–1782) in 1760 derived the following differential equation:

$$\frac{dS}{dt} = -pS(t) + \frac{S(t)}{N(t)}\frac{dN}{dt} + p\frac{S^2(t)}{mN(t)}. \tag{2.9}$$

Multiplying both sides of (2.9) by N/S^2, and arranging the terms, we find

$$\frac{1}{S}\frac{dN}{dt} - \frac{N}{S^2}\frac{dS}{dt} = \frac{pN}{S} - \frac{p}{m},$$

which is the same as

$$\frac{d}{dt}\left(\frac{N}{S}\right) = p\left(\frac{N}{S}\right) - \frac{p}{m}. \tag{2.10}$$

Clearly, (2.10) is a first-order linear differential equation in N/S, and its solution can be written as

$$\frac{N(t)}{S(t)} = Ce^{pt} + \frac{1}{m}, \tag{2.11}$$

where C is an arbitrary constant.

Now since at birth every member of the group is susceptible, $S(0) = N(0)$. Thus, from (2.11) it follows that $C = (m - 1)/m$. Therefore, we have the relation

$$S(t) = \frac{mN(t)}{1 + (m - 1)e^{pt}}. \tag{2.12}$$

For the case of smallpox in Paris in 1760, Bernoulli used the available data $t = 24$ (years), $p = 1/8$ and $m = 8$, to obtain

$$S(24) = \frac{8N(24)}{1 + 7e^3} = 0.056498\, N(24),$$

i.e., only one in about 18 of those 24 years old would not have had smallpox.

Example 2.6 A homogeneous hollow metallic ball of inner radius r_1 and outer radius r_2 is in a stationary thermal state. The temperature on the inner surface is T_1 and on the outer surface T_2. We shall find the temperature T at a distance r from the center of the ball, where $r_1 \le r \le r_2$.

Clearly, from the symmetry it follows that T is a function of r alone. Since the quantity of heat remains invariable between two concentric spheres with centers at the center of the ball (their radii can vary from r_1 to r_2), the same quantity of heat Q flows through each sphere. Hence, the differential equation describing this process can be written as

$$-4\pi k r^2 \frac{dT}{dr} = Q, \tag{2.13}$$

where k is the coefficient of thermal conduction.

In (2.13) separating the variables and then integrating, we obtain

$$4\pi k \int_{T_1}^{T} dT = -Q \int_{r_1}^{r} \frac{dr}{r^2} \quad (\text{recall } T(r_1) = T_1),$$

which gives

$$4\pi k (T - T_1) = Q \left(\frac{1}{r} - \frac{1}{r_1} \right). \tag{2.14}$$

Now we use the condition $T(r_2) = T_2$ in (2.14), to determine the value of Q as

$$Q = \frac{4\pi k (T_2 - T_1)}{\frac{1}{r_2} - \frac{1}{r_1}} = \frac{4\pi k (T_2 - T_1) r_1 r_2}{r_1 - r_2}.$$

Example 2.7 When hydrogen is burnt in oxygen, two atoms of hydrogen combine with one of oxygen to produce water. The rate of the reaction at time t is proportional to the product of the numbers of hydrogen and oxygen atoms present at time t. We shall find the number of water molecules x at time t, and calculate $\lim_{t \to \infty} x(t)$.

The answer depends on the amounts present initially. Let a and b be the number of atoms of hydrogen and oxygen (i.e., $a/2$ molecules of H_2 and b molecules of O). Let x be the number of molecules of H_2O produced at time t. Then,

$$\frac{dx}{dt} = K(a - 2x)(b - x), \tag{2.15}$$

which is a first- order differential equation with variables separable, and hence

$$
\begin{aligned}
Kt &= \int \frac{dx}{(a - 2x)(b - x)} + c \\
&= \int \frac{1}{a - 2b}\left(\frac{1}{b - x} - \frac{2}{a - 2x}\right) dx + c \quad \text{(assuming } a \neq 2b) \\
&= \frac{1}{a - 2b} \ln \frac{a - 2x}{b - x} + c.
\end{aligned}
$$

Since $x = 0$ when $t = 0$, it follows that

$$c = -\frac{\ln a/b}{a - 2b}$$

and hence

$$Kt = \frac{1}{a - 2b} \ln\left(\frac{b(a - 2x)}{a(b - x)}\right),$$

which gives

$$\frac{a - 2x}{b - x} = \frac{a}{b} \exp\left[(a - 2b)Kt\right].$$

Thus, it follows that

$$x = \frac{a\left(\exp[(2b - a)Kt] - 1\right)}{2\exp[(2b - a)Kt] - (a/b)} = \frac{a(1 - \exp[(a - 2b)Kt])}{2 - (a/b)\exp[(a - 2b)Kt]}.$$

Hence, if $a > 2b$, $x \to b$ as $t \to \infty$, and if $a < 2b$, $x \to a/2$ as $t \to \infty$. These results are reasonable, i.e. the reaction runs until one or the other of the components is exhausted.

Now, if $a = 2b$, then we have

$$Kt = \int \frac{dx}{2(b - x)^2} + c = \frac{1}{2(b - x)} + c$$

and hence in view of $x = 0$ when $t = 0$, we find

$$Kt = \frac{1}{2(b - x)} - \frac{1}{2b},$$

which gives

$$x = b - \frac{1}{2Kt + (1/b)}.$$

Hence, $x \to b$ as $t \to \infty$.

A reaction described by the differential equation (2.15) is called a second-order reaction. It can be generalized as follows: Let two chemical substances combine in the ratio $a : b$ to form a new substance x. If $x(t)$ is the amount of new substance at time t, then the amounts of original substances remaining at time t are, respectively,

$$A - \frac{a}{a+b}x \quad \text{and} \quad B - \frac{b}{a+b}x,$$

where A and B are the initial amounts of these substances. Now the rate of formation of the new substance is proportional to the product of the remaining substances; i.e.,

$$\frac{dx}{dt} = k\left(A - \frac{a}{a+b}x\right)\left(B - \frac{b}{a+b}x\right),$$

where k is a constant. Chemists refer to these reactions to as the *law of mass action*.

In an analogous way, an nth-order reaction leads to the nonlinear differential equation

$$\frac{dx}{dt} = k(A_1 - a_1x)(A_2 - a_2x)\cdots(A_n - a_nx),$$

where $a_1 + a_2 + \cdots + a_n = 1$.

Example 2.8 Evangelista Torricelli's (1608–1647) law states that water in an open tank will flow out through a small hole in the bottom with the speed it would acquire in falling freely from the water level to the hole (see Fig. 2.4), i.e., the velocity v with which water flows from the orifice is

$$v = k\sqrt{2gh}, \tag{2.16}$$

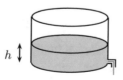

Fig. 2.4

where k is a proportionality constant, g is the gravitational constant, and h is the height of water above the orifice at any time t. For water it has been experimentally determined that $k = 0.6$. Now, if V is the volume of water in the tank at any time t

and A is the cross-sectional area of the orifice, then the rate of change of volume is related to A by

$$\frac{dV}{dt} = -Av \tag{2.17}$$

and hence from (2.16) it follows that

$$\frac{dV}{dt} = -0.6A\sqrt{2gh}. \tag{2.18}$$

Let $V = V(h)$, i.e., the volume we express as a function of height h. Then, we have

$$\frac{dV}{dt} = \frac{dV}{dh} \cdot \frac{dh}{dt} = B(h)\frac{dh}{dt}, \tag{2.19}$$

where $B(h)$ denotes the cross-sectional area of the tank at height h. Finally, combining (2.18) and (2.19) we get the differential equation

$$B(h)\frac{dh}{dt} = -0.6A\sqrt{2gh}. \tag{2.20}$$

We shall find the time t needed to empty a cylindrical tank of radius $6\,\text{in}$ and vertical height of $2\,\text{ft}$ originally full of water and has an orifice of diameter $1/2\,\text{in}$ at the bottom. For this, clearly we have $A = \pi(1/48)^2$, $V = \pi(1/2)^2 h$ and hence $B(h) = \pi/4$ and $g = 32$. Thus the differential equation (2.20) takes the form

$$\frac{\pi}{4}\frac{dh}{dt} = -0.6\pi\left(\frac{1}{48}\right)^2 \sqrt{2 \times 32 \times h},$$

or

$$\frac{dh}{dt} = -\frac{1}{120}\sqrt{h}.$$

Separating the variables in the above equation and then integrating, we find

$$2\sqrt{h} = c - \frac{1}{120}t,$$

where c is a constant. However, since initially the height of the water is $2\,\text{ft}$, i.e., $h(0) = 2$ it follows that $c = 2\sqrt{2}$, and hence

$$t = 240(\sqrt{2} - \sqrt{h}).$$

Now the tank will be empty when $h = 0$, and this gives $t = 240\sqrt{2} \simeq 339\,\text{s} = 5\,\text{m}$ $39\,\text{s}$.

Example 2.9 We shall find the shape of a mirror that reflects, parallel to a given direction, all the rays emanating from a given point.

For simplicity, we locate the given point at the origin, and direct the x-axis as parallel to the given direction. Let a ray fall on the mirror at the point $P = P(x, y)$ (see Fig. 2.5). We assume that the section of the mirror (cut by the xy-plane) passes through the x-axis and the point P. Let PQ be the tangent on the surface of the mirror from the point P. Since the angle of incidence of the ray is equal to the angle of reflection, the triangle PQO is an isosceles triangle. Thus, we have

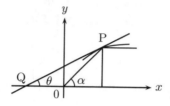

Fig. 2.5

$$\tan(\theta) = \frac{dy}{dx} = \frac{y}{x + \sqrt{x^2 + y^2}} = \frac{y(x - \sqrt{x^2 + y^2})}{x^2 - (x^2 + y^2)} \tag{2.21}$$

and hence

$$x\,dx + y\,dy = \sqrt{x^2 + y^2}\,dx,$$

which is the same as

$$\frac{x\,dx + y\,dy}{\sqrt{x^2 + y^2}} = dx. \tag{2.22}$$

Integrating (2.22), we find

$$\sqrt{x^2 + y^2} = x + c,$$

which is a family of parabolas

$$y^2 = 2cx + c^2.$$

Example 2.10 A rocket of mass M is projected straight up from the surface of the Earth with an initial velocity u_0. We shall calculate its escape velocity. Since Earth's gravitational field varies with altitude, the weight w of the rocket is derived by employing the inverse square law of gravitational attraction, i.e.,

$$w(y) = \frac{c}{(y + R)^2}, \tag{2.23}$$

where y is the altitude above sea level, R Earth's radius, and c is a constant. Since at see level $y = 0$ and $w = Mg$, from (2.23) it is clear that $c = MgR^2$. Hence,

$$w(y) = \frac{MgR^2}{(y+R)^2}.$$

Let the air resistance be negligible, so that $F = M \times$ acceleration. This leads to the following equation of motion

$$M\frac{dv}{dt} = -\frac{MgR^2}{(y+R)^2}.$$

However, since

$$\frac{dv}{dt} = \frac{dv}{dy} \cdot \frac{dy}{dt} = \frac{dv}{dy} \cdot v$$

it follows that

$$v\frac{dv}{dy} = -\frac{gR^2}{(y+R)^2},$$

which is an equation with variables separable, and hence can be solved rather easily to obtain

$$\frac{v^2}{2} = \frac{gR^2}{y+R} + A. \tag{2.24}$$

Finally, since when $y = 0$, $v = u_0$, from (2.24) we find $A = (1/2)u_0^2 - gR$, and hence we have

$$v^2 = u_0^2 - 2gR + \frac{2gR^2}{y+R}. \tag{2.25}$$

Clearly, for the rocket to escape the gravitational pull of the Earth, the velocity v must remain positive for all y. From (2.25) it is clear that this certainly happens if $u_0^2 \geq 2gR$. Hence, the escape velocity v_e is $v_e = (2gR)^{1/2} \simeq 7$ miles/s.

Problems

2.1 A college dormitory houses 100 students each of whom is susceptible to a certain virus infection. A simple model of epidemics assumes that during the course of an epidemic the rate of change with respect to time of the number of infected students I is proportional to the number of infected students and also proportional to the number of uninfected students $(100 - I)$, i.e., $\frac{dI}{dt} = kI(100 - I)$, where k is a constant. If at time $t = 0$ a single student becomes infected, show that the number of infected students at time t is given by

$$I(t) = \frac{100e^{100kt}}{99 + e^{100kt}}.$$

2.2 Male walruses live in a colony with many others, however, each dominates a certain subterritory of the land occupied by the whole colony. Thus, each male member of the colony has under his control a packaged resource. Every new incoming male member, including one born in the colony, must either find an unused package or contend with a present member for his package. Competition of this type restricts the growth of the colony's population, and results in a reduction of growth rate. The rate of change of the population is positive for natural propagation, but negative for the competition. Thus, if $M(t)$ denotes the male walrus population, the resulting differential equation is

$$M' = aM - bM(M - 1),$$

where a is the natural unconstrained growth rate and the constant $b > 0$ is called the competition parameter. Show that $\lim_{t\to\infty} M(t) = (a + b)/b$, i.e., the eventual population is independent of the initial size of the population.

2.3 In a study of the behavior of flood waves in rivers, Stoker [11] obtained the following Bernoulli's differential equation

$$\frac{dc_1}{d\xi} + ac_1 = bc_1^2, \tag{2.26}$$

where

$$a = \frac{gS}{(v_0 + c_0)^2}\left(\frac{1}{v_0} - \frac{2}{3c_0}\frac{gB}{gB + 2c_0^2}\right) \quad \text{and} \quad b = \frac{3}{(v_0 + c_0)^2}$$

and g is the acceleration due to gravity ($g = 32.16$ ft/s^2), B is the width of the river (any cross section of the river perpendicular to the direction of the river flow is assumed to be a rectangle of constant width B, but variable depth y), S represents the slope of the river bed, v_0 represents the initial velocity of the wave, $c_0 = (gy_0)^{1/2}$, where y_0 is the initial depth, c_1 is an unknown function that is related to the depth y of the river by the approximation $c_1(\xi) \simeq (\sqrt{gy} - c_0)/\tau$. Further, $\xi = x$, $\tau = (v_0 + c_0)t - x$, and t is time, x is a space variable whose axis is parallel to the direction of flow of the river. Find the solution of Eq. (2.26).

2.4 A gas is to be ionized in such a way that the number of electrons per unit volume equals the number of positive ions per unit volume. Positive ions and electrons recombine to form neutral molecules at a rate equal to kn^2, where the constant $k > 0$ is called the *constant of recombination*. Let the gas be initially ($t = 0$) unionized, and for $t > 0$, C (a constant) ions per unit volume are produced. This leads to the initial value problem

$$\frac{dn}{dt} = C - kn^2, \quad n(0) = 0. \tag{2.27}$$

Find the solution of the problem (2.27).

2.5 Let a population be growing according to logistic law and being harvested at a constant rate H (like hunting, fishing, or disease), so that the modified differential equation is

$$\frac{dP}{dt} = aP - bP^2 - H. \tag{2.28}$$

Show that if $H \geq a^2/(4b)$ its general solution can be written as

$$P(t) = \frac{1}{2b}\left[a + \sqrt{4bH - a^2}\tan\left(\frac{1}{2}(c-t)\sqrt{4bH - a^2}\right)\right]$$

where c is a constant. What happens ultimately if $H < a^2/(4b)$?

2.6 A modification of the logistic equation is the Benjamin Gompertz (1779–1865) equation

$$\frac{dP}{dt} = P[a - b\ln(P)], \tag{2.29}$$

where a and b are constants. Show that
(1) The general solution of (2.29) can be written as

$$P(t) = e^{a/b}e^{-ce^{-bt}}, \tag{2.30}$$

where c is an arbitrary constant. The graph of the function (2.30) is called a Gompertz curve and is similar to the logistic function. Note that when $b > 0$, $P \to e^{a/b}$ as $t \to \infty$, whereas if $b < 0$, $c > 0$, $P \to 0$ as $t \to \infty$. Functions such as (2.30) are encountered in studies of the growth or decline of certain populations, in the growth of solid tumors, in actuarial predictions, and in the study of growth of revenue in the sale of a commercial product.
(2) If $P(0) = P_0$, then $c = \frac{a}{b} - \ln(P_0)$.
(3) If $0 < P_0 < e^{a/b}$ and $a > 0$, then the point of inflection for the Gompertz curve (2.30) is $\left[(1/b)\ln[a/b - \ln(P_0)], e^{a/b-1}\right]$.

2.7 Often a population must attain a critical size P_C before it can sustain itself and grow. Show that

$$\frac{dP}{dt} = kP(P - P_C)(P_M - P),$$

where k, P_C and P_M are constants with $P_C < P_M$, represents such a situation. Solve this equation with the initial population P_0, i.e., $P(0) = P_0$, where P_0 satisfies (1) $P_0 < P_C$, (2) $P_C < P_0 < P_M$, and (3) $P_0 > P_M$.

2.8 Taking into consideration that death rates are declining much faster than birth rates because of the rapid advances in modern medical science, Foerster, Mora and Amiot [5] postulated for the world population P that $P'/P =$ birth rate $-$ death rate $= kP^n$, where k and n are positive constants. This leads to the initial value problem $P' = kP^{n+1}$, $P(0) = P_0$. Show that $P(t) = \left(P_0^{-n} - nkt\right)^{-1/n}$ and hence $P(t) \to \infty$ as $t \to P_0^{-n}/(nk)$. Discuss the deficiencies of the model that leads to the unbounded population.

2.9 *Aquaculture* is the art of cultivating the natural products of water. A differential equation describing the growth of catfish, say, may be expressed as

$$w' = kw^\alpha, \tag{2.31}$$

where $w(t)$ is the weight of the fish at time t, and k and α are empirically determined growth constants. Biologists often refer to (2.31) as the *allometric equation*.
(1) Solve (2.31) when $\alpha \neq 1$.
(2) The solution obtained in part (1) grows large without bound, but in practice there is some limiting maximum weight w_M for the fish. This limiting weight is included in the differential equation describing growth by inserting a dimensionless variable h that can range between 0 and 1 and involves an empirically determined dimensionless parameter μ. Thus, we assume

$$w' = kw^\alpha h, \tag{2.32}$$

where $h = 1 - (w/w_M)^\mu$. Solve (2.32) when $\mu = 1 - \alpha$.
(3) The differential equation describing the total cost in dollars $c(t)$ of raising a fish for t months appears as

$$c' = a_1 + a_2 w', \tag{2.33}$$

where the constant a_1 represents the cost per month (due to costs such as interest, depreciation, and labor), and the constant a_2 multiplies the growth rate (because the amount of food consumed by the fish is approximately proportional to the growth rate). Solve (2.33) when $w(t)$ is as determined in part (2) with $\mu = 1 - \alpha$.

2.10 In a study of the rhesus monkey psychologists found that the stimulus s and response r satisfy the following differential equation:

$$\frac{dr}{ds} = k\frac{r^n}{s},$$

where k is a positive constant and n is a positive integer. Show that $r^{1-n} = k(1 - n)\ln(s) + c$.

2.11 In an attempt to cast the human eye-positioning mechanism into a mathematical model Cook and Stark [4] derived the following differential equation: for an isometric contraction of the muscle

$$T - \left(\frac{1.25T}{b+x'}\right)x' = K(x-L),$$

where T is muscle tension controlled by the nerve signal, L is the length of the elastic component, $b = 1500$, and $K = 6$. Find its general solution. (Write the equation as $x' = (A - Kx)/(C + Kx)$.)

2.12 In autocatalytic reactions, the rate of reaction is increased by the presence of the substance being produced. Suppose that in an autocatalytic reaction a chemical X is being converted into a chemical Y with x and y as the concentrations of X and Y, respectively. Assume that the rate of change of X is proportional to the product xy. If $x(0) = \alpha$, $y(0) = \beta$ and $\alpha - x(t) = y(t) - \beta$ are known, show that x satisfies the initial value problem

$$x' = -kx(\beta + \alpha - x), \quad x(0) = \alpha$$

where k is a positive constant. Find its solution and show that $\lim_{t\to\infty} x(t) = 0$, $\lim_{t\to\infty} y(t) = \alpha + \beta$.

2.13 The reaction between nitrogen oxide and oxygen to form nitrogen dioxide is given by the balanced chemical equation $2NO + O_2 = 2NO_2$. At high temperatures the dependence of the rate of this reaction on the concentrations of NO, O_2 and NO_2 is complicated. However, at $25\,°C$, the rate at which NO_2 is formed obeys the law of mass action and is given by the rate equation

$$\frac{dx}{dt} = k(a-x)^2\left(b - \frac{x}{2}\right),$$

where $x(t)$ denotes the concentration of NO_2 at time t, k is the rate constant, a is the initial concentration of NO, and b is the initial concentration of O_2. Find its general solution.

2.14 Find the time t needed to empty a cubical tank whose edge is 5 ft originally full of water and has an orifice of diameter 1 in at the bottom.

2.15 Find the time t needed to empty a hemispherical tank of radius 4 ft originally full of water and has an orifice of diameter 1 in at the bottom. (35 m 50 s.)

2.16 Find the time t needed to empty an inverted conical tank with the base radius 3 ft and height 6 ft originally full of water and has an orifice of diameter 1 in at the bottom.

2.17 Deoxyribonucleic acid (DNA) is the complex chemical in the cell nuclei, which is responsible for the translation and transcription of genetic information. All organisms from viruses to human have their own type of DNA. DNA is in the form of two very long molecular strands, which are wound helically around each other and attached through cross-links. If DNA is heated, these links break and the double strands dissociate, and single strands result. When the separated strands are allowed to cool, then they reform double strands. This process of reassociation is found experimentally to involve two different second-order chemical reactions: a very rapid reassociation for some fraction of the DNA and a much slower reassociation for the remaining DNA. Both reassociations obey the differential equation

$$y' = -ky^2,$$

where $y(t)$ is the concentration of the dissociated fractions and k is the same for both reactions.

(1) Show that if α_1 and α_2 are the initial concentrations of each fraction of DNA, and t_1 and t_2 are the half-lifes for the two reactions, then $\alpha_1 = (t_2/t_1)\alpha_2$.

(2) For the calf DNA it is found that $t_2/t_1 = 1000,000$. Show that this implies that initially one fraction of DNA is 100,000 times more concentrated than the remaining portion.

Using other biological knowledge, it can be concluded that the basic genetic information lies in the more concentrated segment and it is repeated in hundreds of thousands of copies. It is believed that this great number of copies may account for the great reliability of genetic transmittal and the slowness of evolution.

2.18 In considering a case of slow selection involving two genes, the following differential equation arises (see [6]):

$$\frac{dy}{dx} = \frac{y^2(1-y)(x^2-a^2)}{x^2(1-x)(y^2-b^2)},$$

where a and b are constants. Find its solution.

2.19 The equation

$$\frac{dy}{dt} = ay - \frac{a}{b}\left(1 + e^{c-dt}\right)y^2$$

arises in one model of the growth process of ownership of a consumer durable (see [9]). Solve this differential equation.

2.20 The Cobb–Douglas growth model, after Charles Wiggins Cobb (1875–1949) and Paul Howard Douglas (1892–1976), in economics simplifies to (see [7])

$$\frac{dy}{dt} = ae^{bt}y^\alpha,$$

where $y(t)$ is the aggregate input of capital stock at time t, and the constants a, b and α depend on the output elasticity of capital, the input of labor, the growth of labor and technology, and the ratio of full employment saving to the gross national product. Show that its solution satisfying the initial condition $y(0) = y_0$ can be written as

$$y(t) = \left(\frac{a}{b}(1-\alpha)(e^{bt} - 1) + y_0^{1-\alpha} \right)^{1/(1-\alpha)}.$$

2.21 A mathematical model describing the state of a learner $y(t)$ at time t while learning a specific task is governed by the differential equation (see [12])

$$\frac{dy}{dt} = \frac{2k}{\sqrt{m}}[y(1-y)]^{3/2},$$

where k and m are positive constants depending on the individual learner and the complexity of the task, respectively. Find its solution.

2.22 According to Josef Stefan's (1835–1893) law of radiation the rate of change of temperature from a body at absolute temperature θ is

$$\frac{d\theta}{dt} = k(\theta^4 - T^4),$$

where T is the absolute temperature of the surrounding medium. Show that the solution of this differential equation can be written as

$$\ln \left| \frac{\theta - T}{\theta + T} \right| - 2 \tan^{-1} \left(\frac{\theta}{T} \right) = 4T^3 kt + c \quad (c \text{ is a constant}).$$

(It can be shown that if $\theta - T$ is small compare to T, then the above equation approximates Newton's law of cooling).

2.23 The nonlinear differential equation

$$\left(\frac{dr}{dt} \right)^2 = \frac{2\mu}{r} + 2h,$$

where μ and h are nonnegative constants arises in the study of the two-body problem of celestial mechanics. Here r represents the distance between the two masses. Find its solution.

2.24 When a viscous fluid flows through a tube, fluid pressure along the tube drops to overcome the frictional resistance at the walls. As a result of this pressure gradient and the elasticity of the tube wall, the radius of the tube R decreases with distance along the axis. In cylindrical coordinates r, θ, z where r is radial and z is the distance along the axis of the tube, scientists have considered the flow across any tube section of radius $R(z)$. If $R(z)$ is constant, the tube is rigid. For a nonrigid tube, such as a blood vessel, complicated expressions (such as the equation of elasticity, the equation of continuity, and the Navier–Stokes equation, after Claude-Louis Navier (1785–

1836) and George Gabriel Stokes (1819–1903), for axially symmetric study, and the laminar flow of an incompressible fluid) have been combined to obtain the differential equation

$$\frac{dR}{dz} = -\frac{8\mu Q}{E\delta\pi\rho R^2},$$

where E is William Henry Young's (1863–1942) modulus, Q is the total mass flow across a tube section, ρ is the density, μ is the coefficient of viscosity, and δ is the wall thickness. Show that

$$R(z) = \left(-\frac{24\mu Qz}{E\delta\pi\rho}\right)^{1/3} + c,$$

where c is a constant. Thus, the radius decreases in proportion to the cube root of the distance z along the tube.

2.25 In a study on accumulation processes in the primitive solar nebula, the following differential equation occurs (see [3]):

$$\frac{dx}{dt} = \frac{ax^{5/6}}{(b - kt)^{3/2}},$$

where a, b and k are constants. Find its solution.

2.26 An equation of the form

$$y' = p(t)y^2 + q(t)y + r(t) \tag{2.34}$$

is called Count Jacopo Francesco Riccati's (1676–1754) equation. In general, this equation cannot be solved in terms of the known functions p, q and r.
(1) A body of mass M falling through a viscous medium encounters a resisting force proportional to the square of its instantaneous velocity. The motion of the body is governed by the initial value problem

$$M\frac{dv}{dt} = Mg - kv^2, \quad v(0) = v_0.$$

Show that its solution can be written as

$$\frac{v(t) + c}{v(t) - c} = \frac{v_0 + c}{v_0 - c}e^{2(\sqrt{gk/M})t},$$

where

$$c = \sqrt{\frac{Mg}{k}} \quad \left(v(t) = c\tanh\left(\sqrt{\frac{gk}{M}}\,t\right), \quad \text{if } v_0 = 0\right).$$

Hence, the limiting velocity of the falling body is $\sqrt{Mg/k}$.

(2) Show that if $y_1(t)$ is a solution of (2.34), then the transformation $y(t) = y_1(t) + 1/z(t)$ reduces it to the first-order differential equation

$$z' + [q(t) + 2y_1(t)p(t)]z = -p(t).$$

(3) In the propagation of a single act in a large population (an act that is performed at most once in the lifetime of an individual, such as suicide), the following Riccati differential equation occurs (see [8, 10]):

$$y' = (1 - y)[S(t) + dy],$$

where $y(t)$ is the fraction of the population who performed the act at time t, $S(t)$ is the external influence or stimulus and the product dy is the imitation component. Show that the solution of this Riccati differential equation in the case $S(t) = \frac{1}{t} - d$, is

$$y(t) = 1 + \frac{1/d}{t[\ln t + c]},$$

where c is a constant.

2.27 A man standing at $(0, 0)$ holds a rope of length a to which a weight is attached at the point $(a, 0)$. The man then walks along the positive y-axis, dragging the weight after him (see Fig. 2.6). Show that the differential equation of the path along which the weight moves is

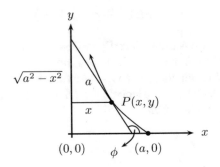

Fig. 2.6

$$\frac{dy}{dx} = -\frac{\sqrt{a^2 - x^2}}{x}$$

and its solution satisfying the initial condition $y(a) = 0$ can be written as

$$y(x) = a \ln\left(\frac{a + \sqrt{a^2 - x^2}}{x}\right) - \sqrt{a^2 - x^2}.$$

This curve is called *tractrix*, and is of considerable importance as the surface obtained by revolving it about y-axis is a model for Nikolai Ivanovich Lobachevsky's (1792–1856) version of non-Euclidean geometry.

2.28 The problem of determining the curve along which a particle slides (without friction) from the point O to the point A in the shortest time (see Fig. 2.7), where gravity is the only acting force is called the *brachistochrone problem* (quickest descent). The curve that solves this problem is a solution of the differential equation

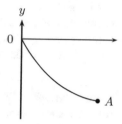

Fig. 2.7

$$y[1 + (y')^2] = c,$$

where c is a constant. Clearly, this equation is the same as

$$\left(\frac{y}{c - y}\right)^{1/2} dy = dx.$$

Use the substitution $y = c \sin^2(\phi)$ and the initial condition $y = 0$ when $x = 0$ to show that the parametric form of the required curve is

$$x = a[\theta - \sin(\theta)], \quad y = a[1 - \cos(\theta)],$$

where $a = c/2$ and $\theta = 2\phi$. These are the standard equations of the *cycloid* which is generated by a point on the circumference of a circle of radius a rolling along the x-axis.

2.29 Let $f(x, y, a) = 0$ and $g(x, y, b) = 0$ be the equations of two families of curves each dependent upon one parameter. When each member of the second family cuts each member of the first family according to a definite law, any curve of either of the families is said to be a *trajectory* of the family. The most important case is that in which curves of the families intersect at a constant angle. The *orthogonal trajectories* of a given family of curves are the curves that cut the given family at right angles. The slopes y_1' and y_2' of the tangents to the curves of the given family and to the sought-for orthogonal trajectories must at each point satisfy the orthogonality condition $y_1' y_2' = -1$. Orthogonal trajectories find applications in wave propagation such as light waves, sound waves and electromagnetic waves, in meteorological maps, fluid flow, and heat flow.

(1) Let $T(x, y)$ represents the temperature at the point (x, y). The curves given by $T(x, y) = c$ (constant) are called *isotherms*. The orthogonal trajectories are curves along which heat flows. Show that the isotherms of the heat flow $y^2 + 2xy - x^2 = a$ are $x^2 + 2xy - y^2 = b$. (Both the families are hyperbolas.)

(2) If the *streamlines* of the flow (paths of the particles of the fluid) in a channel are $\psi(x, y) = xy = a$, show that its orthogonal trajectories (*equipotential lines*) are $y^2 - x^2 = b$. (Both the families are hyperbolas.)

(3) In the electric field between two concentric cylinders the *equipotential lines* (curves of constant potential) are circles given by $U(x, y) = x^2 + y^2 = a$ (constant). Show that their orthogonal trajectories (curves of electric force) are the straight lines $y = bx$.

2.30 A given family of curves is said to be *self-orthogonal* if it has the property that its family of orthogonal trajectories is the same as that of the given family. Show that the family of parabolas $y^2 = 2cx + c^2$ is self-orthogonal.

2.31 A family of curves that intersect another family of curves at a constant angle θ ($\theta \neq \pi/2$) are called *isogonal* (oblique) trajectories of each other. If $y' = f(x, y)$ is the differential equation of the first family of curves, show that

$$y' = \frac{f(x, y) \pm \tan(\theta)}{1 \mp f(x, y) \tan(\theta)}$$

is the governing differential equation of the second isogonal family. Show that the family of curves that intersect the family $x^2 - y^2 = a$ at an angle $\pi/4$ is $x^2 \pm 2xy - y^2 = b$.

2.32 Recall that the angle ψ measured in the counterclockwise direction from the radius vector to the tangent line at a point satisfies the relation $\tan(\psi) = r d\theta/dr$, where r and θ are polar coordinates (see Fig. 2.8). Show that two polar curves $r = f_1(\theta)$ and $r = f_2(\theta)$ are orthogonal at a point of intersection if and only if $\tan(\psi_1) \tan(\psi_2) = -1$. In particular, verify that the orthogonal trajectories of $r = a[1 - \sin(\theta)]$ are $r = b[1 + \sin(\theta)]$.

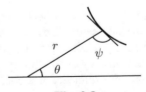

Fig. 2.8

2.33 Find the curves that satisfy each of the following geometric conditions:

(1) the tangent at a point is perpendicular to the line joining the point to the origin, $(x^2 + y^2 = c^2)$

(2) the part of the tangent cut off by the axes is bisected by the point of tangency, $(xy = c)$

(3) the projection of the normal on the x-axis is of constant length, $(y^2 = 2kx + c)$

(4) the projection of the tangent on the x-axis is of constant length, $(y = ce^{kx})$

(5) the tangent makes a constant angle with the radius vector, $(r = ce^{\theta \cot \alpha})$.

2.34 Four flies sit at the corners of a card table, facing inward. They start simultaneously walking at the same rate, each directing its motion steadily toward the fly on its right. Find the path of each fly. $(r = ae^{-\theta/\sqrt{2}}$, a is the side).

2.35 It is observed that a bug moves in a plane containing two lights in such a way that the intensities of the lights on its two sides are the same. Show that the equation of the path is $k_1 \cos(\theta_1) - k_2 \cos(\theta_2) = c$, where k_1 and k_2 are the strengths of the lights, and θ_1 and θ_2 are interior angles at the lights in the triangle whose vertices are the bug and the lights. Show that the bug, if approaching, will enter the first or second light according as $c < k_1 - k_2$, or $c > k_1 - k_2$.

2.36 An equation describing the process of fluid separation from an airfoil surface is

$$y\frac{d^2 y}{d\eta^2} + \nu \left(\frac{dy}{d\eta}\right)^2 = 0, \tag{2.35}$$

where $\nu = (1 - \mu)/\mu$. In (2.35) set $p = dy/d\eta$ to obtain the first-order equation

$$yp\frac{dp}{dy} = -\nu p^2.$$

Hence, show that the solution of (2.35) is $y(\eta) = [(A\eta + B)/\mu]^\mu$.

2.37 In certain cases, a steady-state diffusion process can be represented by the differential equation

$$c\frac{d^2 \phi}{dx^2} - v\frac{d\phi}{dx} = 0,$$

where c is the diffusion constant, and v is the velocity of the process. Show that $\phi(x) = A + Be^{vx/c}$, where A and B are arbitrary constants.

2.38 Stoke's law states that for spherical droplets falling in motionless air and having a diameter $D < 0.762$ cm, the acceleration due to gravity is opposed by an amount proportional to the velocity of the raindrop, specifically by an amount equal to $(0.329 \times 10^{-5}/D^2)dy/dt$. Thus, we can write the differential equation for the rain drop as

$$\frac{d^2y}{dt^2} = 980 - \frac{0.329 \times 10^{-5}}{D^2}\frac{dy}{dt}.$$

Find the solution of this differential equation. The velocity which is an upper bound to how fast the body can go at any time during the fall is called the *terminal velocity*, at this velocity clearly the acceleration $d^2y/dt^2 = 0$. Thus, the terminal velocity of the falling droplets is $dy/dt = 980 \times 10^5 \times D^2/0.329$ cm/s.

2.39 For spherical raindrops falling in still air and having diameter $D > 0.12$ cm, the acceleration due to gravity is opposed by an amount proportional to the square of its velocity, specifically an amount equal to $(0.00046/D)(dy/dt)^2$. This leads to the differential equation

$$\frac{d^2y}{dt^2} = 980 - \frac{0.00046}{D}\left(\frac{dy}{dt}\right)^2.$$

Find the solution of this differential equation, and show that the terminal velocity is $dy/dt = (980D/0.00046)^{1/2}$ cm/s.

2.40 The following differential equation arises in the design of a magnetohydrodynamic power generator

$$((k+1)y')' = 0,$$

where k is a positive constant. Find its solution.

2.41 In the study of astrodynamics, the following differential equation occurs

$$y'' = [1 + (y')^2]f(x, y, y').$$

Show that its solution in the particular case $f(x, y, y') = y'$ can be written as $y(x) = \sin^{-1}(Ae^x) + B$, where A and B are constants.

2.42 Alexis Claude Clairaut's (1713–1765) equation

$$y = ty' + f(y') \tag{2.36}$$

can be solved directly. Indeed differentiating both sides with respect to t, we obtain

$$y' = y' + ty'' + f'(y')y''$$

and hence

$$(t + f'(y'))y'' = 0.$$

In the above equation if $y'' = 0$, then $y' = c$ and the general solution of (2.36) is a collection of straight lines $y = ct + f(c)$. However, if $t + f'(y') = 0$ then $t = -f'(y')$ and Eq. (2.36) can be written as $y = f(y') - y'f'(y')$. Thus, if we let $y' = s$

then we obtain the parameterized curve $t = -f'(s)$, $y = f(s) - sf'(s)$ which is a *singular solution* of (2.36), i.e., it cannot be obtained from the general solution $y = ct + f(c)$.

(1) The differential equation

$$x \left(\frac{dx}{dy} \right)^2 + 2y \frac{dx}{dy} = x, \tag{2.37}$$

where $x = x(y)$ occurs in the steady of optics. This equation describes the type of plane curve that will reflect all incoming light rays to the same point. Show that the substitution $w = x^2$ transforms (2.37) to a Clairaut's equation, and its solution represents a family of parabolas $x^2 = cy + c^2/4$.

(2) Show that the straight lines whose segment between the positive x-axis and y-axis has constant length 1 are solutions of the Clairaut equation $y = xy' - y'/\sqrt{1 + y'^2}$, whose singular solution is the *astroid* $x^{2/3} + y^{2/3} = 1$.

2.43 Let a cable of uniformly distributed weight be suspended between two supports (see Fig. 2.9). Suppose the origin of the coordinate system is at the lowest point of the cable and any point on the cable is denoted as (x, y). The differential equation relating x and y is

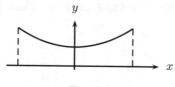

Fig. 2.9

$$a \frac{d^2 y}{dx^2} = \left[1 + \left(\frac{dy}{dx} \right)^2 \right]^{1/2}, \tag{2.38}$$

where a is the ratio of the tension in the cable at the lowest point to the density of the cable. Show that the solution of (2.38) subject to the conditions $y(0) = a$, $y'(0) = 0$ can be written as $y(x) = a \cosh(x/a)$. The graph of this curve is called a *catenary* which is the Latin word for chain (catena). Catenaries arise in a number of other physical problems. The St. Louis arch is in the shape of a catenary. (Note that the substitution $v = y'$ reduces (2.38) to a equation of first order $av' = [1 + v^2]^{1/2}$ in which the variables can be separated.)

2.44 A *curve of pursuit* is a path generated by a point P that always moves in the direction of a second point Q constrained to move along a prescribed path. As an interesting example assume that a cat is chasing a rat as shown in the Fig. 2.10. Initially, the rate is at the origin and the cat is at $(c, 0)$. The rat and the cat run at constant speeds v_r and v_c, where $v_c > v_r$. The rat runs along the y-axis and the cat always runs directly toward the rat.

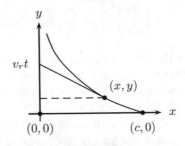

Fig. 2.10

(1) Show that the points (x, y) on the path of the cat satisfy

$$x \frac{dy}{dx} = y - v_r t. \tag{2.39}$$

(2) Differentiate (2.39) with respect to x and use the fact that $\left(\frac{dx}{dt}\right)^2 + \left(\frac{dy}{dt}\right)^2 = v_c^2$, to obtain

$$x \frac{d^2 y}{dx^2} = \frac{v_r}{v_c} \sqrt{1 + \left(\frac{dy}{dx}\right)^2}. \tag{2.40}$$

(3) Solve (2.40) with the required initial conditions $y(c) = 0$, $\left. \frac{dy}{dx} \right|_{x=c} = 0$ to show that

$$y(x) = \frac{c}{2} \left[\frac{(x/c)^{1+v_r/v_c}}{1 + v_r/v_c} - \frac{(x/c)^{1-v_r/v_c}}{1 - v_r/v_c} + \frac{2 v_r v_c}{v_c^2 - v_r^2} \right].$$

(4) Find the time when cat will catch the rat. $(c v_c / (v_c^2 - v_r^2))$.

2.45 The modeling the shape of a *tsunami* (from the Japanese harbor, "tsu", and wave "nami") leads to the differential equation (see [13])

$$\frac{dW}{dx} = W \sqrt{4 - 2W},$$

where $W(x) > 0$ is the height of the wave expressed as a function of its position relative to a point offshore. Find all solutions of this differential equation. $(0, 2, 2\text{sech}^2 (x - c))$

2.46 The initial value problem

$$(1 + \epsilon\theta)\frac{d\theta}{d\tau} + \theta = 0, \quad \theta(0) = 1$$

occurs in cooling of a lumped system. In this problem, we let ϵ as the perturbing parameter and seek the solution in the form $\theta(\tau) = \sum_{n=0}^{\infty} \epsilon^n \theta_n(\tau)$. Show that

$$\theta(\tau) = e^{-\tau} + \epsilon\left(e^{-\tau} - e^{-2\tau}\right) + \epsilon^2\left(e^{-\tau} - 2e^{-2\tau} + \frac{3}{2}e^{-3\tau}\right) + O(\epsilon^3).$$

Compare this approximation with the exact solution $\ln(\theta) + \epsilon(\theta - 1) = -\tau$.

2.47 The initial value problem

$$\frac{d\theta}{d\tau} + \theta + \epsilon\theta^4 = 0, \quad \theta(0) = 1$$

occurs in cooling of a lumped system. In this problem we let ϵ as the perturbing parameter and seek the solution in the form $\theta(\tau) = \sum_{n=0}^{\infty} \epsilon^n \theta_n(\tau)$. Show that

$$\theta(\tau) = e^{-\tau} + \epsilon\frac{1}{3}\left(e^{-4\tau} - e^{-\tau}\right) + \epsilon^2\frac{2}{9}\left(e^{-\tau} - 2e^{-4\tau} + e^{-7\tau}\right) + O(\epsilon^3).$$

Compare this approximation with the exact solution $\frac{1}{3}\ln((1 + \epsilon\theta^3)/[(1 + \epsilon)\theta^3]) = \tau$.

References

1. R.P. Agarwal, D. O'Regan, *An Introduction to Ordinary Differential Equations* (Springer, New York, 2008)
2. R.P. Agarwal, D. O'Regan, *Ordinary and Partial Differential Equations with Special Functions, Fourier Series and Boundary Value Problems* (Springer, New York, 2009)
3. A.G.W. Cameron, Icarus **18**, 407–450 (1973)
4. G. Cook, L. Stark, Bull. Math. Biophys. **29**, 153–174 (1967)
5. H.V. Foerster, P.M. Mora, L.W. Amiot, Science **12**, 1291–1295 (1960)
6. J.B.S. Haldane, *Causes of Evolution* (Cornell University Press, Ithaca, 1966)
7. D. Hamberg, *Models of Economic Growth* (Harper & Row, New York, 1971)
8. J.Z. Hearon, Bull. Math. Biophys. **17**, 7–13 (1962)
9. W.J. Oomens, *The Demand for Consumer Durables* (Tilburg University Press, The Netherlands, 1976)
10. A. Rapoport, Bull. Math. Biophys. **14**, 159–169 (1952)
11. J.J. Stoker, *Water Waves* (Interscience, New York, 1957)
12. L.L. Thurstone, J. Gen. Psychol. **3**, 469–493 (1930)
13. D.G. Zill, *A First Course in Differential Equations with Modeling Applications* (Cengage Learning, Boston, 2016)

Chapter 3
Second- and Higher Order Differential Equations

Consider the second-order linear differential equation

$$p_0(t)y'' + p_1(t)y' + p_2(t)y = r(t), \tag{3.1}$$

where the functions p_0, p_1, p_2, and r are continuous in an open interval $I = (\alpha, \beta)$, and $p_0(t) \neq 0$ for all $t \in I$ [1, 2]. For (3.1) the corresponding homogeneous differential equation

$$p_0(t)y'' + p_1(t)y' + p_2(t)y = 0 \tag{3.2}$$

plays an important role. We summarize some of the well-known results for the Eqs. (3.1) and (3.2) in the following theorems.

Theorem 3.1 *There exist exactly two solutions y_1 and y_2 of (3.2) which are linearly independent (essentially different) in I, i.e., there does not exist a constant c such that $y_1(t) = cy_2(t)$ for all $t \in I$.*

Theorem 3.2 *Two solutions y_1 and y_2 of (3.2) are linearly independent in I if and only if the value of the Wronskian (Jósef Maria Hoëne Wronski (1776–1853)) of y_1 and y_2 defined by*

$$W(t) = \begin{vmatrix} y_1(t) & y_2(t) \\ y_1'(t) & y_2'(t) \end{vmatrix} \tag{3.3}$$

is different from zero for some t in I.

Theorem 3.3 *For the Wronskian defined in (3.3) the following Niels Henrik Abel's (1802–1829) identity holds:*

© Springer Nature Switzerland AG 2019
R. P. Agarwal et al., *500 Examples and Problems of Applied Differential Equations*, Problem Books in Mathematics,
https://doi.org/10.1007/978-3-030-26384-3_3

$$W(t) = W(t_0) \exp\left(-\int_{t_0}^{t} \frac{p_1(s)}{p_0(s)} ds\right), \quad t_0 \in I. \tag{3.4}$$

Thus, the Wronskian is zero at some $t_0 \in I$, if and only if it is zero for all $t \in I$.

Theorem 3.4 *If y_1 and y_2 are solutions of (3.2) and c_1 and c_2 are arbitrary constants, then $(c_1y_1 + c_2y_2)$ is also a solution of (3.2). Further, if y_1 and y_2 are linearly independent, then any solution y of (3.2) can be written as $y = \bar{c}_1 y_1 + \bar{c}_2 y_2$, where $\bar{c}_1$ and $\bar{c}_2$ are suitable constants.*

In contrast with first-order linear differential equations, there is no general method to find a solution of (3.2). However, if one solution y_1 of (3.2) is known, then its second solution y_2 interms of known functions can be written as

$$y_2(t) = y_1(t) \int^{t} \frac{1}{y_1^2(s)} \exp\left(-\int^{s} \frac{p_1(\tau)}{p_0(\tau)} d\tau\right) ds. \tag{3.5}$$

Moreover, if the differential equation (3.2) is with constant coefficients, i.e.,

$$p_0 y'' + p_1 y' + p_2 y = 0 \tag{3.6}$$

then its solution can be obtained rather easily. For this based on our experience, we assume that $y = e^{mt}$ is a solution of (3.6). Thus, it follows that

$$(p_0 m^2 + p_1 m + p_2)e^{mt} = 0,$$

or

$$p_0 m^2 + p_1 m + p_2 = 0. \tag{3.7}$$

Equation (3.7) is called the characteristic equation of the differential equation (3.6). Its significance lies in the fact that if m is a root of (3.7), then e^{mt} is a solution of (3.6). Since Eq. (3.6) has two roots, say, m_1 and m_2, depending on the nature of these roots, three different forms of the general solution occurs:
Case 1. $m_1 \neq m_2$ (real), $y(t) = Ae^{m_1 t} + Be^{m_2 t}$,
Case 2. $m_1 = m_2 = m$, $y(t) = (A + Bt)e^{mt}$,
Case 3. $m_1 = p + iq$, $m_2 = p - iq$ ($i = \sqrt{-1}$, $q \neq 0$), $y(t) = e^{pt}(A \cos qt + B \sin qt)$, where A and B are arbitrary constants.

Theorem 3.5 *For the initial value problem consisting of the differential equation (3.1) and the initial conditions*

$$y(t_0) = y_0, \quad y'(t_0) = y_1, \quad t_0 \in I \tag{3.8}$$

there exists a unique solution y in I. Thus, in particular for the differential equation (3.2) the only solution y satisfying the initial conditions

$$y(t_0) = 0, \quad y'(t_0) = 0, \quad t_0 \in I \tag{3.9}$$

is the trivial solution, i.e., $y(t) = 0$ for all $t \in I$.

Theorem 3.6 *If u is a solution of (3.2) and v is any solution of (3.1), then $y = u + v$ is also a solution of (3.1). Thus, if y_1 and y_2 are linearly independent solutions of (3.2), and v is a solution of (3.1) satisfying the initial conditions (3.9), then the general solution y of (3.1) can be written as*

$$y(t) = c_1 y_1(t) + c_2 y_2(t) + v(t), \quad t \in I. \tag{3.10}$$

Moreover, the solution v in terms of the known functions can be written as

$$v(t) = \int_{t_0}^{t} \begin{vmatrix} y_1(s) & y_2(s) \\ y_1(t) & y_2(t) \end{vmatrix} \Bigg/ \begin{vmatrix} y_1(s) & y_2(s) \\ y_1'(s) & y_2'(s) \end{vmatrix} \frac{r(s)}{p_0(s)} ds. \tag{3.11}$$

Example 3.1 Consider a simple mechanical system consisting of a mass M connected to a spring (see Fig. 3.1). If the mass is displaced from its equilibrium position by a distance x, then by Robert Hooke's (1635–1703) law a force $F = -kx$ is exerted on the mass by the spring, where k is a positive constant. Now, by Newton's second law $F = M \times$ acceleration $= M\frac{d^2x}{dt^2}$. Thus, it follows that

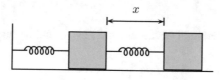

Fig. 3.1

$$M\frac{d^2x}{dt^2} = -kx, \tag{3.12}$$

or

$$x'' + \omega^2 x = 0, \quad \omega^2 = k/M. \tag{3.13}$$

Any mechanical system described by Eq. (3.12) is called a *harmonic oscillator*.

Equation (3.13) has the characteristic equation $m^2 + \omega^2 = 0$ with complex roots given by $m = \pm i\omega$, and hence its solution can be written as

$$x(t) = A\cos(\omega t) + B\sin(\omega t) \tag{3.14}$$

$$= \sqrt{A^2 + B^2} \left[\frac{A}{\sqrt{A^2 + B^2}} \cos(\omega t) + \frac{B}{\sqrt{A^2 + B^2}} \sin(\omega t) \right]$$

$$= \sqrt{A^2 + B^2}\,[\sin(\theta)\cos(\omega t) + \cos(\theta)\sin(\omega t)], \quad \tan(\theta) = A/B$$

$$= C\sin(\omega t + \theta), \quad C = \sqrt{A^2 + B^2}, \quad \theta = \tan^{-1}(A/B). \tag{3.15}$$

Since for all $t \geq 0$, $-1 \leq \sin(\omega t + \theta) \leq 1$ it follows that $-C \leq x(t) \leq C$. The constant C is called the *amplitude* of the oscillation. The argument $(\omega t + \theta)$ is called the *phase angle*. When $t = 0$, the phase angle is θ. Hence, θ is called the *initial phase*. Now, since

$$x\left(t + \frac{2\pi}{\omega}\right) = C\sin(\omega t + 2\pi + \theta) = C\sin(\omega t + \theta),$$

the motion described by x in (3.15) is periodic with period $T = 2\pi/\omega$. Further, the *natural frequency* f of the motion, i.e., the number of complete oscillations per unit time is given by $f = \omega/2\pi$.

If the initial displacement and the initial velocity are given as

$$x(0) = x_0, \quad v(0) = x'(0) = x_1 \tag{3.16}$$

then from (3.14) it follows that $x_0 = A$ and $x_1 = B\omega$. Hence,

$$C = \sqrt{x_0^2 + \frac{x_1^2}{\omega^2}} \quad \text{and} \quad \theta = \tan^{-1}\left(\frac{x_0\omega}{x_1}\right). \tag{3.17}$$

If we multiply the differential equation (3.12) by $\frac{dx}{dt}$ and integrate, we obtain

$$\frac{1}{2}M\left(\frac{dx}{dt}\right)^2 + \frac{1}{2}kx^2 = E. \tag{3.18}$$

The first term in Eq. (3.18) is the kinetic energy of the object, while the second term represents the potential energy. For example, in the spring–mass system, the second term represents the energy stored in the spring. Equation (3.18) is in fact a statement of conservation of energy, i.e., $E =$ total energy.

Example 3.2 (*Paul Anthony Samuelson's (1915–2009) investment model*) Let $C(t)$ and $I(t)$ represent the capital and the investment at time t. We assume that
(i) the investment gives the rate of increase of capital so that

$$\frac{dC}{dt} = I, \tag{3.19}$$

(ii) the deficiency of the capital below a certain equilibrium level, say, C_e leads to an acceleration of the rate of investment proportional to this deficiency, and a surplus of the capital above C_e level leads to a deceleration of the rate of investment, again proportional to the surplus, so that

$$\frac{dI}{dt} = -m[C(t) - C_e].$$ (3.20)

Let $x(t) = C(t) - C_e$, so that (3.19) and (3.20) can be written as

$$\frac{dx}{dt} = I, \quad \frac{dI}{dt} = -mx$$

and hence

$$\frac{d^2x}{dt^2} = \frac{dI}{dt} = -mx,$$

or

$$\frac{d^2x}{dt^2} + mx = 0.$$ (3.21)

Clearly, Eq. (3.21) is the same as (3.13). If $x(0) = x_0$, $x'(0) = I(0) = 0$ then from (3.21) it is immediate that

$$x(t) = x_0 \cos(\sqrt{m}t) \quad \text{and} \quad I(t) = -x_0\sqrt{m} \sin(\sqrt{m}t)$$

so that both $x(t)$ and $I(t)$ oscillate with a time period $2\pi/\sqrt{m}$.

Example 3.3 Consider an elastic string of length L whose one end is fixed and to the other end a particle of mass M is attached. Let the string be stretched to a distance a and then released. Let at time t, $x(t)$ be the extension, then the equation of motion of the particle is (see Fig. 3.2)

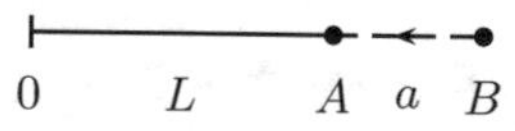

Fig. 3.2

$$M\frac{d^2x}{dt^2} = -\frac{\mu}{L}x = -kx,$$ (3.22)

where $k > 0$ is the elastic constant. Again Eq. (3.22) is the same as (3.12).

Now let the particle move in a resisting medium with resistance proportional to the velocity x'. Then, Eq. (3.22) becomes

$$Mx'' + ax' + kx = 0,$$ (3.23)

where $a > 0$ is called the *damping constant*. The term ax' is called the *damping term* and (3.23) is the equation of free *damped* motion.

In Eq. (3.23) for simplicity of the arguments, we let $2\lambda = a/M$ and $\omega^2 = k/M$ so that the characteristic equation can be written as $m^2 + 2\lambda m + \omega^2 = 0$. The corresponding roots are then $m_1 = -\lambda + \sqrt{\lambda^2 - \omega^2}$ and $m_2 = -\lambda - \sqrt{\lambda^2 - \omega^2}$. We now need to consider the following three cases:

Case 1. $\lambda^2 - \omega^2 > 0$, i.e., $a^2 > 4Mk$ and hence, the damping constant a is large compared to the spring constant k. Since the roots m_1 and m_2 are real, distinct, and negative the solution $x(t) = Ae^{(-\lambda + \sqrt{\lambda^2 - \omega^2})t} + Be^{(-\lambda - \sqrt{\lambda^2 - \omega^2})t} \to 0$ as $t \to \infty$. The motion in this case is called *overdamped*.

Case 2. $\lambda^2 - \omega^2 = 0$, i.e., $a^2 = 4Mk$. Since the roots m_1 and m_2 are real and equal, the solution $x(t) = (A + Bt)e^{-(a/2M)t} \to 0$ as $t \to \infty$. The motion in this case is said to be *critically damped*.

Case 3. $\lambda^2 - \omega^2 < 0$, i.e., $a^2 < 4Mk$ and hence the damping constant a is small compared to the spring constant k. Since the roots m_1 and m_2 are complex, the solution $x(t) = e^{-\lambda t}[A\cos(\sqrt{\omega^2 - \lambda^2}t) + B\sin(\sqrt{\omega^2 - \lambda^2}t)]$ always oscillates and tends to zero (damps out) as $t \to \infty$. The motion in this case is called *underdamped*. As in Example 3.1 the solution $x(t)$ in this case can be written as $x(t) = Ce^{-\lambda t}\sin(\sqrt{\omega^2 - \lambda^2}t + \theta)$, where $C = \sqrt{A^2 + B^2}$ and $\theta = \tan^{-1}(A/B)$. The coefficient $Ce^{-\lambda t}$ is sometimes called the damped amplitude of vibrations. Now since $|\sin(\sqrt{\omega^2 - \lambda^2}t + \theta)| \le 1$ for all t the graph of $x(t)$ lies between the graphs of $Ce^{-\lambda t}$ and $-Ce^{-\lambda t}$. Thus if t_1 is a maximum (minimum) point of $x(t)$, i.e., $x(t_1) = Ce^{-\lambda t_1}(-Ce^{-\lambda t_1})$ then the next maximum (minimum) point of $x(t)$ will be at $t_2 = t_1 + \frac{2\pi}{\sqrt{\omega^2 - \lambda^2}}$. Hence, the period of the oscillatory motion is

$$\frac{2\pi}{\sqrt{\omega^2 - \lambda^2}} = \frac{4M\pi}{\sqrt{4kM - a^2}} = 2\pi\sqrt{\frac{M}{k}}\frac{1}{\sqrt{1 - \frac{a^2}{4kM}}},$$

which is an increasing function of a, i.e., if resistance is increased, the particle will oscillate more slowly. In fact, as a increases toward the critical damped case $a = 2\sqrt{kM}$, then $\left(1 - \frac{a^2}{4kM}\right)$ approaches zero, and so the period grows without bound. Finally, we note that the ratio between two consecutive maximum (minimum) displacements $x(t_1)$ and $x(t_2)$ is a constant given by

$$\frac{x(t_1)}{x(t_2)} = e^{2\pi\lambda/\sqrt{\omega^2 - \lambda^2}}.$$

The number $\delta = \ln\left(x(t_1)/x(t_2)\right) = 2\pi\lambda/\sqrt{\omega^2 - \lambda^2}$ is called the *logarithmic decrement*.

Example 3.4 (*Modified Samuelson's investment model*) If in Example 3.2, the rate of investment slowed not only by excess capital, but it is also slowed by a high investment level, then Eq. (3.19) remains the same, whereas (3.20) is modified to

$$\frac{dI}{dt} = -m[C(t) - C_e] - nI. \tag{3.24}$$

Again the substitution $x(t) = C(t) - C_e$ in (3.19) and (3.24) leads to the equation

$$x'' + nx' + mx = 0,$$

which is exactly the same as (3.23).

Equation (3.23) also occurs in modeling several other real-world applications. However, in many situations, instead of linear damping term ax' in (3.23) some form of nonlinear damping leads to better results. For example, the equation

$$Mx'' + ax'|x'| + kx = 0,$$

which implies that the damping force is always directed opposite to the direction of the motion, gives more accurate results when an airplane tires from wet snow or slush. Stewart [19] points out that drag from only four inches of slush was enough to cause the 1958 crash during take off of the plane carrying the Manchester United soccer team. Large airplanes are now allowed to take off and land in no more than one-half inch of wet snow or slush.

Example 3.5 If an external force $F(t)$ is applied to the spring–mass system discussed in Example 3.1, then the resulting differential equation becomes nonhomogeneous. Here we shall discuss the undamped motion governed by the differential equation

$$M\frac{d^2x}{dt^2} + kx = F(t), \tag{3.25}$$

or

$$x'' + \omega^2 x = f(t), \quad \omega^2 = k/M \quad \text{and} \quad f(t) = F(t)/M. \tag{3.26}$$

Clearly, in view of (3.10) and (3.11) the general solution of (3.26) can be written as

$$
\begin{aligned}
x(t) &= A\cos(\omega t) + B\sin(\omega t) + \int_{t_0}^{t} \frac{\begin{vmatrix} \cos(\omega s) & \sin(\omega s) \\ \cos(\omega t) & \sin(\omega t) \end{vmatrix}}{\begin{vmatrix} \cos(\omega s) & \sin(\omega s) \\ -\omega\sin(\omega s) & \omega\cos(\omega s) \end{vmatrix}} f(s)\,ds \\
&= A\cos(\omega t) + B\sin(\omega t) + \frac{1}{\omega}\int_{t_0}^{t} \sin[\omega(t - s)]f(s)\,ds.
\end{aligned}
\tag{3.27}
$$

In particular, if $F(t) = P$ (constant), then (3.27) reduces to

$$
\begin{aligned}
x(t) &= A\cos(\omega t) + B\sin(\omega t) + \frac{1}{\omega}\frac{P}{M}\int_{t_0}^{t} \sin[\omega(t - s)]ds \\
&= A\cos(\omega t) + B\sin(\omega t) + \frac{P}{k}(1 - \cos[\omega(t - t_0)]).
\end{aligned}
\tag{3.28}
$$

Now we assume that the system is initially at rest in its equilibrium position before $F(t)$ is applied. This means we impose the initial conditions

$$x(0) = 0, \quad x'(0) = 0. \tag{3.29}$$

In this case solutions (3.27) and (3.28), respectively reduce to

$$x(t) = \frac{1}{\omega} \int_0^t \sin[\omega(t - s)] f(s) ds \tag{3.30}$$

and

$$x(t) = \frac{P}{k}(1 - \cos(\omega t)). \tag{3.31}$$

Thus, when $F(t) = P$ the mass oscillates at the natural frequency of the spring–mass system between the points 0 and $2P/k$.

Next, we consider the case when the forcing function is a simple periodic function $F(t) = P \cos \omega_0 t$ (*sinusoidal force*) where $\omega \neq \omega_0$. Hence, we need to solve the initial value problem

$$x'' + \omega^2 x = \frac{P}{M} \cos(\omega_0 t), \quad x(0) = 0, \quad x'(0) = 0. \tag{3.32}$$

In view of (3.30) the solution of (3.32) can be written as

$$\begin{aligned} x(t) &= \frac{1}{\omega} \int_0^t \sin[\omega(t - s)] \frac{P}{M} \cos(\omega_0 s) ds \\ &= \frac{1}{\omega^2} \frac{P}{M} [\cos(\omega_0 t) - \cos(\omega t)] + \frac{\omega_0^2}{\omega^2} x(t) \end{aligned}$$

and hence

$$x(t) = \frac{P}{M(\omega^2 - \omega_0^2)} [\cos(\omega_0 t) - \cos(\omega t)], \tag{3.33}$$

i.e., the motion consists of the superposition of two modes of vibrations—the *natural mode* at frequency ω and the *forced mode* at frequency ω_0.

Now in view of the trigonometric identity

$$\cos(A) - \cos(B) = 2 \sin\left(\frac{B - A}{2}\right) \sin\left(\frac{A + B}{2}\right)$$

the solution (3.33) can be written as

$$x(t) = \frac{2P}{M(\omega^2 - \omega_0^2)} \sin\left(\frac{\omega - \omega_0 t}{2}\right) \sin\left(\frac{\omega + \omega_0 t}{2}\right). \qquad (3.34)$$

Thus if the forcing frequency ω_0 is close to the natural frequency ω, i.e., $0 < |\omega - \omega_0| \ll 1$, then the period of the sine wave $\sin\left(\frac{(\omega - \omega_0)t}{2}\right)$ is large compared with the period of the sine wave $\sin\left(\frac{(\omega + \omega_0)t}{2}\right)$. Hence, the motion described by (3.34) can be visualized as a rapid oscillation with angular frequency $(\omega + \omega_0)/2$, but with a slowly varying sinusoidal amplitude known as the *envelope* (see Fig. 3.3). An oscillatory motion possessing a periodic variation of amplitude describes the phenomenon of *beats*. In the case of two tuning forks, the variation in the amplitude of the sound is easily heard. In applications to electric circuits, the slowly varying amplitude is usually called an amplitude modulation.

The solution for the case $\omega_0 = \omega$ can be obtained from (3.33) by using Guillaume Francois Antoine de L'Hôpital's (1661–1704) rule. Indeed, we have

$$x(t) = \lim_{\omega_0 \to \omega} \left[\frac{P}{M(\omega^2 - \omega_0^2)} [\cos(\omega_0 t) - \cos(\omega t)] \right] = \frac{P}{2M\omega} t \sin(\omega t). \qquad (3.35)$$

The same solution can be obtained directly from (3.30) as follows:

$$\begin{aligned}
x(t) &= \frac{1}{\omega} \int_0^t \sin[\omega(t - s)] \frac{P}{M} \cos(\omega s) ds \\
&= \frac{P}{2\omega M} \int_0^t [\sin(\omega t) + \sin(\omega t - 2\omega s)] ds \\
&= \frac{P}{2\omega M} \left[t \sin(\omega t) + \frac{\cos \omega[(t - 2s)]}{2\omega} \bigg|_0^t \right] = \frac{P}{2\omega M} t \sin(\omega t).
\end{aligned}$$

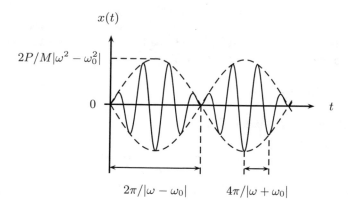

$x(t)$

$2P/M|\omega^2 - \omega_0^2|$

0

t

$2\pi/|\omega - \omega_0|$ $\qquad$ $4\pi/|\omega + \omega_0|$

Fig. 3.3

A sketch of this solution is depicted in Fig. 3.4. Thus, when $\omega_0 = \omega$ the forced response is not a periodic function. It is a growing oscillatory function and the amplitude becomes unbounded as $t \to \infty$. We refer to this phenomenon as *resonance*, and the value $\omega_0 = \omega$ is called the *resonant circular frequency*. Resonance occurs to situations such as vibrations in an aircraft wing, a skyscraper, a glass, or a bridge. The excitation that leads to the vibrations of these structures include strong winds, unbalanced rotating devices, and moving vehicles. Naturally, if the amplitude becomes too large, the system falls apart. This situation actually cased the Tacoma Narrows bridge at Puget Sound (State of Washington) to collapse on November 7, 1940 only 4 months after its grand opening. In designing such structures, it is very important to make natural frequency of the structure different (if possible) from the frequency of any probable forcing function. Resonance also suggest the reason that soldiers should not march in steps across bridges.

$x(t)$

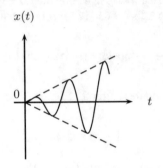

Fig. 3.4

Example 3.6 Consider the nonhomogeneous differential equation with the damping term of the type

$$Mx'' + ax' + kx = F(t). \tag{3.36}$$

From Theorem 3.6 its general solution x can be written as $x = u + v$, where u is the general solution of the homogeneous equation

$$x'' + 2\lambda x' + \omega^2 x = 0, \quad 2\lambda = a/M, \quad \omega^2 = k/M \tag{3.37}$$

and v is any solution of the equation

$$x'' + 2\lambda x' + \omega^2 x = F(t)/M. \tag{3.38}$$

In Example 3.3 we have seen that u assumes the form

$$u(t) = \begin{cases} e^{-\lambda t}\left(Ae^{\alpha t} + Be^{-\alpha t}\right), & \alpha = \sqrt{\lambda^2 - \omega^2}, \quad \lambda^2 - \omega^2 > 0 \\ e^{-\lambda t}(A + Bt), & \lambda^2 - \omega^2 = 0 \\ e^{-\lambda t}[A\cos(\mu t) + B\sin(\mu t)], & \mu = \sqrt{\omega^2 - \lambda^2}, \quad \omega^2 - \lambda^2 > 0. \end{cases}$$

Clearly, $u(t) \to 0$ as $t \to \infty$. Thus, u contributes only initial effects to the motion of the system. For this reason, in physical applications u is often referred to as a *transient solution*. Once the initial effects diminish the particular solution v dominates the response of the system. We call v as the *steady-state solution* (only that part of v which does not vanish as $t \to \infty$). Now since the transient solution does not have long term effects, it suffices to find the steady-state solution v of the differential equation (3.38) ignoring completely the initial conditions.

Again, for simplicity we consider the forcing function $F(t) = P \cos(\omega_0 t)$ (sinusoidal force), i.e., we shall solve the differential equation

$$x'' + 2\lambda x' + \omega^2 x = \frac{P}{M} \cos(\omega_0 t). \tag{3.39}$$

A solution of (3.39), which does not tend to zero and is independent of the sign of $(\lambda^2 - \omega^2)$ can be obtained by using (3.11), and after several manipulations appears as

$$v(t) = \frac{P}{(k - M\omega_0^2)^2 + \omega_0^2 a^2} \left[(k - M\omega_0^2) \cos(\omega_0 t) + \omega_0 a \sin(\omega_0 t) \right]. \tag{3.40}$$

(The same solution can be obtained rather easily by employing the method of undetermined coefficients, or the method of operators.)

Clearly, the steady-state solution v obtained in (3.40) is the sum of two sinusoids of the same frequency, and hence it can be written as

$$v(t) = R \cos(\omega_0 t - \phi), \tag{3.41}$$

where

$$R = \frac{P}{\sqrt{(k - M\omega_0^2)^2 + \omega_0^2 a^2}} \quad \text{and} \quad \tan(\phi) = \frac{\omega_0 a}{k - M\omega_0^2}. \tag{3.42}$$

From (3.41) and (3.42), it follows that like the forcing function $F(t) = P \cos(\omega_0 t)$, $v(t)$ is a sinusoid with the same angular frequency ω_0; however, $v(t)$ is out of phase with $F(t)$ by the angle ϕ, and its magnitude is different by the factor

$$G(\omega_0) = \frac{1}{\sqrt{(k - M\omega_0^2)^2 + \omega_0^2 a^2}} = \frac{1/M}{\sqrt{\left(\frac{k}{M} - \omega_0^2\right)^2 + \left(\frac{a}{M}\right)^2 \omega_0^2}}. \tag{3.43}$$

The graph of $G(\omega_0)$ is called the *frequency response curve*, or *resonance curve*.

From (3.43) it is clear that $G(0) = 1/k$ and $G(\omega_0) \to 0$ as $\omega_0 \to \infty$. Further, since

$$G'(\omega_0) = \frac{-\left(\frac{2\omega_0}{M}\right)\left[\omega_0^2 - \left(\frac{k}{M} - \frac{a^2}{2M^2}\right)\right]}{\left[\left(\frac{k}{M} - \omega_0^2\right)^2 + \left(\frac{a}{M}\right)^2 \omega_0^2\right]^{3/2}},$$

$G'(\omega_0) = 0$ if and only if

$$\omega_0 = 0, \quad \text{or} \quad \omega_0 = \omega_1 = \left(\frac{k}{M} - \frac{a^2}{2M^2}\right)^{1/2}. \tag{3.44}$$

Thus, if $a^2 > 2Mk$, ω_1 is imaginary, and hence $G'(\omega_0) = 0$ only when $\omega_0 = 0$. Hence, as ω_0 increases from zero to infinity, $G(\omega_0)$ decreases from $1/k$ to 0. If $a^2 < 2Mk$, ω_1 is real and positive, and $G(\omega_0)$ attains its maximum at ω_1 and its maximum value is

$$G(\omega_1) = \frac{1}{a\sqrt{\frac{k}{M} - \frac{a^2}{4M^2}}}. \tag{3.45}$$

The value $\omega_1/(2\pi)$ is called the *resonance frequency* for the system, and when the external force is applied at this frequency, the system is said to be at *resonance*.

Example 3.7 An electrical circuit contains an inductance L, a resistance R, a capacitance C, a current I, and a generator which delivers E volts. These quantities satisfy the differential equation

$$L\frac{d^2 I}{dt^2} + R\frac{dI}{dt} + \frac{1}{C}I = \frac{dE}{dt}, \tag{3.46}$$

which is obtained by differentiating the equation

$$L\frac{d^2 q}{dt^2} + R\frac{dq}{dt} + \frac{1}{C}q = E(t), \quad (q \text{ is the charge}) \tag{3.47}$$

and using the relation $I = \frac{dq}{dt}$. Equations (3.46) and (3.47) are called LRC-circuit systems. If $E = E_0 \sin(\gamma t)$ we shall find the steady-state solution $I_p(t)$ of (3.46).

Since $E = E_0 \sin \gamma t$ we need to solve the differential equation

$$L\frac{d^2 I}{dt^2} + R\frac{dI}{dt} + \frac{1}{C}I = E_0 \gamma \cos(\gamma t). \tag{3.48}$$

As earlier if $R^2 < 4L/C$ the general solution of (3.48) can be written as

$$\begin{aligned}
I(t) = {} & e^{\alpha t}[A \cos(\beta t) + B \sin(\beta t)] \\
& + \frac{E_0 \gamma}{\left(\frac{1}{C} - \gamma^2 L\right)^2 + \gamma^2 R^2} \left[\left(\frac{1}{C} - \gamma^2 L\right) \cos(\gamma t) + \gamma R \sin(\gamma t)\right],
\end{aligned} \tag{3.49}$$

where

$$\alpha = -\frac{R}{2L} < 0 \quad \text{and} \quad \beta = \frac{\left(\frac{4L}{C} - R^2\right)^{1/2}}{2L}.$$

Clearly, the first term of $I(t)$ tends to zero as $t \to \infty$, and hence the steady-state solution $I_p(t)$ of (3.46) is

$$I_p(t) = \frac{E_0\gamma}{\sqrt{\left(\frac{1}{C} - \gamma^2 L\right)^2 + \gamma^2 R^2}} \cos(\gamma t - \phi), \quad \tan(\phi) = \frac{\gamma R}{(1/C) - \gamma^2 L}.$$

Hence, at the steady-state the amplitude is $E_0/\sqrt{[(1/C\gamma) - \gamma L]^2 + R^2}$, which assumes its largest value when $1/(C\gamma) - \gamma L = 0$, i.e., $\gamma = 1/\sqrt{LC}$. The maximum amplitude is E_0/R. This (roughly) happens when one tunes a radio.

Example 3.8 A particle is attracted toward a fixed point O with a force inversely proportional to its instantaneous distance from O. If the particle is released from rest, we shall find the time for it to reach O.

Let O be the origin, and the particle be located on the x-axis. If at time $t = 0$ it is at distance a, then by Newton's second law, we have

$$M\frac{d^2x}{dt^2} = -\frac{k}{x}, \quad x(0) = a, \quad x'(0) = 0,$$

where M is the mass of the particle and $k > 0$ is a constant of proportionality. Multiplying the differential equation by dx/dt and integrating, we get

$$\frac{M}{2}\left(\frac{dx}{dt}\right)^2 = k\ln\left(\frac{a}{x}\right),$$

or

$$\frac{dx}{dt} = -\sqrt{\frac{2k}{M}}\sqrt{\ln\left(\frac{a}{x}\right)},$$

where the negative sign is taken because as t increases x decreases. Thus the required time T is given by

$$T = \sqrt{\frac{M}{2k}}\int_0^a \frac{dx}{\sqrt{\ln(a/x)}} = a\sqrt{\frac{M}{2k}}\int_0^\infty u^{-1/2}e^{-u}\,du$$
$$= a\sqrt{\frac{M}{2k}}\,\Gamma\left(\frac{1}{2}\right) = a\sqrt{\frac{\pi M}{2k}}.$$

Next for a moving particle, we shall need the components of the velocity and acceleration along radial and transverse directions. If a particle moves from a point P to a point Q, then the displacement along the radius vector is (see Fig. 3.5)

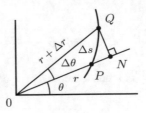

Fig. 3.5

$$ON - OP = (r + \Delta r)\cos(\Delta\theta) - r$$

and hence the radial component u of velocity is

$$u = \lim_{\Delta t \to 0} \frac{(r + \Delta r)\cos(\Delta\theta) - r}{\Delta t} = \lim_{\Delta t \to 0} \frac{\Delta r}{\Delta t} = \frac{dr}{dt} = r'. \qquad (3.50)$$

Similarly, the displacement perpendicular to the radius vector is $(r + \Delta r)\sin(\Delta\theta)$, and the transverse component v of velocity is

$$v = \lim_{\Delta t \to 0} \frac{(r + \Delta r)\sin(\Delta\theta)}{\Delta t} = \lim_{\Delta t \to 0} r\frac{\sin(\Delta\theta)}{\Delta\theta}\frac{\Delta\theta}{\Delta t} = r\frac{d\theta}{dt} = r\theta'. \quad (3.51)$$

Now the change in the velocity along the radius vector is (see Fig. 3.6)

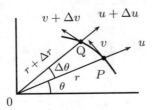

Fig. 3.6

$$(u + \Delta u)\cos(\Delta\theta) - (v + \Delta v)\sin(\Delta\theta) - u$$

and hence the radial component of acceleration is

$$\lim_{\Delta t \to 0} \frac{(u + \Delta u)\cos(\Delta\theta) - (v + \Delta v)\sin(\Delta\theta) - u}{\Delta t}$$
$$= \lim_{\Delta t \to 0} \frac{\Delta u - v\Delta\theta}{\Delta t} = \frac{du}{dt} - v\frac{d\theta}{dt} \qquad (3.52)$$
$$= \frac{d}{dt}(r') - r\theta'\theta' = r'' - r\theta'^2.$$

Similarly, the transverse component of acceleration is

$$\lim_{\Delta t \to 0} \frac{(u + \Delta u)\sin(\Delta\theta) + (v + \Delta v)\cos(\Delta\theta) - v}{\Delta t}$$

$$= \lim_{\Delta t \to 0} \frac{u\,\Delta\theta + \Delta v}{\Delta t} = u\frac{d\theta}{dt} + \frac{dv}{dt} = r'\theta' + \frac{d}{dt}(r\theta') = \frac{1}{r}(r^2\theta')'. \tag{3.53}$$

Example 3.9 Any object that swings back and forth is called a *physical pendulum*. A *simple pendulum* is a special case of physical pendulums. It consists of a string to which a mass is attached at one end. The swinging bob in a grandfather's clock and a child's swing are examples of pendulums. In describing the motion of a simple pendulum, we shall assume that the mass of the string is negligible and there is no other external force on the system.

Let a particle of mass M be suspended from the end of a string of length L. Assume that the motion is in a circle of radius $r = L$ (see Fig. 3.7). Thus, it follows that the radial component of the velocity is $r' = 0$, the transverse component of velocity is $v = r\theta' = L\theta'$, the radial component of acceleration is $r'' - r\theta'^2 = -L\theta'^2$, and the transverse component of acceleration is $(1/r)(r^2\theta')' = (1/L)(L^2\theta')' = L\theta''$. Thus, from Newton's second law the equations of motion can be written as

$$ML\theta'' = -Mg\sin(\theta), \tag{3.54}$$

$$-ML\theta'^2 = Mg\cos(\theta) - F, \tag{3.55}$$

where F is the tension toward the origin exerted by the string to hold the mass M. We shall see that it has a nonconstant magnitude.

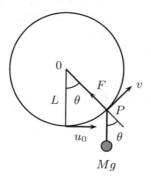

Fig. 3.7

If θ is small, i.e., the angular displacement is small, then $\sin\theta \simeq \theta$, and the Eq. (3.54) takes the form

$$\theta'' + \omega^2\theta = 0, \quad \omega^2 = g/L, \tag{3.56}$$

which has exactly the same structure as the differential equation (3.13).

If θ is not necessarily small, we multiply (3.54) by $2\theta'$ and integrate, to obtain

$$L\theta'^2 = 2g\cos(\theta) + c,$$

which is the same as

$$\frac{v^2}{L} = 2g\cos(\theta) + c, \tag{3.57}$$

where c is a constant. Now, let the particle be projected from the lowest point with velocity u_0, i.e., when $\theta = 0$, $v = L\theta' = u_0$. Then, from (3.57) it follows that $c = (u_0^2/L) - 2g$, and hence

$$v^2 = u_0^2 - 2gL[1 - \cos(\theta)], \tag{3.58}$$

or

$$\frac{1}{2}Mv^2 = \frac{1}{2}Mu_0^2 - MgL(1 - \cos(\theta)) = \frac{1}{2}Mu_0^2 - Mgh, \tag{3.59}$$

where h is the vertical distance traveled by the particle.

Now, from Eqs. (3.55) and (3.58), we get

$$F = ML\theta'^2 + Mg\cos(\theta) = M\frac{v^2}{L} + Mg\cos(\theta) = M\frac{u_0^2}{L} - 2Mg + 3Mg\cos(\theta). \tag{3.60}$$

At the highest point $\theta = \pi$ and $F = Mu_0^2/L - 5Mg$. Thus, if $u_0^2 \geq 5Lg$, the particle will move in a complete circle again and again. However, if $u_0^2 < 5Lg$, F will vanish before the particle reaches the highest point. When F vanishes, the particle begins to move freely under gravity and describes a parabolic path till the string again becomes tight and the circular motion is started again.

Equation (3.54) can be made more complicated by assuming that the pendulum of length L and mass M is constrained to move in a plane rotating with angular velocity Ω about a vertical line through the pivot point. If $\theta(t)$ denotes the angular deviation of the pendulum from the vertical, the equation of motion takes the form

$$I\theta'' - M\Omega^2 L^2 \sin(\theta)\cos(\theta) + MgL\sin(\theta) = 0,$$

where I is the moment of inertia of the pendulum. If θ is small, the above equation reduces to

$$I\theta'' + ML(g - \Omega^2 L)\theta = 0.$$

When $\Omega^2 L > g$, the solution of this reduced equation does not display swinging motion.

Example 3.10 If a particle of mass M moves on the inside of a circular smooth wire of radius a, the equations of motion are (see Fig. 3.8)

$$Ma\theta'' = -Mg\sin(\theta) \tag{3.61}$$

$$Ma\theta'^2 = R - Mg\cos(\theta). \tag{3.62}$$

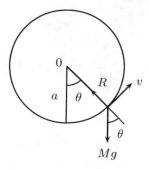

Fig. 3.8

Clearly, (3.61) and (3.62) are exactly the same as (3.54) and (3.55), where L is replaced by a, and F is replaced by the normal reaction R. Thus, if the initial velocity $u_0^2 \geq 5ag$, the particle will make an infinite number of complete rounds of the circular wire. And if $u_0^2 < 5ag$, the reaction R vanishes before the particle reaches the highest point, the particle leaves the curve, describes a parabolic path till it meets the circular wire, and then describes a circular path. This continues again and again.

If the particle moves on the outside of a circular smooth wire from the point A ($\theta = 0$), the equations of motion are (see Fig. 3.9)

$$Ma\theta'' = Mg\sin(\theta) \tag{3.63}$$

$$Ma\theta'^2 = -R + Mg\cos(\theta). \tag{3.64}$$

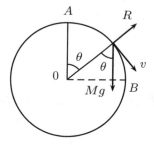

Fig. 3.9

Integrating (3.63), we find

$$a\theta'^2 = \frac{u_0^2}{a} + 2g[1 - \cos(\theta)]$$

and hence from (3.64) it follows that

$$R = 3Mg\cos(\theta) - M\frac{u_0^2}{a} - 2Mg.$$

Thus, at the point A,

$$\theta = 0, \quad R = Mg - M\frac{u_0^2}{a}, \tag{3.65}$$

whereas at the point B,

$$\theta = \frac{\pi}{2}, \quad R = -M\frac{u_0^2}{a} - 2Mg < 0. \tag{3.66}$$

From (3.65) and (3.66) it is clear that if $u_0^2 > ag$, the particle leaves the contact with the wire immediately and describes a parabolic path. If $u_0^2 < ag$, the particle remains in contact for some distance, but must leave contact when R vanishes, i.e., before it reaches the point B and then it describes a parabolic path.

If the particle moves on the inside of rough vertical circular wire, then there is an additional functional force μR along the tangent opposing the motion. Thus, the Eq. (3.62) remains the same, whereas (3.61) is modified to

$$Ma\theta'' = -Mg\sin(\theta) - \mu R. \tag{3.67}$$

Eliminating R between Eqs. (3.67), (3.62) we get the nonlinear differential equation

$$a\theta'' + \mu a\theta'^2 = -g\sin(\theta) - \mu g\cos(\theta). \tag{3.68}$$

Using the substitution $\theta' = w$ so that $\theta'' = w\frac{dw}{d\theta}$, Eq. (3.68) becomes

$$aw\frac{dw}{d\theta} + \mu aw^2 = -g\sin(\theta) - \mu g\cos(\theta). \tag{3.69}$$

Now letting $w^2 = z$ in (3.69), we obtain

$$a\frac{dz}{d\theta} + 2\mu az = -2g\sin(\theta) - 2\mu g\cos(\theta), \tag{3.70}$$

which is a first-order linear differential equation, and can be solved rather easily. Knowing z, we can find θ, and then determine the value of θ when R vanishes.

Similarly, for the rough circular wire, the Eq. (3.64) remains the same, whereas (3.63) is modified to

$$Ma\theta'' = Mg\sin(\theta) - \mu R. \tag{3.71}$$

Example 3.11 A model representing the angular displacement $\theta(t)$ of a swinging door is (see Fig. 3.10)

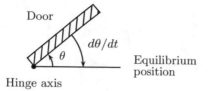

Door

$d\theta/dt$

θ

Equilibrium position

Hinge axis

Fig. 3.10

$$I\theta'' = -k_1\theta - k_2\theta', \tag{3.72}$$

where θ is the angle measured from the equilibrium position of the door, I is the moment of inertia of the door with respect to the hinge axis, k_1 is the constant of proportionality of a spring that acts to close the door, and k_2 is the constant of proportionality of a hydraulic mechanism that acts as a damper opposing the movement of the door.

Clearly, (3.72) is exactly the same as (3.23) except some different terminology.

Example 3.12 The differential equation governing the torsional motion of a weight suspended from the end of an elastic shaft is (see Fig. 3.11)

$$I\theta'' + a\theta' + k\theta = T(t), \tag{3.73}$$

where I is the moment of inertia, a is the damping constant, k is the shaft constant, $\theta(t)$ represents the amount of twist of the weight at any time, and $T(t)$ is the applied torque.

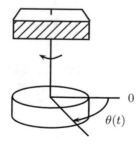

0

$\theta(t)$

Fig. 3.11

On comparing Eqs. (3.36), (3.47) and (3.73) it is clear that with the exception of terminology, there is no difference in mathematics.

Example 3.13 Assume that the force acting on a particle of mass M is $MF(r)$ and it is directed toward the origin, then the equations of motion are given by

$$M \left(r'' - r(\theta')^2 \right) = -MF(r) \tag{3.74}$$

and

$$\frac{M}{r} \frac{d}{dt} (r^2 \theta') = 0. \tag{3.75}$$

From these equations, we respectively have

$$r'' - r(\theta')^2 = -F(r) \tag{3.76}$$

and

$$r^2 \theta' = h, \tag{3.77}$$

where h is an arbitrary constant.

Now let $u = 1/r$, so that (3.77) is the same as $\theta' = hu^2$, and since

$$r' = \frac{dr}{dt} = \frac{dr}{du} \frac{du}{d\theta} \frac{d\theta}{dt} = -\frac{1}{u^2} \frac{du}{d\theta} hu^2 = -h \frac{du}{d\theta},$$

$$r'' = \frac{d}{dt} \left(-h \frac{du}{d\theta} \right) = \frac{d}{d\theta} \left(-h \frac{du}{d\theta} \right) \frac{d\theta}{dt} = -h \frac{d^2 u}{d\theta^2} hu^2 = -h^2 u^2 \frac{d^2 u}{d\theta^2}$$

the Eq. (3.76) can be written as

$$-h^2 u^2 \frac{d^2 u}{d\theta^2} - \frac{1}{u} h^2 u^4 = -F(1/u),$$

which is the same as

$$\frac{d^2 u}{d\theta^2} + u = \frac{F(1/u)}{h^2 u^2}. \tag{3.78}$$

Clearly, Eq. (3.78) is independent of the time t, and its integration provides the path described by a particle moving under a central force F per unit mass.

If the central force per unit mass is K/r^2 or Ku^2, where K is a constant, then Eq. (3.78) is the same as

$$\frac{d^2 u}{d\theta^2} + u = \frac{K}{h^2}. \tag{3.79}$$

The solution of Eq. (3.79) can be written as

$$u(\theta) = C \cos(\theta - \phi) + \frac{K}{h^2},$$

or

$$\frac{L}{r} = 1 + e\cos(\theta - \phi),\tag{3.80}$$

where $h^2 = KL$ and $e = LC$.

Equation (3.80) represents a conic section (ellipse if $e < 1$, parabola if $e = 1$, and hyperbola if $e > 1$) with a focus at the center of force. Thus, we can conclude that if a particle moves under a central force K/r^2 per unit mass, then the path is a conic section with a focus at the center. Conversely, if we know that the path is a conic section with a focus at the center of force, i.e., (3.80), then the force per unit mass is given by

$$F = h^2 u^2 \left(\frac{d^2 u}{d\theta^2} + u\right) = h^2 u^2 \left(\frac{-e\cos(\theta - \phi)}{L} + \frac{1 + \cos(\theta - \phi)}{L}\right) = \frac{K}{r^2}.$$

Thus, the central force follows the inverse square law. Now since all planets are observed to move in elliptic orbits with the Sun at one focus, it follows that the law of attraction between different planets and Sun must be the inverse square law.

Remark 3.1 Following a long period of observations of planetary motions by his predecessors specially of Nicolaus Copernicus (1473–1543) and Tycho Brahe (1546–1601), Johannes Kepler (1571–1630) formulated three laws of motion empirically. These laws can be deduced from Example 3.13 as follows:

1. Every planet describes an ellipse with Sun at one focus: From Example 3.13 it is clear that under the inverse square law, the path has to be a conic section and this includes elliptic orbits.

2. The radius vector from the Sun to a planet describes equal areas in equal intervals of time: Since $r^2\theta' = h$, we have

$$\lim_{\Delta t \to 0} \frac{1}{2}\frac{r^2 \Delta\theta}{\Delta t} = \frac{1}{2}h.\tag{3.81}$$

Now from Fig. 3.6 the area ΔA bounded by radius vectors OP and OQ and the arc PQ is $(1/2)r^2 \sin(\Delta\theta)$, so that (3.81) gives

$$\frac{dA}{dt} = \lim_{\Delta t \to 0}\frac{\Delta A}{\Delta t} = \frac{1}{2}r^2 \lim_{\Delta\theta \to 0}\frac{\sin(\Delta\theta)}{\Delta\theta}\lim_{\Delta t \to 0}\frac{\Delta\theta}{\Delta t} = \frac{1}{2}r^2\frac{d\theta}{dt} = \frac{1}{2}h.$$

Thus, the rate of description of sectorial area is constant, and hence equal areas are described in equal intervals of time.

3. The squares of periodic time of planets are proportional to the cubes of the semimajor axes of the orbits of the planets: Let a and b be the semimajor and semiminor axes of the ellipse, so that its total area is πab. Since the areal velocity is $h/2$, the periodic time T is given by

$$T = \frac{\pi ab}{h/2} = \frac{2\pi ab}{h}.\tag{3.82}$$

Now since the majoraxis of the ellipse is the sum of greatest and least distances from the focus, from (3.80) it follows that

$$2a = \frac{L}{1-e} + \frac{L}{1+e} = \frac{2L}{1-e^2} = \frac{2h^2}{K(1-e^2)},$$

which from the well-known relation $b^2 = a^2(1-e^2)$ gives

$$b^2 = \frac{h^2 a}{K}. \tag{3.83}$$

From (3.82) and (3.83), we have the required relation

$$T^2 = \frac{4\pi^2 a^2 b^2}{h^2} = \frac{4\pi^2}{K} a^3.$$

Remark 3.2 If the central force per unit mass is K/r^n, then Eq. (3.78) is the same as

$$\frac{d^2 u}{d\theta^2} + u = K u^{n-2}.$$

To find the radius r_0 of a circular orbit, we note that $u = u_0 = 1/r_0$ must be a constant, and hence $u_0 = K u_0^{n-2}$, or $u_0^{3-n} = K$, or $u_0 = K^{1/(n-3)}$ ($n \neq 3$). For $n = 3$ circular orbits do not exist unless $K = 1$. Now suppose the planet is given a small "push" so that $u = u_0 + \epsilon(\theta)$, where ϵ is small. Then, we have

$$\frac{d^2 \epsilon}{d\theta^2} + u_0 + \epsilon = K(u_0 + \epsilon)^{n-2} \simeq K u_0^{n-2} + K(n-2) u_0^{n-3} \epsilon.$$

Using $u_0 = K u_0^{n-2}$ and $K u_0^{n-3} = 1$ it follows that

$$\frac{d^2 \epsilon}{d\theta^2} + \epsilon \simeq (n-2)\epsilon,$$

or

$$\frac{d^2 \epsilon}{d\theta^2} \simeq (n-3)\epsilon,$$

which can be solved to obtain

$$\epsilon(\theta) \simeq \begin{cases} A \cosh(\sqrt{n-3}\theta) + B \sinh(\sqrt{n-3}\theta), & n > 3 \\ A\theta + B, & n = 3 \\ A \cos(\sqrt{3-n}\theta) + B \sin(\sqrt{3-n}\theta), & n < 3. \end{cases}$$

So ϵ is bounded in general only if $n < 3$, i.e., only in this case does a small disturbance remains small. Hence, the trajectories are circular orbits for $1/r$, $1/r^2$ force laws, but become degenerate conics for $1/r^3$, $1/r^4$ etc.

Example 3.14 Using clever substitutions some particular second-order differential equations can be transformed to integrable first-order equations. In particular, we consider the equation for the capillary curve (see [18])

$$\frac{d^2y}{dx^2} = \frac{4y}{C^2}\left[1 + \left(\frac{dy}{dx}\right)^2\right]^{3/2}, \quad C^2 = \frac{4T}{\rho g}, \tag{3.84}$$

where T is the surface tension, ρ is the density of the liquid, and the physical situation is depicted in Fig. 3.12.

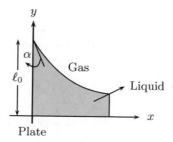

Fig. 3.12

Thus, it follows that

$$y(0) = \ell_0, \quad \frac{dx}{dy}(0) = -\cot(\alpha). \tag{3.85}$$

In (3.84) we set $p = y'$, $y'' = p\,dp/dy$, to obtain

$$\frac{p\,dp}{(1 + p^2)^{3/2}} = \frac{4y\,dy}{C^2},$$

which yields

$$-(1 + p^2)^{-1/2} = \frac{2y^2}{C^2} + c_1. \tag{3.86}$$

Since $p = y' = 0$ when $y = 0$ it follows that $c_1 = -1$, and hence Eq. (3.86) is the same as

$$(1 + p^2)^{-1/2} = \frac{C^2 - 2y^2}{C^2}. \tag{3.87}$$

Using (3.85) in (3.87), we get the expression

$$\ell_0 = C\sqrt{\frac{1 - \sin(\alpha)}{2}} = C\sin\left(\frac{\pi}{4} - \frac{\alpha}{2}\right).$$

Now from (3.87), we find

$$p = \frac{dy}{dx} = -\frac{2y(C^2 - y^2)^{1/2}}{C^2 - 2y^2},$$

where we have taken the negative sign since the physical slope of the curve in Fig. 3.12 is negative for a "wetting" fluid. Separating the variables, we obtain

$$dx = \frac{ydy}{(C^2 - y^2)^{1/2}} - \frac{C^2}{2}\frac{dy}{y(C^2 - y^2)^{1/2}},$$

which yields

$$x + c_2 = -(C^2 - y^2)^{1/2} + \frac{C}{2}\mathrm{sech}^{-1}\left(\frac{y}{C}\right).$$

However, since $y(0) = \ell_0$ it follows that

$$c_2 = -(C^2 - \ell_0^2)^{1/2} + \frac{C}{2}\mathrm{sech}^{-1}\left(\frac{\ell_0}{C}\right).$$

Example 3.15 Consider a pipe of infinite length laid along the x-axis. Assume that the outer and inner radii of the pipe are r_1 and r_2, respectively. Thus, the thickness of the wall of the pipe is $T = r_1 - r_2$. Let the average of the internal and external radii be $R = (r_1 + r_2)/2$. Suppose that a liquid being pumped through the pipe exerts a constant pressure P on the wall of the pipe. This pressure causes a slight radial deflection of the wall of the pipe. To strengthen the walls of the pipe, a reinforcing ring is placed around the pipe at $x = 0$ (for a partial cross-sectional view see Fig. 3.13). The radial deflection of the pipe wall at a distance x from the reinforcing ring $w(x)$ is governed by the differential equation

$$\frac{d^4w}{dx^4} + \frac{12}{R^2T^2}w = \frac{12}{ET^3}P, \tag{3.88}$$

where E is the modulus of elasticity of the pipe. At the ring we assume that there is no deflection.

We shall find the general solution of Eq. (3.88). For this, note that the corresponding homogeneous equation of (3.88) is

$$\frac{d^4w}{dx^4} + \frac{12}{R^2T^2}w = 0. \tag{3.89}$$

Since for the characteristic equation

$$m^4 + a^4 = 0, \quad a = (12/R^2T^2)^{1/4},$$

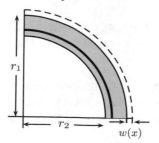

Fig. 3.13

the roots are $a(\pm 1 \pm i)/\sqrt{2}$, four linearly independent solutions of (3.89) are

$$e^{ax/\sqrt{2}} \cos\left(\frac{ax}{\sqrt{2}}\right), \quad e^{ax/\sqrt{2}} \sin\left(\frac{ax}{\sqrt{2}}\right), \quad e^{-ax/\sqrt{2}} \cos\left(\frac{ax}{\sqrt{2}}\right), \quad e^{-ax/\sqrt{2}} \sin\left(\frac{ax}{\sqrt{2}}\right).$$

It is also easy to verify that $R^2 P/ET$ is a particular solution of (3.88). Thus, the general solution of (3.88) can be written as

$$w(x) = e^{ax/\sqrt{2}} \left[c_1 \cos\left(\frac{ax}{\sqrt{2}}\right) + c_2 \sin\left(\frac{ax}{\sqrt{2}}\right)\right]$$
$$+ e^{-ax/\sqrt{2}} \left[c_3 \cos\left(\frac{ax}{\sqrt{2}}\right) + c_4 \sin\left(\frac{ax}{\sqrt{2}}\right)\right] + \frac{R^2 P}{ET}.$$

Problems

3.1 A solid cylinder is placed on the top of an inclined plane and released from rest. If it rolls without slipping and if a frictional force acts in the opposite direction to its motion, the differential equation of motion is

$$\frac{d^2 s}{dt^2} = kg \sin(\alpha),$$

where s is the distance displaced, $k > 0$ is a constant, and α is the angle of the inclined plane. Find its solution.

3.2 The relation between the national income and the burden of debt incurred by the citizens is governed by the differential equation (see [6])

$$\frac{d^2 D}{dt^2} - kD = 0,$$

where D is the total public debt, and the income is assumed to grow at a constant relative rate k $(0 < k < 1)$. Show that its general solution can be written as $D(t) = A\cosh(\sqrt{k}t) + B\sinh(\sqrt{k}t)$.

3.3 If a floating body is depressed vertically from its equilibrium position by an amount y, the resulting motion due to Archimedes' (287–212 BC) principle and Newton's second law of motion is governed by the differential equation

$$\frac{d^2y}{dt^2} + \frac{\rho A}{M}y = 0,$$

where ρ is the density of the liquid, A the cross-sectional area of the body, and M is its mass. Find its general solution.

3.4 Before using a metal in spacecrafts, a torsional stiffness test is applied to a slender vertical rod of the test metal to which a solid metal cylinder is attached. If r is the radius of the cylinder and k is a torsional stiffness constant, then

$$\frac{Mr^2}{2}\frac{d^2\theta}{dt^2} + k\theta = 0$$

is the differential equation that is solved for the angle θ through which the cylinder is twisted about an axis through the vertical rod. Here M is the mass of the system. Find its general solution.

3.5 In the study of a one-dimensional reactor, the following differential equation is encountered (see [15]):

$$E\frac{d^2C}{dz^2} - \overline{U}\frac{dC}{dz} - kC = 0,$$

where $C = C(z)$ is the reactant concentration, z the axial coordinate, E the axial dispersion coefficient, $\overline{U}$ the average velocity of the reaction mixture, and k the reaction constant. Find its general solution.

3.6 The differential equation that describes the motion of a spring–mass system with a single degree of freedom excited by the force Px' is

$$Mx'' + ax' + kx = Px'.$$

Show that if $P > a$ the motion of the system diverges, i.e., dynamically unstable, if $P = a$ the solution is the solution for a free undamped system, and if $P < a$ the solution is a free damped system.

3.7 When a shell is fired from a large artillery gun, the barrel recoils on a well-lubricated guide and its rapid motion is braked by a battery of heavy springs, which first stop the recoil and then push the barrel back into firing position. The return motion is also braked by a hydraulic cylinder, called a dashpot, consisting of a closed cylinder within which slides a piston. Assuming that the firing is horizontal and friction in the guide is negligible, the following differential equations describe the motion of the barrel:

$$\frac{d^2x}{dt^2} + \frac{k}{M}x = 0, \quad x(0) = x_0, \quad x'(0) = v_0 \quad \text{(recoil)} \qquad (3.90)$$

$$\frac{d^2x}{dt^2} + \frac{a}{M}\frac{dx}{dt} + \frac{k}{M}x = 0, \quad x(t_1) = x_1, \quad x'(t_1) = 0, \quad \text{(return)} \qquad (3.91)$$

where k is the spring constant of the breaking system, M is the mass of the moving system, a is the dashpot coefficient, and x_1 is the maximum displacement from rest, occurring at time t_1. Find the general solutions of the Eqs. (3.90) and (3.91). In particular, show that for the consistent constants $M = 1$, $k = 16$, $x_1 = 15$, $a = 10$ and $t_1 = 0$ the solution of (3.91) is $20e^{-2t} - 5e^{-8t}$.

3.8 The equation

$$a_2 x^2 y'' + a_1 xy' + a_0 y = 0, \quad x > 0, \quad ' = \frac{d}{dx},$$

where a_0, a_1 and a_2 are constants is called *Leonhard Euler's* (1707–1783) *equidimensional equation*. Show that the change of independent variable $x = e^t$ transforms it into the equation with constant coefficients

$$a_2 \ddot{y} + (a_1 - a_2)\dot{y} + a_0 y = 0, \quad \cdot = \frac{d}{dt}.$$

Hence, if m_1 and m_2 are the roots of the characteristic equation $a_2 m^2 + (a_1 - a_2)m + a_0 = 0$, then the solution of Euler's equation can be written as

$$y(x) = \begin{cases} Ax^{m_1} + Bx^{m_2} & \text{if } m_1 \neq m_2 \text{ are real} \\ [A + B\ln(x)]x^m & \text{if } m_1 = m_2 = m \\ x^p[A\cos(q\ln(x)) + B\sin(q\ln(x))] & \text{if } m_1 = p + iq, \; m_2 = p - iq. \end{cases}$$

3.9 Physiologists model the circulation of blood in blood vessels by studying the velocity of wave propagation in elastic tubes. If y is the radial displacement of the artery wall, then the equation determined by Mirsky [14] is

$$r^2 \frac{d^2y}{dr^2} + r\frac{dy}{dr} - m^2 y = 0,$$

where the constant m is derived from stress and strain components. Show that the solution of this differential equation is $y(r) = Ar^m + Br^{-m}$.

3.10 The following differential equation occurs in Astronomy (see [10]):

$$v\frac{d^2\psi}{dv^2} + 4\frac{d\psi}{dv} = 0, \quad v > 0.$$

Show that its general solution is $\psi(v) = c_1 + c_2 v^{-3}$.

3.11 The following differential equation arises in the study of peristaltic flow in tubes (see [3])

$$\frac{d^2\rho}{dv^2} + \frac{1}{v}\frac{d\rho}{dv} = 0, \quad v > 0.$$

Show that its general solution is $\rho(v) = c_1 + c_2 \ln(v)$.

3.12 Lewis [11] model for the investment growth is

$$\frac{d^2k}{dt^2} + c^2k = c^2k_1,$$

where k is the capital stock, k_1 the desired level of capital stock, and c a positive constant. Show that the capital stock oscillates about the desired level k_1.

3.13 In a study of riots by oppressed groups Rashevsky [17] obtained the differential equation

$$\frac{d^2y}{dt^2} - ay = b,$$

where $a > 0$ and b is an arbitrary constant. Solve this equation.

3.14 In an effort to describe experimental data recorded in a simple avoidance situation using rats, the following initial value problem occurs (see [4])

$$\frac{d^2y}{dt^2} + \frac{1}{\tau}\frac{dy}{dt} + by = \frac{\beta}{\tau}, \quad y(0) = \alpha, \quad y'(0) = \beta$$

where $y(t)$ represents the value of the learning curve of a rat at time t, b and τ are positive constants represent the characteristic of the experimental situation, and α and β are initial values. Find its solution.

3.15 In mining certain minerals, the cost rises linearly with cumulative production. The optimum process for mining this type of minerals is governed by the differential equation (see [9])

$$\frac{d^2x}{dt^2} - r\frac{dx}{dt} - \frac{r\beta}{b}x = -\frac{rK}{b},$$

where r, β, b and K are nonzero constants. Solve this equation.

3.16 In the study of the gravitational equilibrium of a star, an application of Newton's law of gravity and of the Stefan–Boltzmann (Ludwig Eduard Boltzmann (1844–1906)) law of gases leads to the equilibrium equation

$$\frac{1}{r^2}\frac{d}{dr}\left(\frac{r^2}{\rho}\frac{dP}{dr}\right) = -4\pi\rho G,$$

where P is the sum of the gas kinetic pressure and the radiation pressure, r is the distance from the center of the star, ρ is the density of matter (assumed to be a constant), and G is the gravitational constant. Find its general solution.

3.17 Consider the spring–mass system governed by

$$x'' + \omega^2 x = \frac{P}{M}\sin(\omega_0 t), \quad \omega^2 = k/M.$$

Show that its general solution can be written as

$$x(t) = A\cos(\omega t) + B\sin(\omega t) + \begin{cases} \dfrac{P}{M(\omega^2 - \omega_0^2)}\sin(\omega_0 t), & \omega_0 \neq \omega \\ -\dfrac{P}{2M\omega}t\cos(\omega t), & \omega_0 = \omega. \end{cases}$$

Further, if $x(0) = x'(0) = 0$ show that this solution reduces to

$$x(t) = \begin{cases} \dfrac{P}{M(\omega^2 - \omega_0^2)\omega}(\omega\sin(\omega_0 t) - \omega_0\sin(\omega t)], & \omega_0 \neq \omega \\ \dfrac{P}{2M\omega^2}[\sin(\omega t) - t\omega(\cos\omega t)], & \omega_0 = \omega. \end{cases}$$

3.18 Consider the spring–mass system governed by

$$x'' + 2ax' + (\omega^2 + a^2)x = \sin(\beta t), \quad x(0) = 0, \quad x'(0) = 0.$$

show that its solution can be written as

$$x(t) = -\left[A\cos(\omega t) + \frac{B\beta + Aa}{\omega}\sin(\omega t)\right]e^{-at} + A\cos(\beta t) + B\sin(\beta t),$$

$$A = -2a\beta/C, \quad B = (a^2 + \omega^2 - \beta^2), \quad C = (a^2 + \omega^2 - \beta^2)^2 + (2a\beta)^2.$$

Further, show that the amplitude of the steady-state solution is $C^{-1/2}$, and the value of β which maximizes this amplitude is $\sqrt{\omega^2 - a^2}$.

3.19 Show that the steady-state solution of the spring–mass system

$$Mx'' + ax' + kx = P\sin(\omega_0 t)$$

can be written as $v(t) = R\sin(\omega_0 t - \phi)$, where R and ϕ are the same as in (3.42).

3.20 Consider the LRC-circuit equation

$$L\frac{d^2q}{dt^2} + R\frac{dq}{dt} + \frac{1}{C}q = E_0 e^{i\gamma t}, \quad i = \sqrt{-1}.$$

(1) Show that its steady-state solution is

$$q_p(t) = \frac{E_0}{(1/c) - \gamma^2 L + i\gamma R} e^{i\gamma t}.$$

(2) Show that its steady-state current is

$$I_p(t) = \frac{E_0}{R + i[\gamma L - (1/\gamma C)]} e^{i\gamma t}.$$

3.21 In a model of foreign exchange speculation under floating exchange, it is assumed that the excess demand at time t caused by speculators is of the form $[a_0 + a_1 R(t) + B \cos(wt)]$, where $a_0 > 0$, $a_1 < 0$ and $B > 0$ are constants, $R(t)$ is the rate of exchange at time t, and $B \cos wt$ represents external factors, such as seasonal influences. It is also assumed that the expected rate of exchange at time t (a value that is expected to be realized at some given time in the future) is of the form $[b_0 R(t) + b_1 R'(t) + b_2 R''(t)]$, where b_0, b_1 and b_2 are constants whose signs indicate the attitude of the speculators. Based on these assumptions Gondolfo [8] obtained the following differential equation

$$b_2 R'' + b_1 R' + [a_1 - (1 - b_0)]R = -a_0 - B \cos(wt).$$

Find the solution of this equation for the values $a_0 = 200$, $a_1 = -10$, $B = 10$, $w = 4\pi$, $b_0 = 1$, $b_1 = -0.4$, $b_2 = -0.2$ considered in [8], satisfying the initial conditions $R(0) = 0$, $R'(0) = 1$.

3.22 An LC-circuit is subjected to an impressed voltage $E(t) = P[\sin t + \sin at]$, where a is not an integer (E is a quasi-periodic function). Then, the differential equation for the charge q in the circuit is

$$\frac{d^2q}{dt^2} + \omega^2 q = \frac{P}{L}[\sin(t) + \sin(at)], \quad \omega^2 = 1/(LC).$$

Find the forced response and discuss the following cases: $a \ll \omega^2$, $a \ll 1$, $1 \ll a \ll \omega^2$, a near 1, $a = 1$ and $a = \omega$.

3.23 Find the forced response of a pendulum of length L that starts from rest, $\theta(0) = 0$, $\theta'(0) = 0$, with a forcing function $F(t) = \sum_{n=1}^{N} P_n \times \sin n\pi t$, where P_n are constants, i.e., solve the initial value problem

$$\theta'' + \omega^2\theta = \sum_{n=1}^{N} P_n \sin(n\pi t), \quad \theta(0) = 0, \quad \theta'(0) = 0, \quad \omega^2 = \frac{g}{L}.$$

3.24 The frequency of blowing across the open top of a flask/bottle to produce a foghorn type of sound depends on the size of the air cavity in the flask. A flask with an air space that is used to produce sound is called an *acoustic resonator* or a Hermann von *Helmholtz* (1821–1894) *resonator*. Consider a flask with an interior cavity of volume V_0 that contains a gas of density ρ_0 at some ambient pressure p_0. The neck of the flask has length ℓ and a cross-sectional area A. If our blowing across the flask produces a dynamic force $F(t)$, the mass of gas $m = \rho_0 A \ell$ in the neck moves to the right by a distance x, then the mass m vibrates against the stiffness $k = \gamma p_0 A^2 / V_0$ (here γ is the ratio of the specific heats, for air $\gamma = 1.4$) in accordance with the differential equation (see [7])

$$\rho_0 A \ell \frac{d^2 x}{dt^2} + \gamma \frac{p_0 A^2}{V_0} x = F(t).$$

Find the general solution of this differential equation.

3.25 In recording an earthquake, the response of the accelerogram is governed by the equation

$$x'' + 2\eta \omega x' + \omega^2 x = -f(t),$$

where $x(t)$ is the instrument response recorded on the accelerogram, $f(t)$ the ground acceleration, η the damping ratio of the instrument, and ω the undamped natural frequency of the instrument (see [13]). Find the general solution of this differential equation.

3.26 In a study of wavelike motions of the flagellum (taillike part of an organism such as a bacterium or protozoan) leads to the following differential equation (see [5]) for the curvature $C(s)$ of the flagellum at a distance s along the flagellum

$$\frac{d^2 C}{ds^2} - a^2 C = \begin{cases} -a^2 C_0 & \text{if } s < 0 \\ 0 & \text{if } s > 0, \end{cases}$$

where C_0 is the curvature that results if there are only contractive forces, and $a > 0$ is a constant which depends on physical properties of the flagellum. Find the solution of this differential equation subject to the conditions that C and dC/ds are continuous, $\lim_{s \to -\infty} C(s) = C_0$, and $\lim_{s \to \infty} C(s) = 0$.

3.27 The differential equation

$$x'' + x = \begin{cases} 1 - t^2, & 0 \le t \le 1 \\ 0, & t > 1 \end{cases}$$

is an undamped system in which a force acts only during the time $0 \le t \le 1$. This occurs, for example, in a gun barrel when a shell is fired. The barrel is braked with heavy springs. Show that its solution satisfying the initial conditions $x(0) = 0$, $x'(0) = 0$ can be written as

$$x(t) = \begin{cases} 3 - t^2 - 3\cos t, & 0 \le t \le 1 \\ \cos(t)[2\cos(1) + 2\sin(1) - 3] + \sin(t)[2\sin(1) - 2\cos(1)], & t > 1. \end{cases}$$

3.28 Suppose a straight, light, elastic rod has one end clamped in the chunk of a lathe, and the other end is attached with an imperfectly balanced disk of mass M. Let $h \ne 0$ be the distance from the center of mass of the disk to the axis of the rod. We will determine the motion of the disk when the rod is rotated. This system has two degrees of freedom x and y in the plane in which the rotating disk vibrates. We assume that a force kz is required to deflect the rod a distance z in the direction z. If the rod is rotated with a constant angular velocity w, then the horizontal x, and vertical y deflections of the disk satisfy the equations (see [12]),

$$M\frac{d^2x}{dt^2} + kx = hw^2 \cos(wt)$$

$$M\frac{d^2y}{dt^2} + ky = hw^2 \sin(wt).$$

Show that the deflections $x(t)$ and $y(t)$ given that the disk begins at rest, i.e., $x(0) = x'(0) = y(0) = y'(0) = 0$ are as follows:

If $w^2 \ne k/M$, then

$$x(t) = c_1 \cos\left(\sqrt{\frac{k}{M}}t\right) + c_2 \sin\left(\sqrt{\frac{k}{M}}t\right) + \frac{hw^2}{k - Mw^2} \cos wt$$

$$y(t) = c_3 \cos\left(\sqrt{\frac{k}{M}}t\right) + c_4 \sin\left(\sqrt{\frac{k}{M}}t\right) + \frac{hw^2}{k - Mw^2} \sin(wt).$$

If $w^2 = k/M$, then

$$x(t) = c_1 \cos\left(\sqrt{\frac{k}{M}}t\right) + c_2 \sin\left(\sqrt{\frac{k}{M}}t\right) + \frac{hw^2}{2k}t \sin(wt)$$

$$y(t) = c_3 \cos\left(\sqrt{\frac{k}{M}}t\right) + c_4 \sin\left(\sqrt{\frac{k}{M}}t\right) + \frac{hw^2}{2k}t \cos(wt).$$

3.29 The differential equation

$$y'' + \delta(xy' + y) = 0$$

arises in the study of the turbulent flow of a uniform stream past a circular cylinder. Verify that $y_1(x) = \exp(-\delta x^2/2)$ is one solution. Show that its general solution is

$$y(x) = c_1 e^{-\delta x^2/2} \int^x e^{\delta t^2/2} dt + c_2 e^{-\delta x^2/2}.$$

3.30 Let a piston of mass M be placed at the midpoint of a closed cylinder of cross-sectional area A and length $2L$, as depicted in Fig. 3.14. Assume that the pressure p on either side of the piston satisfies Robert Boyle's (1627–1691) law, (i.e., the pressure times the volume is constant), and let p_0 be the pressure on both sides when

$x = 0$. If the piston is disturbed from its equilibrium position $x = 0$, show that the governing equation of motion is

$$Mx'' + 2p_0AL\frac{x}{L^2 - x^2} = 0.$$

When the amplitude of oscillation is small compared to L, this equation can be linearized to

$$Mx'' + \frac{2p_0A}{L}x = 0,$$

which is the same as (3.13) with $\omega^2 = 2p_0A/(LM)$.

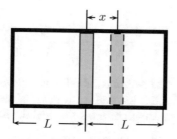

Fig. 3.14

3.31 Consider a rod of length L with a point mass M at its end, where the mass of the rod is negligible compared to M. The rod is welded at a right angle to another, which rotates without friction about an axis that is tilted by an angle of α with respect to the vertical (see Fig. 3.15). Let θ denote the angle of rotation of the pendulum, with respect to its equilibrium position (where M is at its lowest possible point, namely, in the plane of the paper). Show that the governing equation of motion is

$$\theta'' + \frac{g}{L}\sin(\alpha)\sin(\theta) = 0.$$

When the amplitude of oscillation is small this equation can be linearized to

$$\theta'' + \frac{g}{L}\sin(\alpha)\theta = 0,$$

which is the same as (3.13) with $\omega^2 = (g/L)\sin(\alpha)$.

3.32 Consider the equation of simple pendulum (3.54) together with the initial conditions

$$\theta(0) = \alpha, \quad \theta'(0) = 0.$$

Show that

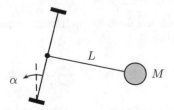

Fig. 3.15

(1) $\left(\dfrac{d\theta}{dt}\right)^2 = \dfrac{2g}{L}[\cos(\theta) - \cos(\alpha)]$.

(2) The period of the pendulum (the time for the pendulum to go from α to $(-\alpha)$ and back to α) is

$$T = 2\sqrt{\dfrac{2L}{g}} \int_0^\alpha \dfrac{d\theta}{\sqrt{\cos(\theta) - \cos(\alpha)}} = 4\sqrt{\dfrac{L}{g}} \int_0^{\pi/2} \dfrac{d\phi}{\sqrt{1 - k^2 \sin^2(\phi)}},$$

where $k = \sin(\alpha/2)$. The last integral is called an *elliptic integral of the first kind* and is denoted by $F(k, \pi/2)$. For this integral ready made tables are available. Obviously, the period T depends on the length of the string and the initial displacement α.

3.33 A spherical body of radius R is in equilibrium when half of it is submerged in a liquid. If the body is pushed down and then released, its motion is governed by the initial value problem

$$\dfrac{d^2y}{dt^2} = -\dfrac{g}{2}\left[3\dfrac{y}{R} - \left(\dfrac{y}{R}\right)^3\right], \quad y(0) = \sqrt{6}R, \quad y'(0) = 0$$

where y is the distance displaced from equilibrium, and g is the acceleration due to gravity. Show that $y(t) = \sqrt{6}R \sec\left(\sqrt{\dfrac{3g}{2R}}t\right)$.

3.34 The *Emden differential equation* $y'' - t^p y^q = 0$ arises in various astrophysical problems, including the study of the density of stars. Show that if $q \neq 1$ for each pair (p, q) there is a unique pair of (A, n) so that $y = At^n$ is a solution. What happens if $q = 1$?

3.35 The following fourth-order differential equations occur in applications as indicated:
(1) $y'''' - k^4 y = 0$ (vibration of a beam)
(2) $y'''' + 4k^4 y = 0$ (beam on an elastic foundation)
(3) $y'''' - 2k^2 y'' + k^4 y = 0$, (bending of an elastic plate)
where $k \neq 0$ is a constant. Find their general solutions.
(Solutions): (1) $(c_1 e^{kx} + c_2 e^{-kx} + c_3 \cos(kx) + c_4 \sin(kx))$ (2) $(e^{kx}[c_1 \cos(kx) + c_2 \sin(kx)] + e^{-kx}[c_3 \cos(kx) + c_4 \sin(kx)])$ (3) $(e^{kx}(c_1 + c_2 x) + e^{-kx}(c_3 + c_4 x))$.

3.36 A vibration absorber consists of a gyrostat suspended about an axis perpendicular to that of the torsional vibration it is supposed to suppress. The gyrostat is attached to springs and a viscous damping mechanism that limits the vibration about its axis. If a disturbing torque $T = \sin pt$ is applied to this vibrating absorber, rotating it through an angle θ, then the oscillation about the equilibrium satisfies (see [11]),

$$\frac{d^4\theta}{dt^4} + 43517.65\frac{d^2\theta}{dt^2} + 156960\,\theta = (1.57 - 0.04p^2)\sin(pt).$$

Find the general solution of this differential equation for $p = 1$.
$(c_1\cos(208.6t) + c_2\sin(208.6t) + c_3\cos(1.9t) + c_4\sin(1.9t) + 1.35 \times 10^{-5}\sin(t))$.

3.37 The deflection y of a suspension bridge under a load W distributed along the bridge is governed by the differential equation

$$EI\frac{d^4y}{dx^4} - (H+h)\frac{d^2y}{dx^2} = W - q\frac{h}{H},$$

where E is Young's modulus, I is the moment of inertia of the cross section of the bridge, H is the horizontal tension in the cables due to the load W, h is the tension in the cables when there is no load on the bridge, and q is the weight per unit length of the bridge. The product EI is known as the *flexural rigidity* of the beam and it is assumed to be a constant. Find the general solution of this equation for the special cases

(1) $W - q(h/H) = 0$ $(c_1 + c_2x + c_3e^{\lambda x} + c_4e^{-\lambda x}$, where $\lambda = \sqrt{(H+h)/(EI)}$
(2) $W - q(h/H) = x(L - x)$, where L is the length of the bridge.

3.38 Consider a uniform cylindrical rod whose one end $(x = 0)$ is clamped and the other end $(x = L)$ is supported in a bearing and is free to rotate. It can be shown that the deflection y of the rotating rod satisfies the differential equation

$$EI\frac{d^3y}{dx^3} + \frac{1}{2}Ww^2y = 0,$$

where W is the load at the midpoint, and w is the angular velocity. Find the general solution of this equation.

3.39 The deflection y of an elastically supported uniform beam with a constant axial force p satisfies the differential equation

$$EI\frac{d^4y}{dx^4} + p\frac{d^2y}{dx^2} + ky = 0,$$

where $k > 0$ is the constant of proportionality. Show that if $p^2 < 4kEI$ then the general solution of this equation involves only trigonometric functions.

3.40 Consider the following system of first-order chemical reactions

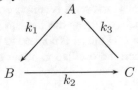

so that

$$\frac{dA}{dt} = k_3 C - k_1 A$$

$$\frac{dB}{dt} = k_1 A - k_2 B$$

$$\frac{dC}{dt} = k_2 B - k_3 C.$$

Show that

$$\frac{d^3 A}{dt^3} + (k_1 + k_2 + k_3)\frac{d^2 A}{dt^2} + (k_1 k_3 + k_1 k_2 + k_2 k_3)\frac{dA}{dt} = 0.$$

Also, find the general expressions for $A(t)$, $B(t)$ and $C(t)$ in the special case $k_1 = k_3 = k_2/4$.

3.41 In a study which returns the economy to its equilibrium the following third-order initial value problem occurs (see [16]),

$$\frac{d^3 P}{dt^3} + (\alpha L + \beta)\frac{d^2 P}{dt^2} + \alpha\beta(L + f_1)\frac{dP}{dt} + \alpha\beta f_2 P = 0, \quad t > 0$$
$$P(0) = 0, \quad P'(0) = -\alpha, \quad P''(0) = \alpha^2 L,$$

where $P(t)$ represents the production at time t, and α, β, f_1, f_2 and L are positive constants. Find the solution of this problem for the constants $\alpha = 4$, $\beta = 8$, $f_1 = f_2 = 1/2$ and $L = 1/4$ used in [16]. $\quad \left(-\frac{28}{9}e^{-t} + \left(\frac{28}{9} + \frac{16}{3}t\right)e^{-4t} \right).$

3.42 Show that the generalized harmonic motion equation $y^{(n)} + -w^n y = 0$ has a general solution which is unbounded as $t \to \infty$, $n \geq 3$. ($\lambda^n + w^n = 0$ gives $\lambda = w \times$ (n-th roots of (-1)). For $n = 2$, n-th roots of -1 have no real part, whereas for $n > 3$, n-th roots of (-1) have negative as well as positive real parts. For example, for $n = 3$ the solution is $y(t) = Ae^{-wt} + e^{wt/2}\left[a\cos(\sqrt{3}wt/2) + b\sin(\sqrt{3}wt/2) \right]$, which is unbounded whether $w > 0$ or $w < 0$).

3.43 Consider the case of dropping a stone from the height h. Let $r = r(t)$ denote the distance of the stone from the surface at time t. Then, the equation of motion is

$$\frac{d^2 r}{dt^2} = -\frac{\gamma M}{(R + r)^2}, \quad r(0) = h, \quad r'(0) = 0 \qquad (3.92)$$

where R and M are the radius and the mass of the earth. Let $\epsilon = 1/R$ in (3.92), to obtain

$$\frac{d^2r}{dt^2} = -\frac{\gamma M \epsilon^2}{(1 + \epsilon r)^2}, \quad r(0) = h, \quad r'(0) = 0. \tag{3.93}$$

In (3.93) use the expansion $r(t) = \sum_{i=0}^{4} \epsilon^i r_i(t)$ to show that

$$r(t) = h - \frac{\gamma M}{R^2}\left(1 - \frac{2h}{R}\right)\frac{t^2}{2} + O\left(\frac{1}{R^4}\right).$$

3.44 Consider the *satellite equation*

$$\frac{d^2y}{dt^2} + y = ky^2$$

together with the initial conditions $y(0) = A, \ y'(0) = 0$. Show that

$$y(t) = A\cos(t) + kA^2\left(\frac{1}{2} - \frac{1}{3}\cos(t) - \frac{1}{6}\cos 2t\right) +$$
$$+ k^2 A^3\left(-\frac{1}{3} + \frac{29}{144}\cos(t) + \frac{5}{12}t\sin t + \frac{1}{9}\cos(2t) + \frac{1}{48}\cos(3t)\right) + O(k^3).$$

References

1. R.P. Agarwal, D. O'Regan, *An Introduction to Ordinary Differential Equations* (Springer, New York, 2008)
2. R.P. Agarwal, D. O'Regan, *Ordinary and Partial Differential Equations with Special Functions, Fourier Series and Boundary Value Problems* (Springer, New York, 2009)
3. C. Barton, S. Raynor, Bull. Math. Biophys. **30**, 663–680 (1968)
4. J.P. Brady, C. Marmasse, Bull. Math. Biophys. **26**, 77–81 (1964)
5. C.J. Brokow, Nature **209**, 161–163 (1966)
6. E.D. Domar, Am. Econ. Rev. **798–827**, (1944)
7. C.L. Dym, E.S. Ivey, *Principles of Mathematical Modeling* (Academic, New York, 1980)
8. G. Gondolfo, *Mathematical Models in Economic Dynamics* (North Holland, Amsterdam, 1971)
9. O.C. Herfindahl, A.V. Kneese, *Economic Theory for Natural Resources* (Charles E. Merrill, Columbus, 1974)
10. Z. Kopal, Icarus **2**, 376–395 (1963)
11. J.P. Lewis, *Introduction to Mathematics for Students of Economics* (Macmillan, New York, 1969)
12. H. McCallion, *Vibration of Linear Mechanical Systems* (Halsted Press, New York, 1973)
13. G.A. McLennan, Bull. Seismol. Soc. Am. **59**, 1591–1598 (1969)
14. I. Mirsky, Bull. Math. Biophys. **29**, 311–318 (1967)
15. E. Petersen, *Chemical Reaction Analysis* (Prentice-Hall, Englewood Cliffs, 1965)

16. A.W. Phillips, Econ. J. **64**, 290–323 (1954)
17. N. Rashevsky, Bull. Math. Biophys. **30**, 510–518 (1968)
18. H.W. Reddick, F.H. Miller, *Advanced Mathematics for Engineers*, 3rd edn. (Wiley, New York, 1956)
19. S. Stewart, *Air Disasters* (Barnes & Noble, New York, 1986)

Chapter 4
Power Series Solutions

Generally, second-order differential equations with variable coefficients cannot be solved in terms of the known functions. However, there is a fairly large class of differential equations whose solutions can be expressed either in terms of power series, or as simple combination of power series and elementary functions [1–3]. It is this class of differential equations that we shall study in this chapter. For this, it is convenient to let the independent variable be x instead of t. We begin by introducing some basic concepts.

Power series. A *power series* is a series of functions of the form

$$\sum_{m=0}^{\infty} c_m (x - x_0)^m = c_0 + c_1(x - x_0) + c_2(x - x_0)^2 + \cdots + c_m(x - x_0)^m + \cdots$$

in which the coefficients c_m, $m = 0, 1, \cdots$ and the point x_0 are independent of x. The point x_0 is called the *point of expansion* of the series.

A function $f(x)$ is said to be *analytic* at $x = x_0$ if it can be expanded in a power series in powers of $(x - x_0)$ in some interval of the form $|x - x_0| < \mu$, where $\mu > 0$. If $f(x)$ is analytic at $x = x_0$, then

$$f(x) = \sum_{m=0}^{\infty} c_m (x - x_0)^m, \quad |x - x_0| < \mu$$

where $c_m = f^{(m)}(x_0)/m!$, $m = 0, 1, \cdots$ which is the same as Taylor's expansion of $f(x)$ at $x = x_0$.

The following properties of power series are well known.

© Springer Nature Switzerland AG 2019
R. P. Agarwal et al., *500 Examples and Problems of Applied Differential Equations*, Problem Books in Mathematics, https://doi.org/10.1007/978-3-030-26384-3_4

1. A power series $\sum_{m=0}^{\infty} c_m(x - x_0)^m$ is said to *converge* at a point x if $\lim_{n\to\infty} \sum_{m=0}^{n} c_m(x - x_0)^m$ exists. It is clear that the series converges at $x = x_0$; it may converge for all x, or it may converge for some values of x and not for others.

2. A power series $\sum_{m=0}^{\infty} c_m(x - x_0)^m$ is said to converge *absolutely* at a point x if the series $\sum_{m=0}^{\infty} |c_m(x - x_0)^m|$ converges. If the series converges absolutely, then the series also converges, however, the converse is not necessarily true.

3. If the series $\sum_{m=0}^{\infty} c_m(x - x_0)^m$ converges absolutely for $|x - x_0| < \mu$ and diverges for $|x - x_0| > \mu$, then μ is called the *radius of convergence*. For a series that converges nowhere except at x_0, we define μ to be zero; for a series that converges for all x, we say μ is infinite.

4. The derivative of a power series is obtained by term-by-term differentiation, i.e., if $f(x) = \sum_{m=0}^{\infty} c_m(x - x_0)^m$, then

$$
f'(x) = c_1 + 2c_2(x - x_0) + 3c_3(x - x_0)^2 + \cdots
$$
$$
= \sum_{m=1}^{\infty} m c_m(x - x_0)^{m-1} = \sum_{m=0}^{\infty} (m + 1)c_{m+1}(x - x_0)^m.
$$

Further, the radii of convergence of these two series are the same. Similarly, the second derivative of $f(x)$ can be written as

$$
f''(x) = 2c_2 + 3.2c_3(x - x_0) + \cdots = \sum_{m=0}^{\infty} (m + 1)(m + 2)c_{m+2}(x - x_0)^m.
$$

Gamma and Beta functions. It is possible to write long expressions in very compact form using Gamma and Beta functions which we shall define now. The *Gamma function*, denoted by $\Gamma(x)$, is defined by

$$
\Gamma(x) = \int_0^{\infty} t^{x-1} e^{-t} dt, \quad x > 0. \tag{4.1}
$$

This improper integral converges only for $x > 0$, thus the Gamma function is defined by this formula only for the positive values of its arguments. It is easy to show that $\Gamma(1) = 1$, $\Gamma(1/2) = \sqrt{\pi}$, and the *recurrence formula*

$$
\Gamma(x + 1) = x\Gamma(x). \tag{4.2}
$$

From (4.2) it is immediate that for any nonnegative integer n the function $\Gamma(n + 1) = n!$, and hence the Gamma function can be considered as a generalization of the factorial function.

The *Beta function* $B(x, y)$ is defined as

$$
B(x, y) = \int_0^1 t^{x-1}(1 - t)^{y-1} dt, \tag{4.3}
$$

which converges for $x > 0$, $y > 0$. Gamma and Beta functions are related as follows

$$B(x, y) = \frac{\Gamma(x)\, \Gamma(y)}{\Gamma(x + y)}. \qquad (4.4)$$

Ordinary and singular points. Consider the differential equation

$$y'' + p_1(x)y' + p_2(x)y = 0. \qquad (4.5)$$

If at a point $x = x_0$ the functions $p_1(x)$ and $p_2(x)$ are analytic, then the point x_0 is said to be an *ordinary point* of the differential equation (4.5). Further, if at $x = x_0$ the functions $p_1(x)$ and/or $p_2(x)$ are not analytic, then x_0 is said to be a *singular point* of (4.5). Thus, if $p_1(x)$ and $p_2(x)$ are constants, then every point is an ordinary point for (4.5); for the equation $y'' + xy = 0$ also every point is an ordinary point; however, for Euler's equation $x^2 y'' + a_1 xy' + a_2 y = 0$ the point $x = 0$ is a singular point, but every other point is an ordinary point.

A singular point x_0 at which the functions $p(x) = (x - x_0)p_1(x)$ and $q(x) = (x - x_0)^2 p_2(x)$ are analytic is called a *regular singular point* of the differential equation (4.5). Thus, a second- order differential equation with a regular singular point x_0 has the form

$$y'' + \frac{p(x)}{(x - x_0)} y' + \frac{q(x)}{(x - x_0)^2} y = 0, \qquad (4.6)$$

where the functions $p(x)$ and $q(x)$ are analytic at $x = x_0$. Hence, for Euler's equation $x^2 y'' + a_1 xy' + a_2 y = 0$, the point $x = 0$ is a regular singular point. If a singular point x_0 is not a regular singular point, then it is called an *irregular singular point*.

Theorem 4.1 *Let the functions $p_1(x)$ and $p_2(x)$ be analytic at $x = x_0$, and hence these can be expressed as power series in $(x - x_0)$ in some interval $|x - x_0| < \mu$. Then, the differential equation (4.5) together with the initial conditions*

$$y(x_0) = c_0, \quad y'(x_0) = c_1 \qquad (4.7)$$

possesses a unique solution $y(x)$ that is analytic at x_0, and hence can be expressed as

$$y(x) = \sum_{m=0}^{\infty} c_m (x - x_0)^m \qquad (4.8)$$

in some interval $|x - x_0| < \mu$. The coefficients c_m, $m \geq 2$ in (4.8) can be obtained in terms of c_0 and c_1 by substituting it in the differential equation (4.5) directly.

Theorem 4.2 (Ferdinand Georg Frobenius (1849–1917)) *Let in the differential equation (4.6) the functions $p(x)$ and $q(x)$ be analytic at $x = x_0$, and hence these can be expressed as power series*

$$p(x) = \sum_{m=0}^{\infty} p_m (x - x_0)^m \quad \text{and} \quad q(x) = \sum_{m=0}^{\infty} q_m (x - x_0)^m$$

for $|x - x_0| < \mu$. Further, let r_1 and r_2 be the roots (called exponents) of the indicial equation

$$r(r - 1) + p_0 r + q_0 = 0. \tag{4.9}$$

Then,

(i)   if $Re(r_1) \geq Re(r_2)$ and $(r_1 - r_2)$ is not a nonnegative integer, then the two linearly independent solutions of the differential equation (4.6) are

$$y_1(x) = |x - x_0|^{r_1} \sum_{m=0}^{\infty} c_m (x - x_0)^m \tag{4.10}$$

and

$$y_2(x) = |x - x_0|^{r_2} \sum_{m=0}^{\infty} \overline{c}_m (x - x_0)^m, \tag{4.11}$$

(ii)   if the roots of the indicial equation are equal, i.e., $r_2 = r_1$ then the two linearly independent solutions of the differential equation (4.6) are (4.10) and

$$y_2(x) = y_1(x) \ln|x - x_0| + |x - x_0|^{r_1} \sum_{m=1}^{\infty} d_m (x - x_0)^m, \tag{4.12}$$

(iii) if the roots of the indicial equation are such that $r_1 - r_2 = n$ (a positive integer) then the two linearly independent solutions of the differential equation (4.6) are (4.10) and

$$y_2(x) = c y_1(x) \ln|x - x_0| + |x - x_0|^{r_2} \sum_{m=0}^{\infty} e_m (x - x_0)^m, \tag{4.13}$$

where the coefficients c_m, $\overline{c}_m$, d_m, e_m and the constant c can be determined by substituting the form of the series for $y(x)$ in the Eq. (4.6). The constant c may turn out to be zero, in which case there is no logarithmic term in the solution (4.13). Each of the solutions given in (4.10)–(4.13) converges at least for $0 < |x - x_0| < \mu$.

Example 4.1 Solutions of Sir George Biddell Airy's (1801–1892) differential equation

$$y'' - xy = 0 \tag{4.14}$$

known as *Airy functions* find applications in physical optics, radiative transfer, fluid mechanics, nonlinear wave propagation, propagation of light, electromagnetic waves, electromagnetic diffraction, radiowave propagation, fluid dynamics, theory of elasticity, and quantum mechanics. Clearly, for (4.14) hypotheses of Theorem 4.1 are satisfied for all x, and hence its solutions have power series expansion about any point $x = x_0$. In the case $x_0 = 0$, we assume that $y(x) = \sum_{m=0}^{\infty} c_m x^m$ is a solution of (4.14). A direct substitution of this in (4.14) gives

$$\sum_{m=0}^{\infty} (m+1)(m+2)c_{m+2}x^m - x \sum_{m=0}^{\infty} c_m x^m = 0,$$

which is the same as

$$2c_2 + \sum_{m=1}^{\infty} [(m+1)(m+2)c_{m+2} - c_{m-1}]x^m = 0.$$

Hence, it follows that

$$c_2 = 0, \quad c_m = \frac{1}{m(m-1)} c_{m-3}, \quad m \geq 3. \tag{4.15}$$

If $m = 3k + 2$, then (4.15) becomes

$$c_{3k+2} = \frac{1}{(3k+2)(3k+1)} c_{3k-1} = \frac{1 \cdot 2 \cdot 3 \cdot 6 \cdot 9 \cdots (3k)}{(3k+2)!} c_2 = 0, \quad k = 1, 2, \cdots.$$

If $m = 3k + 1$, then (4.15) is the same as

$$c_{3k+1} = \frac{1}{(3k+1)(3k)} c_{3k-2} = \frac{2 \cdot 5 \cdots (3k-1)}{(3k+1)!} c_1, \quad k = 1, 2, \cdots.$$

If $m = 3k$, then (4.15) reduces to

$$c_{3k} = \frac{1}{(3k)(3k-1)} c_{3k-3} = \frac{1 \cdot 4 \cdot 7 \cdots (3k-2)}{(3k)!} c_0, \quad k = 1, 2, \cdots.$$

Since

$$y(x) = c_0 + c_1 x + \sum_{k=1}^{\infty} c_{3k} x^{3k} + \sum_{k=1}^{\infty} c_{3k+1} x^{3k+1} + \sum_{k=1}^{\infty} c_{3k+2} x^{3k+2},$$

Airy functions are given by

$$y(x) = c_0 \left[1 + \sum_{k=1}^{\infty} \frac{1.4 \cdots (3k-2)}{(3k)!} x^{3k} \right] + c_1 \left[x + \sum_{k=1}^{\infty} \frac{2.5 \cdots (3k-1)}{(3k+1)!} x^{3k+1} \right]$$
$$= c_0 y_1(x) + c_1 y_2(x).$$
(4.16)

Finally, since $y_1(0) = 1$, $y_1'(0) = 0$, and $y_2(0) = 0$, $y_2'(0) = 1$ functions $y_1(x)$ and $y_2(x)$ are linearly independent solutions of Airy's equation (cf. Theorem 3.2).

Example 4.2 Charles *Hermite's* (1822–1901) *differential equation*

$$y'' - 2xy' + 2ay = 0 \tag{4.17}$$

is used in quantum mechanics to study the spatial position of a moving particle that undergoes simple harmonic motion in time. In quantum mechanics the exact position of a particle at a given time cannot be predicted, as in classical mechanics. It is possible to determine only the probability of the particle being at a given location at a given time. The unknown function $y(x)$ in (4.17) is then related to the probability of finding the particle at the position x. The constant a is related to the energy of the particle. This equation also arises in the solution of Pierre Simon de Laplace's (1749–1827) equation in parabolic coordinates. The solutions of (4.17) are called *Hermite functions*. Clearly, for (4.17) also hypotheses of Theorem 4.1 are satisfied for all x, and hence its solutions have power series expansion about any point $x = x_0$. In the case $x_0 = 0$, we again assume that $y(x) = \sum_{m=0}^{\infty} c_m x^m$ is a solution of (4.17), and obtain the recurrence relation

$$c_m = \frac{2(m-2-a)}{m(m-1)} c_{m-2}, \quad m = 2, 3, \cdots. \tag{4.18}$$

From (4.18) it is easy to find

$$c_{2m} = \frac{(-1)^m 2^{2m} \Gamma\left(\frac{1}{2}a + 1\right)}{(2m)! \, \Gamma\left(\frac{1}{2}a - m + 1\right)} c_0, \quad m = 0, 1, \cdots$$

and

$$c_{2m+1} = \frac{(-1)^m 2^{2m+1} \Gamma\left(\frac{1}{2}a + \frac{1}{2}\right)}{2(2m+1)! \, \Gamma\left(\frac{1}{2}a - m + \frac{1}{2}\right)} c_1, \quad m = 0, 1, \cdots.$$

Hence, Hermite functions can be written as

$$\begin{aligned} y(x) = {} & c_0 \, \Gamma\left(\frac{1}{2}a + 1\right) \sum_{m=0}^{\infty} \frac{(-1)^m (2x)^{2m}}{(2m)! \, \Gamma\left(\frac{1}{2}a - m + 1\right)} \\ & + c_1 \frac{1}{2} \, \Gamma\left(\frac{1}{2}a + \frac{1}{2}\right) \sum_{m=0}^{\infty} \frac{(-1)^m (2x)^{2m+1}}{(2m+1)! \, \Gamma\left(\frac{1}{2}a - m + \frac{1}{2}\right)} \\ & = c_0 y_1(x) + c_1 y_2(x). \end{aligned} \tag{4.19}$$

Obviously, $y_1(x)$ and $y_2(x)$ are linearly independent solutions of Hermite's equation.

If in (4.17), a is an even integer $2n$, then from (4.18) it is clear that $c_{2n+2} = c_{2n+4} = \cdots = 0$, i.e., $y_1(x)$ reduces to a polynomial of degree $2n$ involving only even powers of x. Similarly, if $a = 2n + 1$ then $y_2(x)$ reduces to a polynomial of degree $(2n + 1)$, involving only odd powers of x. Since $y_1(x)$ and $y_2(x)$ are themselves solutions of (4.17) we conclude that Hermite's differential equation has a polynomial solution for each nonnegative integer value of the parameter a. The interest is now to obtain these polynomials in descending powers of x. For this we note that the recurrence relation (4.18) can be written as

$$c_s = \frac{(s + 2)(s + 1)}{2(s - n)} c_{s+2}, \quad s \le n - 2.$$

This relation with $c_n = 2^n$ gives

$$c_{n-2m} = \frac{(-1)^m \, n! \, 2^{n-2m}}{(n - 2m)! \, m!}.$$

Thus, Hermite polynomials of degree n represented as $H_n(x)$ appear as

$$H_n(x) = \sum_{m=0}^{[n/2]} \frac{(-1)^m \, n!}{m! \, (n - 2m)!} (2x)^{n-2m}. \tag{4.20}$$

From (4.20), we find

$$\begin{aligned}
H_0(x) &= 1 \\
H_1(x) &= 2x \\
H_2(x) &= 4x^2 - 2 \\
H_3(x) &= 8x^3 - 12x \\
H_4(x) &= 16x^4 - 48x^2 + 12 \\
H_5(x) &= 32x^5 - 160x^3 + 120x.
\end{aligned}$$

Hermite polynomials find applications in probability, combinatorics, numerical analysis, physics, systems theory, and random matrix theory.

Example 4.3 The Pafnuty Lvovich *Chebyshev* (1821–1894) *differential equation*

$$(1 - x^2)y'' - xy' + a^2 y = 0, \tag{4.21}$$

where a is a real constant (parameter) finds applications in approximation theory and numerical computation of solutions to ordinary and partial differential equations. Since the functions

$$p_1(x) = -\frac{x}{1 - x^2} \quad \text{and} \quad p_2(x) = \frac{a^2}{1 - x^2}$$

are analytic for $|x| < 1$, $x = x_0 = 0$ is an ordinary point. Thus, Theorem 4.1 ensures that its series solution $y(x) = \sum_{m=0}^{\infty} c_m x^m$ converges for $|x| < 1$. To find this solution, we substitute it directly in (4.21), to find the recurrence relation

$$c_{m+2} = \frac{(m^2 - a^2)}{(m+2)(m+1)} c_m, \quad m \geq 0, \tag{4.22}$$

which can be solved to obtain

$$c_{2m} = \frac{(-a^2)(2^2 - a^2) \cdots ((2m-2)^2 - a^2)}{(2m)!} c_0, \quad m \geq 1$$

$$c_{2m+1} = \frac{(1^2 - a^2)(3^2 - a^2) \cdots ((2m-1)^2 - a^2)}{(2m+1)!} c_1, \quad m \geq 1.$$

Hence, the solution of (4.21) can be written as

$$y(x) = c_0 \left[1 + \sum_{m=1}^{\infty} \frac{(-a^2)(2^2 - a^2) \cdots ((2m-2)^2 - a^2)}{(2m)!} x^{2m} \right]$$

$$+ c_1 \left[x + \sum_{m=1}^{\infty} \frac{(1^2 - a^2)(3^2 - a^2) \cdots ((2m-1)^2 - a^2)}{(2m+1)!} x^{2m+1} \right] \tag{4.23}$$

$$= c_0 y_1(x) + c_1 y_2(x).$$

It is easy to verify that $y_1(x)$ and $y_2(x)$ are linearly independent solutions of Chebyshev's equation.

From the recurrence relation (4.22) it is clear that in the case $a = n$ one of the solutions of (4.21) reduces to a polynomial of degree n. To find this solution, in the differential equation (4.21) with $a = n$, we use the substitution $x = \cos\theta$, to obtain

$$\frac{d^2 y}{d\theta^2} + n^2 y = 0.$$

Since for this differential equation $\sin(n\theta)$ and $\cos(n\theta)$ are the solutions, the solutions of (4.21) with $a = n$ are $\sin(n \cos^{-1}(x))$ and $\cos(n \cos^{-1}(x))$. The solution $T_n(x) = \cos(n \cos^{-1}(x))$ is a polynomial in x of degree n, and is called the *Chebyshev polynomial of the first kind*. To find its explicit representation, we note the following obvious recurrence relation

$$T_{n+1}(x) = 2x T_n(x) - T_{n-1}(x), \quad n \geq 1$$

which gives

$$T_n(x) = \frac{n}{2} \sum_{m=0}^{[n/2]} (-1)^m \frac{(n-m-1)!}{m!(n-2m)!} (2x)^{n-2m}, \quad n \geq 1. \tag{4.24}$$

Since $T_0(x) = \cos(0) = 1$, $T_1(x) = \cos(\cos^{-1}(x)) = x$ from (4.24) it immediately follows that

$$T_2(x) = 2x^2 - 1$$
$$T_3(x) = 4x^3 - 3x$$
$$T_4(x) = 8x^4 - 8x^2 + 1$$
$$T_5(x) = 16x^5 - 20x^3 + 5x.$$

The nth degree *Chebyshev polynomial of the second kind* is defined by the relation

$$U_n(x) = \frac{1}{n+1} T'_{n+1}(x) = \sum_{m=0}^{[n/2]} (-1)^m \frac{(n-m)!}{m!\,(n-2m)!} (2x)^{n-2m}.$$

Example 4.4 The Adrien-Marie *Legendre* (1752–1833) *differential equation*

$$(1 - x^2)y'' - 2xy' + a(a+1)y = 0, \tag{4.25}$$

where a is a real constant (parameter), arises in problems such as the flow of an ideal fluid past a sphere, the determination of the electric field due to a charged sphere, and the determination of the temperature distribution in a sphere given its surface temperature. Since the functions

$$p_1(x) = -\frac{2x}{1-x^2} \quad \text{and} \quad p_2(x) = \frac{a(a+1)}{1-x^2}$$

are analytic for $|x| < 1$, $x = x_0 = 0$ is an ordinary point for (4.25). Thus, Theorem 4.1 ensures that its series solution $y(x) = \sum_{m=0}^{\infty} c_m x^m$ converges for $|x| < 1$. To find this solution, we substitute it directly in (4.25), to obtain

$$(1 - x^2) \sum_{m=0}^{\infty} (m+1)(m+2)c_{m+2}x^m - 2x \sum_{m=0}^{\infty} (m+1)c_{m+1}x^m$$

$$+ a(a+1) \sum_{m=0}^{\infty} c_m x^m = 0,$$

which is the same as

$$\sum_{m=0}^{\infty} \left[(m+1)(m+2)c_{m+2} - \{(m-1)m + 2m - a(a+1)\}\, c_m \right] x^m = 0,$$

or

$$\sum_{m=0}^{\infty} \left[(m+1)(m+2)c_{m+2} + (a+m+1)(a-m)c_m \right] x^m = 0.$$

But, this is possible if and only if

$$(m + 1)(m + 2)c_{m+2} + (a + m + 1)(a - m)c_m = 0, \quad m = 0, 1, \cdots$$

or

$$c_{m+2} = -\frac{(a + m + 1)(a - m)}{(m + 1)(m + 2)}c_m, \quad m = 0, 1, \cdots, \tag{4.26}$$

which is the required *recurrence relation*.

Now a little computation gives

$$
\begin{aligned}
c_{2m} &= \frac{(-1)^m (a + 2m - 1)(a + 2m - 3) \cdots (a + 1)a(a - 2) \cdots (a - 2m + 2)}{(2m)!}c_0 \\
&= (-1)^m \frac{\Gamma\left(\frac{1}{2}a + 1\right)\Gamma\left(\frac{1}{2}a + m + \frac{1}{2}\right)2^{2m}}{\Gamma\left(\frac{1}{2}a + \frac{1}{2}\right)\Gamma\left(\frac{1}{2}a - m + 1\right)(2m)!}c_0, \quad m = 1, 2, \cdots
\end{aligned}
\tag{4.27}
$$

and

$$
\begin{aligned}
c_{2m+1} &= \frac{(-1)^m (a + 2m)(a + 2m - 1) \cdots (a + 2)(a - 1)(a - 3) \cdots (a - 2m + 1)}{(2m + 1)!}c_1 \\
&= (-1)^m \frac{\Gamma\left(\frac{1}{2}a + \frac{1}{2}\right)\Gamma\left(\frac{1}{2}a + m + 1\right)2^{2m+1}}{2\,\Gamma\left(\frac{1}{2}a + 1\right)\Gamma\left(\frac{1}{2}a - m + \frac{1}{2}\right)(2m + 1)!}c_1, \quad m = 1, 2, \cdots.
\end{aligned}
\tag{4.28}
$$

Thus, the series solution of (4.25) can be written as

$$
\begin{aligned}
y(x) &= c_0 \left[1 - \frac{(a + 1)a}{2!}x^2 + \frac{(a + 3)(a + 1)a(a - 2)}{4!}x^4 - \cdots\right] \\
&\quad + c_1 \left[x - \frac{(a + 2)(a - 1)}{3!}x^3 + \frac{(a + 4)(a + 2)(a - 1)(a - 3)}{5!}x^5 - \cdots\right] \\
&= c_0 y_1(x) + c_1 y_2(x).
\end{aligned}
\tag{4.29}
$$

It is clear that $y_1(x)$ and $y_2(x)$ are linearly independent solutions of Legendre's equation.

Exactly as for Hermite's differential equation (4.17) if in (4.25), a is an even integer $2n$, then from (4.26) it is clear that $c_{2n+2} = c_{2n+4} = \cdots = 0$, i.e., $y_1(x)$ reduces to a polynomial of degree $2n$ involving only even powers of x. Similarly, if $a = 2n + 1$ then $y_2(x)$ reduces to a polynomial of degree $(2n + 1)$, involving only odd powers of x. Since $y_1(x)$ and $y_2(x)$ are themselves solutions of (4.25) we conclude that Legendre's differential equations has a polynomial solution for each nonnegative integer value of the parameter a. The interest is now to obtain these polynomials in descending powers of x. For this we note that the recurrence relation (4.26) can be written as

$$c_s = -\frac{(s+1)(s+2)}{(n-s)(n+s+1)}c_{s+2}, \quad s \leq n-2. \tag{4.30}$$

With the help of (4.30) we can express all nonvanishing coefficients in terms of the coefficient c_n of the highest power of x. It is customary to choose

$$c_n = \frac{(2n)!}{2^n(n!)^2} = \frac{1.3.5.\cdots(2n-1)}{n!} \tag{4.31}$$

so that the polynomial solution of (4.25) will have the value 1 at $x = 1$.

From (4.30) and (4.31) it is easy to obtain

$$c_{n-2m} = (-1)^m \frac{(2n-2m)!}{2^n\, m!\, (n-m)!\, (n-2m)!} \tag{4.32}$$

as long as $n - 2m \geq 0$.

The resulting solution of (4.25) is called the *Legendre polynomial of degree n* and is denoted as $P_n(x)$. From (4.32) this solution can be written as

$$P_n(x) = \sum_{m=0}^{\left[\frac{n}{2}\right]} (-1)^m \frac{(2n-2m)!}{2^n\, m!\, (n-m)!\, (n-2m)!} x^{n-2m}. \tag{4.33}$$

From (4.33), we easily obtain

$$P_0(x) = 1$$
$$P_1(x) = x$$
$$P_2(x) = \tfrac{1}{2}(3x^2 - 1)$$
$$P_3(x) = \tfrac{1}{2}(5x^3 - 3x)$$
$$P_4(x) = \tfrac{1}{8}(35x^4 - 30x^2 + 3)$$
$$P_5(x) = \tfrac{1}{8}(63x^5 - 70x^3 + 15x).$$

The other nonpolynomial solution of (4.25) is usually denoted as $Q_n(x)$, and this solution from (3.5) can be written as

$$Q_n(x) = P_n(x) \int^x \frac{dt}{(1-t^2)[P_n(t)]^2}. \tag{4.34}$$

In physics, Legendre polynomials often used in the study of Charles-Augustin de Coulomb's (1736–1806), gravitational, and magnetic fields. In applications involving either the Laplace or the Hermann Ludwig Ferdinand von Helmholtz (1821–1894) equation in spherical, oblate spheroidal, or prolate spheroidal coordinates the following *associated Legendre equation* arises

$$(1 - x^2)y'' - 2xy' + \left[n(n+1) - \frac{m^2}{1-x^2} \right] y = 0. \tag{4.35}$$

If $m = 0$ it reduces to (4.25) with $a = n$. When m and n are nonnegative integers the general solution of (4.35) can be written as

$$y(x) = A P_n^m(x) + B Q_n^m(x),$$

where $P_n^m(x)$ and $Q_n^m(x)$ are called *associated Legendre's functions of the first and second kinds* respectively, and in terms of $P_n(x)$ and $Q_n(x)$ are given by

$$P_n^m(x) = (1 - x^2)^{m/2} \frac{d^m}{dx^m} P_n(x), \quad Q_n^m(x) = (1 - x^2)^{m/2} \frac{d^m}{dx^m} Q_n(x).$$

Example 4.5 The Edmond Nicolas *Laguerre* (1834–1886) *differential equation*

$$xy'' + (a + 1 - x)y' + by = 0, \tag{4.36}$$

where a and b are real constants (parameters) arises in quantum mechanics. Clearly, in this equation $p(x) = (a + 1 - x)$ and $q(x) = bx$ are analytic for all x, and hence the point $x = 0$ is a regular singular point. Since $p_0 = a + 1$, $q_0 = 0$ the indicial equation is $r(r - 1) + (a + 1)r = 0$, and therefore, the exponents are $r_1 = 0$ and $r_2 = -a$. Thus, if a is not zero or an integer, a direct substitution of the form of solutions in Theorem 4.2(i) leads to the recurrence relations

$$m(m + a)c_m = (m - 1 - b)c_{m-1}$$

and

$$m(m - a)\bar{c}_m = (m - 1 - a - b)\bar{c}_{m-1}.$$

Hence, the solutions of (4.36) appear as

$$
\begin{aligned}
y_1(x) &= 1 - \frac{b}{a+1}x + \frac{b(b-1)}{2! \, (a+1)(a+2)}x^2 - \cdots \\
&= \sum_{m=0}^{\infty} \frac{(-1)^m \, \Gamma(a+1) \, \Gamma(b+1)}{m! \, \Gamma(m+a+1) \, \Gamma(b+1-m)} x^m
\end{aligned}
\tag{4.37}
$$

and

$$
\begin{aligned}
y_2(x) &= |x|^{-a} \left(1 - \frac{a+b}{1-a}x + \frac{1}{2!} \frac{(a+b)(a+b-1)}{(1-a)(2-a)}x^2 - \cdots \right) \\
&= |x|^{-a} \sum_{m=0}^{\infty} \frac{(-1)^m \, \Gamma(a+b+1) \, \Gamma(1-a)}{m! \, \Gamma(a+b+1-m) \, \Gamma(m+1-a)} x^m.
\end{aligned}
\tag{4.38}
$$

Clearly, both of these solutions converge at least for $0 < |x| < \infty$. Further, the general solution of (4.36) appears as $y(x) = Ay_1(x) + By_2(x)$, where A and B are arbitrary constants.

For $b = n$, the solution $y_1(x)$ obtained in (4.37) reduces to a polynomial of degree n. This solution multiplied by the constant $\Gamma(n + a + 1)/[n!\,\Gamma(a + 1)]$ is called the *Laguerre polynomial* $L_n^{(a)}(x)$ of degree n and it can be written as

$$L_n^{(a)}(x) = \Gamma(n + a + 1) \sum_{m=0}^{n} \frac{(-1)^m x^m}{m!\,(n - m)!\,\Gamma(m + a + 1)}. \qquad (4.39)$$

In the particular case $a = 0$, Laguerre polynomial of degree n is simply represented as $L_n(x)$ and reduces to

$$L_n(x) = \sum_{m=0}^{n} \frac{1}{m!}\binom{n}{m}(-x)^m. \qquad (4.40)$$

It follows that

$$L_n^{(p)}(x) = (-1)^p \frac{d^p}{dx^p}[L_{n+p}(x)], \quad p = 0, 1, 2, \cdots .$$

From (4.40), we have

$$
\begin{aligned}
L_0(x) &= 1 \\
L_1(x) &= -x + 1 \\
L_2(x) &= \tfrac{1}{2!}(x^2 - 4x + 2) \\
L_3(x) &= \tfrac{1}{3!}(-x^3 + 9x^2 - 18x + 6) \\
L_4(x) &= \tfrac{1}{4!}(x^4 - 16x^3 + 72x^2 - 96x + 24) \\
L_5(x) &= \tfrac{1}{5!}(-x^5 + 25x^4 - 200x^3 + 600x^2 - 600x + 120).
\end{aligned}
$$

Example 4.6 The Friedrich Wilhelm *Bessel* (1784–1846) *differential equation*

$$x^2 y'' + xy' + (x^2 - a^2)y = 0, \qquad (4.41)$$

with $a = n$ first appeared in the works of Daniel Bernoulli in 1732 and Euler in 1764; whereas *Bessel functions* which are the solutions of (4.41), also sometimes termed as *cylindrical functions*, were introduced by Bessel, in 1824, in the discussion of a problem in dynamical astronomy. Clearly, in (4.41) the functions

$$xp_1(x) = x\left(\frac{x}{x^2}\right) = 1 \quad \text{and} \quad x^2 p_2(x) = x^2\left(\frac{x^2 - a^2}{x^2}\right) = x^2 - a^2$$

are analytic for all x, and hence the origin is a regular singular point. Since $p_0 = 1$, $q_0 = -a^2$ the indicial equation is $r^2 - a^2 = 0$, and therefore the exponents are $r_1 = a$ and $r_2 = -a$. Thus, if $r_1 - r_2 = 2a$ is not an integer, a direct substitution of the form of solutions in Theorem 4.2(i) leads to the recurrence relations

$$m(m + 2a)c_m = -c_{m-2}, \quad m = 2, 3, \cdots \tag{4.42}$$

and

$$m(m - 2a)c_m = -c_{m-2}, \quad m = 2, 3, \cdots, \tag{4.43}$$

respectively. From these relations we easily obtain two linearly independent solutions $y_1(x)$ and $y_2(x)$, which appear as

$$y_1(x) = \left[1 - \frac{1}{2^2(1+a)1!}x^2 + \frac{1}{2^4(1+a)(2+a)2!}x^4 - \cdots \right] x^a c_0$$

and

$$y_2(x) = \left[1 - \frac{1}{2^2(1-a)1!}x^2 + \frac{1}{2^4(1-a)(2-a)2!}x^4 - \cdots \right] x^{-a} c_0^*.$$

In the above solutions we take the constants

$$c_0 = \frac{1}{2^a \Gamma(1+a)} \quad \text{and} \quad c_0^* = \frac{1}{2^{-a}\Gamma(1-a)},$$

to obtain

$$y_1(x) = \sum_{m=0}^{\infty} \frac{(-1)^m}{m! \, \Gamma(m+1+a)} \left(\frac{x}{2}\right)^{2m+a} \tag{4.44}$$

and

$$y_2(x) = \sum_{m=0}^{\infty} \frac{(-1)^m}{m! \, \Gamma(m+1-a)} \left(\frac{x}{2}\right)^{2m-a}. \tag{4.45}$$

These solutions are analytic for $|x| > 0$. The function $y_1(x)$ is called the *Bessel function* of order a of the *first kind* and is denoted by $J_a(x)$; $y_2(x)$ is the Bessel function of order $-a$ and is denoted by $J_{-a}(x)$. The general solution of the differential equation (4.41) is given by $y(x) = A J_a(x) + B J_{-a}(x)$, where A and B are arbitrary constants.

Similarly, for the case when $r_1 - r_2 = 2a$ is a positive odd integer, i.e., $2a = 2n + 1$ two solutions are $y_1(x) = J_{n+1/2}(x)$ and $y_2(x) = J_{-n-1/2}(x)$. Thus, in this case the general solution is $y(x) = A J_{n+1/2}(x) + B J_{-n-1/2}(x)$, where A and B are arbitrary constants.

If we take $2a$ as a negative odd integer, i.e., $2a = -2n - 1$, then two solutions are $y_1(x) = J_{-n-1/2}(x)$ and $y_2(x) = J_{n+1/2}(x)$. Thus, in this case the general solution is $y(x) = A J_{-n-1/2}(x) + B J_{n+1/2}(x)$, where A and B are arbitrary constants.

Next we consider the case when $a = 0$. In this case the exponents are $r_1 = r_2 = 0$. Here the first solution is

$$y_1(x) = J_0(x) = \sum_{m=0}^{\infty} \frac{(-1)^m}{(m!)^2} \left(\frac{x}{2}\right)^{2m} \tag{4.46}$$

and second solution denoted as $J^0(x)$, as an application of Theorem 4.2(ii), appears as

$$y_2(x) = J^0(x) = J_0(x) \ln |x| - \sum_{m=1}^{\infty} \frac{(-1)^m}{(m!)^2} \left(\sum_{k=1}^{m} \frac{1}{k}\right) \left(\frac{x}{2}\right)^{2m}. \tag{4.47}$$

The general solution in this case can be written as $y(x) = A J_0(x) + B J^0(x)$, where A and B are arbitrary constants.

Now we shall consider the remaining case, namely, when $r_1 - r_2$ is an even integer, i.e., $2a = 2n$. For $r_1 = n$, there is no difficulty and the first solution can be written as $y_1(x) = J_n(x)$. However, for the second solution corresponding to the exponent $r_2 = -n$ we need to use Theorem 4.2(iii), to obtain

$$y_2(x) = \frac{2}{\pi} \left[\left(\gamma + \ln \left|\frac{x}{2}\right|\right) J_n(x) - \frac{1}{2} \sum_{k=0}^{n-1} \frac{(n-k-1)!}{k!} \left(\frac{x}{2}\right)^{2k-n} \right.$$
$$\left. + \frac{1}{2} \sum_{m=0}^{\infty} (-1)^{m+1} \frac{\phi(m) + \phi(n+m)}{m!\,(n+m)!} \left(\frac{x}{2}\right)^{2m+n} \right], \tag{4.48}$$

where $\phi(0) = 0$, $\phi(m) = \sum_{k=1}^{m} k^{-1}$ and γ is the *Euler constant* defined by

$$\gamma = \lim_{m \to \infty} [\phi(m) - \ln(m)] \approx 0.5772157.$$

This solution $y_2(x)$ is known as Heinrich Friedrich *Weber's* (1843–1912) *Bessel function of the second kind* and is denoted by $Y_n(x)$. In the literature some authors also call $Y_n(x)$ the Karl Gottfried *Neumann* (1832–1925) *function*.

Thus, the general solution in this case can be written as $y(x) = A J_n(x) + B Y_n(x)$.

The *Bessel functions of the third kind* also known as Hermann *Hankel* (1839–1873) *functions* are the complex solutions of the differential equation (4.41) for $a = n$ and are defined by the relations

$$H_n^{(1)}(x) = J_n(x) + i Y_n(x) \tag{4.49}$$

and

$$H_n^{(2)}(x) = J_n(x) - i Y_n(x). \tag{4.50}$$

The *modified Bessel function of the first kind of order n* is defined as

$$I_n(x) = i^{-n} J_n(ix) = e^{-n\pi i/2} J_n(ix). \tag{4.51}$$

If n is an integer, $I_{-n}(x) = I_n(x)$; but if n is not an integer, $I_n(x)$ and $I_{-n}(x)$ are linearly independent.

The *modified Bessel function of the second kind of order n* is defined as

$$K_n(x) = \begin{cases} \dfrac{\pi}{2} \dfrac{I_{-n}(x) - I_n(x)}{\sin(n\pi)}, & n \neq 0, 1, 2, \cdots \\[3mm] \lim\limits_{p \to n} \left(\dfrac{\pi}{2} \dfrac{I_{-p}(x) - I_p(x)}{\sin(p\pi)} \right), & n = 0, 1, 2, \cdots . \end{cases} \tag{4.52}$$

These functions are the solutions of the *modified Bessel differential equation*

$$x^2 y'' + xy' - (x^2 + n^2)y = 0. \tag{4.53}$$

Example 4.7 Hypergeometric differential equation

$$x(1 - x)y'' + [c - (a + b + 1)x]y' - aby = 0, \tag{4.54}$$

where a, b and c are parameters, finds applications in several problems of mathematical physics, quantum mechanics, and fluid dynamics. Its solutions are known as *hypergeometric functions*. The term hypergeometric series was first used by John Wallis (1616–1703) in 1655 and used by Euler, but the first full systematic treatment was given by Karl Friedrich Gauss (1777–1855) in 1813. The fundamental treatment of hypergeometric function by means of the differential equation is due to George Friedrich Bernhard Riemann (1826–1866) in 1857. It is clear that $x = 0$ and 1 are regular singular points of (4.54), whereas all other points are ordinary points. Also, in the neighborhood of zero we have $p_0 = c$, $q_0 = 0$, and the indicial equation $r(r - 1) + cr = 0$ has the roots $r_1 = 0$ and $r_2 = 1 - c$. On substituting directly $y(x) = \sum_{m=0}^{\infty} c_m x^m$ in the equation (4.54), we obtain the first solution

$$y_1(x) = F(a, b, c, x) = \left(\sum_{m=0}^{\infty} \frac{\Gamma(a + m)\, \Gamma(b + m)}{\Gamma(c + m)\, m!} x^m \right) \frac{\Gamma(c)}{\Gamma(a)\, \Gamma(b)}. \tag{4.55}$$

The second solution with the exponent $r_2 = 1 - c$ when c is neither zero nor a negative integer can be obtained as follows: In the differential equation (4.54) using the substitution $y = x^{1-c} w$, we obtain

$$x(1 - x)w'' + [c_1 - (a_1 + b_1 + 1)x]w' - a_1 b_1 w = 0, \tag{4.56}$$

where $c_1 = 2 - c$, $a_1 = a - c + 1$, $b_1 = b - c + 1$. This differential equation has a series solution $w(x) = F(a_1, b_1, c_1, x)$, and hence the second solution of the differential equation (4.54) is

$$y_2(x) = x^{1-c}F(a-c+1, b-c+1, 2-c, x). \tag{4.57}$$

The general solution of the differential equation (4.54) in the neighborhood of $x = 0$ is a linear combination of the two solutions (4.55) and (4.57).

Solutions of the differential equation (4.54) at the singular point $x = 1$ can be obtained directly or may be deduced from the preceding solutions by a change of independent variable $t = 1 - x$. Indeed, with this substitution differential equation (4.54) reduces to

$$t(1-t)\frac{d^2y}{dt^2} + [c_1 - (a+b+1)t]\frac{dt}{dt} - aby = 0, \tag{4.58}$$

where $c_1 = a + b - c + 1$.

Thus, we have the solutions

$$y_1(x) = F(a, b, a+b-c+1, 1-x) \tag{4.59}$$

and

$$y_2(x) = (1-x)^{c-a-b}F(c-b, c-a, c-a-b+1, 1-x) \tag{4.60}$$

provided $(c - a - b)$ is not a positive integer.

Problems

4.1 Show that $x_0 = 0$ is a regular singular point of the *Riccati–Bessel equation* which arises in the problem of scattering of electromagnetic waves by a sphere

$$x^2y'' - (x^2 - k)y = 0, \quad -\infty < k < \infty.$$

4.2 Show that $x_0 = 0$ is a regular singular point of the *Coulomb wave equation*

$$x^2y'' + [x^2 - 2\ell x - k]y = 0, \quad \ell \text{ fixed}, \quad -\infty < k < \infty.$$

4.3 Show that the substitution $x = 1/t$ transforms the differential equation (4.5) into the form

$$\frac{d^2y}{dt^2} + \left(\frac{2}{t} - \frac{1}{t^2}p_1\left(\frac{1}{t}\right)\right)\frac{dy}{dt} + \frac{1}{t^4}p_2\left(\frac{1}{t}\right)y = 0. \tag{4.61}$$

Thus, the nature of the point $x = \infty$ of (4.5) is the same as the nature of the point $t = 0$ of (4.61). Use this to examine the nature of the point at infinity for the following: differential equations

(1) Airy's differential equation (4.14)   (irregular singular),
(2) Hermite's differential equation (4.17)   (irregular singular),
(3) Chebyshev's differential equation (4.21)   (regular singular),
(4) Legendre's differential equation (4.25)   (regular singular),
(5) Laguerre's differential equation (4.36)   (irregular singular),
(6) Bessel's differential equation (4.41)   (irregular singular point), and
(7) Hypergeometric differential equation (4.54)   (regular singular).

4.4 The Erwin *Schrödinger* (1887–1961) *wave equation* for a simple harmonic oscillator is

$$-\frac{h^2}{8\pi^2 m}\frac{d^2\psi}{dz^2} + \frac{K}{2}z^2\psi = E\psi, \tag{4.62}$$

where h is Max Planck's (1858–1947) constant, E, K and m are positive real numbers, and $\psi(x)$ is the Schrödinger wave function. Show that the change to dimensionless coordinate $x = \alpha z$ reduces (4.62) to

$$\frac{d^2\psi}{dx^2} + (2a + 1 - x^2)\psi = 0, \tag{4.63}$$

where $\alpha^4 = 4\pi^2 m K/h^2$ and $2a + 1 = (4\pi E/h)\sqrt{m/K}$. Further, show that the second change of variables $\psi = ye^{-x^2/2}$ reduces (4.63) to the Hermite equation (4.17).

4.5 In a work on trigonometric components of a frequency-modulated wave, Cambi [4] in 1948 studied the differential equation

$$[1 + a\cos(2x)]y'' + \lambda y = 0, \quad a \neq -1, \tag{4.64}$$

where a and λ are physical parameters. Find first few terms of the series solution $y(x) = \sum_{m=0}^{\infty} a_m x^m$ of (4.64) in terms of a, λ, a_0 and a_1.

4.6 The differential equation

$$\frac{d^2y}{dt^2} + [a - 2b\cos(2t)]y = 0,$$

where a and b are constants was introduced in 1868 by Émile Léonard Mathieu (1835–1890) in a discussion of the vibration of elliptic membrane. Find first few terms of the power series solution about $t = 0$ of this equation subject to the conditions $y(0) = 1$, $y(0) = 0$.

4.7 In building design it is sometimes useful to use supporting columns that are special geometrical designs. In studying the buckling of columns of varying cross sections, we obtain the following differential equation:

$$x^n y'' + k^2 y = 0, \tag{4.65}$$

where $k > 0$ and n is a positive integer. In particular, if $n = 1$, the column is rectangular with one dimension constant, whereas if $n = 4$, the column is a truncated pyramid or cone. Show that for the case $n = 1$, the point $x = 0$ is regular singular with exponents 0 and 1. Also, find the series solution at $x = 0$ for the exponent

1. $\left(c_0 \left[x - \frac{k^2}{2!} x^2 + \frac{k^4}{2!3!} x^3 - \frac{k^6}{3!4!} x^4 + \frac{k^8}{4!5!} x^5 - \cdots \right] \right)$

4.8 A supply of hot air can be obtained by passing the air through a heated cylindrical tube. It can be shown that the temperature T of the air in the tube satisfies the differential equation

$$\frac{d^2 T}{dx^2} - \frac{upC}{kA} \frac{dT}{dx} + \frac{2\pi rh}{kA} (T_w - T) = 0, \tag{4.66}$$

where $x =$ distance from intake end of the tube, $u =$ flow rate of air, $p =$ density of air, $C =$ heat capacity of air, $k =$ thermal conductivity, $A =$ cross-sectional area of the tube, $r =$ radius of the tube, $h =$ heat transfer coefficient of air (nonconstant), $T_w =$ temperature of the tube. For the parameters Jenson and Jefferys [5] have taken, the differential equation (4.66) becomes

$$\frac{d^2 T}{dx^2} - 26200 \frac{dT}{dx} - 11430x^{-1/2}(T_w - T) = 0. \tag{4.67}$$

(1) Show that the substitution $y = T_w - T$, $x = z^2$ transforms (4.67) into

$$z \frac{d^2 y}{dz^2} - (1 + 52400z^2) \frac{dy}{dz} - 45720z^2 y = 0 \tag{4.68}$$

for which $z = 0$ is a regular singular point with exponents 0 and 2.

(2) Find first few terms of the series solution of (4.68) at $z = 0$ for the exponent
2. $\left(c_0 \left[z^2 + 13100z^4 + 3048z^5 + \frac{343220000}{3} z^6 + \cdots \right] \right)$

4.9 A large-diameter pipe such as the 30-ft diameter pipe used in the construction of Hoover Dam is strengthened by a device called a *stiffener ring*. To cut down the stress on the stiffener ring, a fillet insert device is used. In determining the radial displacement of the fillet insert due to internal water pressure, one encounters the fourth-order equation

$$x^2 y^{(4)} + 6x y''' + 6y'' + y = 0, \quad x > 0. \tag{4.69}$$

Here y is proportional to the radial displacement and x is proportional to the distance measured along an inside element of the pipe shell from some fixed point. Find series solution of (4.69) at $x = 0$ for which the limit as $x \to 0$ exists. $\left(c_0 \sum_{m=0}^{\infty} \frac{(-1)^m x^{2m}}{((2m)!)^2 (2m+1)} \right.$

$\left. + c_1 \sum_{m=0}^{\infty} \frac{(-1)^m x^{2m+1}}{((2m+1)!)^2 (m+1)} \right)$

4.10 The differential equation

$$\frac{d(y^2 y')}{dx} + xy' = 0 \tag{4.70}$$

arises in the application of high-temperature superconductors. Find first few terms of the power series solution about $x = 0$ of (4.70) subject to the conditions $y(0) = 1$, $y'(0) = 0$.

4.11 In recent years Paul *Painlevé's* (1863–1933) *equations* (*transcendent*) have occurred in modern geometry and statistical mechanics

$$\begin{aligned} y'' &= 6y^2 + \lambda x, \\ y'' &= 2y^3 + xy + \mu, \end{aligned}$$

where λ and μ are constants. Find first few terms of the power series solution about $x = 0$ of the above equations subject to the conditions $y(0) = 1$, $y'(0) = 0$.

4.12 Albert Einstein (1879–1955) in his study of the orbital motion of planets under the assumptions of general relativity (the problem of the perihelion shift) led to the differential equation

$$y'' + y = a + by^2, \tag{4.71}$$

where a and b are constants. Find first few terms of the power series solution about $x = 0$ of (4.71) subject to the conditions $y(0) = 1$, $y'(0) = 0$.

4.13 Paul Gerber (1854–1909) in 1898 in an investigation of the velocity of gravitation was led to correct perihelion shift for Mercury by solving the differential equation

$$(1 + \gamma y)y'' + y = \alpha - \beta(y')^2, \tag{4.72}$$

where α, β and γ are constants. Find first few terms of the power series solution about $x = 0$ of (4.72) subject to the conditions $y(0) = 1$, $y'(0) = 0$.

4.14 Balthasar *van der Pol's* (1889–1959) *equation*

$$y'' + \mu(y^2 - 1)y' + y = 0 \tag{4.73}$$

finds applications in physics and electrical engineering. It first arose as an idealized description of a spontaneously oscillating circuit. Find first few terms of the power series solution about $x = 0$ of (4.73) with $\mu = 1$ subject to the conditions $y(0) = 0$, $y'(0) = 1$. $(x + \frac{1}{2}x^2 - \frac{1}{8}x^4 - \frac{1}{8}x^5)$

4.15 John William Strutt, Lord *Rayleigh's* (1842–1919) *equation*

$$my'' + ky = ay' - b(y')^3 \tag{4.74}$$

models the oscillation of a clarinet reed. Find first few terms of the power series solution about $x = 0$ of (4.74) with $m = k = a = 1$, $b = 1/3$ subject to the conditions $y(0) = 1$, $y'(0) = 0$. $(1 - \frac{1}{2}x^2 - \frac{1}{6}x^3 + \frac{1}{40}x^5)$

4.16 Georg *Duffing's* (1861–1944) *equation*

$$y'' + \delta y' + \alpha y + \beta y^3 = \gamma \cos(\omega x) \tag{4.75}$$

describes the motion of a damped oscillator with a more complex potential than in simple harmonic motion, in physical terms, it models, for example, a spring pendulum whose spring's stiffness does not exactly obey Hooke's law. In (4.75), δ controls the amount of damping, α controls the linear stiffness, β controls the amount of nonlinearity in the restoring force, γ is the amplitude of the periodic driving force, and ω is the angular frequency of the periodic driving force. In particular, when $\beta = -1$ and $\gamma = 0$, it models the swaying of a tall building with P-Delta effect (in structural engineering, P-Delta effect refers to the abrupt changes in ground shear, overturning moment, and/or the axial force distribution at the base of a sufficiently tall structure or structural component when it is subject to a critical lateral displacement). For the values of these parameters $\alpha = 1$, $\beta = 5$, $\delta = 0.02$, $\gamma = 8$ and $\omega = 0.5$, Jules Henri Poincaré (1854–1912) suggested *chaotic behavior* (highly sensitive to initial conditions). With the same values of these parameters find first few terms of the power series solution about $y = 0$ of (4.75) subject to the conditions $y(0) = 1$, $y'(0) = 0$.

4.17 Consider a spherical cloud of gas and denote its total pressure at a distance r from the center by $p(r)$. The total pressure is due to the usual gas pressure and a contribution from radiation,

$$p = \frac{1}{3}aT^4 + \frac{RT}{v},$$

where a, T, R and v are respectively the radiation constant, the absolute temperature, the gas constant, and the volume. Pressure and density $\rho = v^{-1}$ vary with r and $p = K\rho^\gamma$ where γ and K are constants. Let m be the mass within a sphere of radius r and G the constant of gravitation. The equilibrium equations for the configuration are

$$\frac{dp}{dr} = -\frac{Gm\rho}{r^2} \quad \text{and} \quad \frac{dm}{dr} = 4\pi r^2 \rho.$$

Eliminating m yields

$$\frac{1}{r^2}\frac{d}{dr}\left(\frac{r^2}{\rho}\frac{dp}{dr}\right) + 4\pi G\rho = 0.$$

Now let $\gamma = 1 + \mu^{-1}$ and set $\rho = \lambda\phi^\mu$ so

$$p = K\rho^{1+\mu^{-1}} = K\lambda^{1+\mu^{-1}}\phi^{\mu+1}.$$

Thus, it follows that

$$\frac{1}{r^2}\frac{d}{dr}\left(r^2\frac{d\phi}{dr}\right) + k^2\phi^\mu = 0$$

with

$$k^2 = \frac{4\pi G\lambda^{1-\mu^{-1}}}{(\mu+1)K}.$$

Now with $x = kr$, we have

$$\frac{d^2\phi}{dx^2} + \frac{2}{x}\frac{d\phi}{dr} + \phi^\mu = 0.$$

If we let $\lambda = \rho_0$, the density at $r = 0$, then we may take $\phi = 1$ at $x = 0$. By symmetry the other condition is $d\phi/dx = 0$ when $x = 0$. A solution of the differential equation satisfying these initial conditions is called a Lane–Emden function (after Jonathan Homer Lane (1819–1880) and Jacob Robert Emden (1862–1940)), of index $\mu = (\gamma - 1)^{-1}$. The differential equation

$$y'' + \frac{2}{x}y' + g(y) = 0$$

was first studied by Emden when he examined the thermal behavior of spherical clouds of gas acting on gravitational equilibrium and subject to the laws of thermodynamics. The usual interest is in the case $g(y) = y^n$, $n \geq 1$, i.e., in the equation

$$xy'' + 2y' + xy^n = 0, \tag{4.76}$$

which was treated by Subramanyan Chandrasekhar (1910–1995) in his study of stellar structure (see [6] for details),. Show that John Robinson Airey's (1868–1937) series solution of (4.76), with the natural initial conditions $y(0) = 1$, $y'(0) = 0$ is

$$y_n(x) = 1 - \frac{x^2}{3!} + n\frac{x^4}{5!} + (5n - 8n^2)\frac{x^6}{3\cdot 7!} + (70n - 183n^2 + 122n^3)\frac{x^8}{9\cdot 9!}$$
$$+ (3150n - 1080n^2 + 12642n^3 - 5032n^4)\frac{x^{10}}{45\cdot 11!} + \cdots.$$

Thus, in particular deduce that

$$y_0(x) = 1 - \frac{x^2}{6}, \quad y_1(x) = \frac{\sin(x)}{x}, \quad y_2(x) = \frac{\sqrt{3}}{\sqrt{3 + x^2}}.$$

4.18 The *Lane–Emden equation*

$$xy'' + 2y' + xe^y = 0 \tag{4.77}$$

appears in a study of isothermal gas spheres. Show that the series solution of (4.77) with the initial conditions $y(0) = y'(0) = 0$ can be written as

$$y(x) = -\frac{1}{6}x^2 + \frac{1}{5 \cdot 4!}x^4 - \frac{8}{21 \cdot 6!}x^6 + \frac{122}{81 \cdot 8!}x^8 - \frac{61 \cdot 67}{495 \cdot 10!}x^{10} + \cdots .$$

4.19 Sir Owen Willans *Richardson's* (1879–1959) *equation* (see Example 9.5)

$$xy'' + 2y' + xe^{-y} = 0 \tag{4.78}$$

appears in the theory of thermionic currents when one seeks to determine the density and electric force of an electron gas in the neighborhood of a hot body in thermal equilibrium. Find first few terms of the series expansion of the solution of (4.78) satisfying the initial conditions $y(0) = y'(0) = 0$. (If $y(x)$ is a solution of (4.77) then $-y(ix)$ a solution of (4.78), $y(x) = -\frac{1}{6}x^2 - \frac{1}{5\cdot4!}x^4 - \frac{8}{21\cdot6!}x^6 - \frac{122}{81\cdot8!}x^8 - \frac{61\cdot67}{495\cdot10!}x^{10} \cdots)$

4.20 The *White-dwarf* equation

$$xy'' + 2y' + x(y^2 - C)^{3/2} = 0 \tag{4.79}$$

was introduced by Chandrasekhar in his study of gravitational potential of the degenerate (white-dwarf) stars. This equation for $C = 0$ is the same as (4.76) with $n = 3$. Show that the series solution of (4.79) with the initial conditions $y(0) = 1, \; y'(0) = 0$ in terms of $q^2 = 1 - C$ can be written as

$$y(x) = 1 - \frac{q^3}{6}x^2 + \frac{q^4}{40}x^4 - \frac{q^5}{7!}(5q^2 + 14)x^6 + \frac{q^6}{3 \cdot 9!}(339q^2 + 280)x^8$$
$$+ \frac{q^7}{5 \cdot 11!}(1425q^4 + 11436q^2 + 4256)x^{10} + \cdots .$$

4.21 The Irving *Langmuir* (1881–1957) *equation*

$$3yy'' + 4yy' + (y')^2 - 1 + y^2 = 0 \tag{4.80}$$

appears in the theory of currents limited by a space charge between coaxial cables. Find first few terms of the power series solution about $x = 0$ of (4.80) subject to the conditions $y(0) = 1, \; y'(0) = 0$.

4.22 A nontrivial solution of the differential equation

$$y'' + q(x)y = 0 \tag{4.81}$$

is said to be *oscillatory* if it has no last zero, i.e., if $y(x_1) = 0$ then there exists a $x_2 > x_1$ such that $y(x_2) = 0$. Equation (4.81) itself is said to be oscillatory if every solution of (4.81) is oscillatory. A solution which is not oscillatory is called *nonoscillatory*. For example, the differential equation $y'' + y = 0$ is oscillatory, whereas $y'' - y = 0$ is nonoscillatory in $J = [0, \infty)$. It is known that if the function $q(x)$ is continuous in $J = (0, \infty)$, and $\int^\infty q(x)dx = \infty$, then (3.8) is oscillatory in J. Show that the solutions of Bessel's differential equation (4.41) for all a are oscillatory.   (Use the substitution $y(x) = z(x)/\sqrt{x}$ to obtain the differential equation $z'' + \left(1 + \frac{1-4a^2}{4x^2}\right)z = 0$)

4.23 In a study of planetary motion Bessel encountered the integral

$$y(x) = \frac{1}{\pi} \int_0^\pi \cos(n\theta - x\sin(\theta))d\theta.$$

Show that $y(x) = J_n(x)$.

4.24 Show that for any $a \geq 0$, the function

$$y(x) = x^a \int_0^\pi \cos(x\cos(\theta))\sin^{2a}(\theta)d\theta$$

is a solution to the Bessel differential equation (4.41).

4.25 Show that the transformation $x = \alpha t^\beta$, $y = t^\gamma w$, where α, β and γ are constants, converts Bessel's differential equation (4.41) to

$$t^2\frac{d^2w}{dt^2} + (2\gamma + 1)t\frac{dw}{dt} + (\alpha^2\beta^2 t^{2\beta} + \gamma^2 - a^2\beta^2)w = 0. \tag{4.82}$$

For $\beta = 1$, $\gamma = 0$, $a = n$ (nonnegative integer) (4.82) reduces to

$$t^2\frac{d^2w}{dt^2} + t\frac{dw}{dt} + (\alpha^2 t^2 - n^2)w = 0. \tag{4.83}$$

(1)   Show that the solution of the differential equation

$$\frac{d^2w}{dt^2} + t^m w = 0, \quad m > 0$$

can be expressed as

$$w(t) = \sqrt{t}\left[AJ_{1/(m+2)}\left(\frac{2}{m+2}t^{(m+2)/2}\right) + BJ_{-1/(m+2)}\left(\frac{2}{m+2}t^{(m+2)/2}\right)\right].$$

Thus, in particular, the solution of Airy's differential equation (4.14) can be written as

$$y(t) = \sqrt{t}\left(AJ_{1/3}\left(\frac{2}{3}t^{3/2}\right) + BJ_{-1/3}\left(\frac{2}{3}t^{3/2}\right)\right),$$

where $t = -x$.

(2) Express the solution of differential equation

$$\frac{d^2w}{dt^2} + w = 0$$

in terms of Bessel functions.

4.26 The differential equation

$$y' = x^2 + y^2,$$

was first considered unsuccessfully by John Bernoulli (1667–1748) in 1694. About 9 years later in 1703 his brother James Bernoulli obtained a solution in the form $y = F(x)/G(x)$, where $F(x)$ and $G(x)$ are power series in x. Dividing $F(x)$ by $G(x)$ he obtained the following series solution

$$y(x) = \frac{1}{3}x^3 + \frac{1}{3^2 \cdot 7}x^7 + \frac{2}{3^3 \cdot 7 \cdot 11}x^{11} + \frac{13}{3^4 \cdot 5 \cdot 7^2 \cdot 11}x^{15} + \cdots.$$

Show that the general solution of this differential equation can be written as

$$y(x) = x\frac{J_{-3/4}\left(\frac{1}{2}x^2\right) + cJ_{3/4}\left(\frac{1}{2}x^2\right)}{cJ_{-1/4}\left(\frac{1}{2}x^2\right) - J_{1/4}\left(\frac{1}{2}x^2\right)},$$

where c is an arbitrary constant.   (The transformation $y = -u'/u$ converts the differential equation $y' = x^2 + y^2$ to $u'' + x^2u = 0$)

4.27 Consider a long, flat triangular piece of metal whose ends are joined by an inextensible piece of string of length slightly less than that of the piece of metal (see Figure 4.1). The line of the string is taken as the x-axis, with the left end as the origin. The deflection $y(x)$ of the piece of metal from horizontal at x satisfies the differential equation

$$EIy'' + Ty = 0,$$

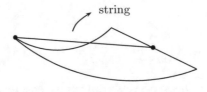

Fig. 4.1

where T is the tension in the string, E is Young's modulus, and I is the moment of inertia of a cross section of the piece of metal. Since the metal is triangular, $I = \alpha x$ for some constant $\alpha > 0$. The above equation thus can be written as

$$xy'' + k^2 y = 0,$$

where $k^2 = T/E\alpha$. Find the general solution of the above equation. $\left(y(x) = \sqrt{x} \left[A J_1 \left(2k\sqrt{x} \right) + B J_{-1} \left(2k\sqrt{x} \right) \right] \right)$

4.28 Consider a vertical column of length ℓ, such as a telephone pole (see Fig. 4.2). In certain cases it leads to the differential equation

$$\frac{d^2 y}{dx^2} + \frac{P}{EI} e^{kx/\ell}(y - a) = 0,$$

where E is the modulus of elasticity, I the moment of inertia at the base of the column about an axis perpendicular to the plane of bending, and P, k and a are constants.

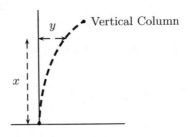

Fig. 4.2

Find the general solutions of the above equation. (Use $z = y - a$, $2\mu = k/\ell$, $\nu = \sqrt{P/EI}$ to convert the differential equation as $\frac{d^2 z}{dx^2} + \nu^2 e^{2\mu x} z = 0$. Now use $t = \nu e^{\mu x}$ to obtain $t^2 \frac{d^2 z}{dt^2} + t \frac{dz}{dt} + \frac{t^2}{\mu^2} z = 0$. Finally, compare this with (4.82).

4.29 A simple pendulum initially has the length ℓ_0 and makes an angle θ_0 with the vertical line. It is then released from this position. If the length of the pendulum increases with time t according to $\ell = \ell_0 + \epsilon t$, where ϵ is a small constant, then the angle θ that the pendulum makes with the vertical line, assuming that the oscillations are small, is the solution of the initial value problem

$$(\ell_0 + \epsilon t)\frac{d^2\theta}{dt^2} + 2\epsilon\frac{d\theta}{dt} + g\theta = 0, \quad \theta(0) = \theta_0, \quad \frac{d\theta}{dt}(0) = 0,$$

where g is the acceleration due to gravity. Find the solution of the above initial value problem.

$(\theta(t)) = \frac{1}{\sqrt{\ell_0 + \epsilon t}}\left[AJ_1\left(\frac{2\sqrt{g}}{\epsilon}\sqrt{\ell_0 + \epsilon t}\right) + BJ_{-1}\left(\frac{2\sqrt{g}}{\epsilon}\sqrt{\ell_0 + \epsilon t}\right)\right]$ where $A = \frac{\sqrt{\ell_0}J'_{-1}(c) - (\epsilon/2\sqrt{g})J_{-1}(c)}{J_1(c)J'_{-1}(c) - J_{-1}(c)J'_1(c)}\theta_0$, $B = \frac{(\epsilon/2\sqrt{g})J_1(c) - \sqrt{\ell_0}J'_1(c)}{J_1(c)J'_{-1}(c) - J_{-1}(c)J'_1(c)}\theta_0$ and $c = \frac{2\sqrt{g\ell_0}}{\epsilon}$)

4.30 The differential equation

$$x^2\frac{d^2E}{dx^2} - \mu x\frac{dE}{dx} - k^2x^\nu E = 0$$

occurs in the study of the flow of current through electrical transmission lines; here μ, ν are positive constants and E represents the potential difference (with respect to one end of the line) at a point a distance x from that end of the line. Find its general solution. (The transformation $x = \alpha t^\beta$, $y = t^\gamma w$, where α, β and γ are constants, converts (4.53) to

$$t^2\frac{d^2w}{dt^2} + (2\gamma + 1)t\frac{dw}{dt} + (-\alpha^2\beta^2 t^{2\beta} + \gamma^2 - n^2\beta^2)w = 0. \tag{4.84}$$

$E(x) = x^{(1+\mu)/2}\left[AI_{(1+\mu)/\nu}\left(\frac{2k}{\nu}x^{\nu/2}\right) + BK_{(1+\mu)/\nu}\left(\frac{2k}{\nu}x^{\nu/2}\right)\right])$

4.31 Consider the wedge-shaped cooling fin as in Fig. 4.3. The differential equation which describes the heat flow through and from this wedge is

$$x^2\frac{d^2y}{dx^2} + x\frac{dy}{dx} - \mu xy = 0,$$

where x is distance from tip to fin, T temperature of fin at x, T_0 constant temperature of surrounding air, $y = T - T_0$, h heat transfer coefficient from outside surface of fin to the surrounding air, k thermal conductivity of fin material, ℓ length of fin, w thickness of fin at its base, θ one-half the wedge angle, and $\mu = 2h\sec(\ell\theta)/kw$.

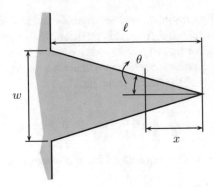

Fig. 4.3

Find the general solutions of the above equation.   (Compare with (4.84) to get
$y(x) = A I_0 \left(2\sqrt{\mu}x^{1/2}\right) + B K_0 \left(2\sqrt{\mu}x^{1/2}\right))$

4.32 Consider a horizontal beam of length 2ℓ supported at both ends, and laterally
fixed at its left end. The beam carries a uniform load w per unit length and is subjected
to a constant end-pull P from the right (see Fig. 4.4). Suppose that the moment of
inertia of a cross section of the beam at a distance s from its left end is $I = 2(s + 1)$.
If origin is the middle of the bar, then the vertical deflection $y(x)$ at x is governed
by the nonhomogeneous differential equation

$$2E(x + 1 + \ell)\frac{d^2 y}{dx^2} - Py = \frac{1}{2}w(x + \ell)^2 - w(x + \ell),$$

where E is Young's modulus.

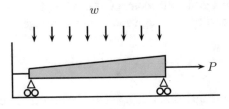

Fig. 4.4

Find the general solutions of the above equation.
$(y(x) = x^{1/2}\left[A I_1 \left(\sqrt{\frac{2P}{E}}x^{1/2}\right) + B K_1 \left(\sqrt{\frac{2P}{E}}x^{1/2}\right)\right] - \frac{w}{2P}(x + \ell)^2 + \frac{w}{P^2}(P - 2E)$
$(x + \ell) - \frac{2E}{P^2}w.)$

4.33 Show that Carl Guslov Jacob *Jacobi's* (1804–1851) *differential equation*

$$x(1-x)y'' + [a - (1+b)x]y' + n(b+n)y = 0$$

is a hypergeometric equation, and hence its general solution near $x = 0$ can be written as

$$y(x) = c_1 F(n+b, -n, a, x) + c_2 x^{1-a} F(n+b-a+1, 1-n-a, 2-a, x).$$

4.34 In the analysis of a deformation due to a uniform load on a certain circular plate the following equation occurs

$$x^2[1 - (\epsilon x)^k]\phi'' + x[1 - (1+3k)(\epsilon x)^k]\phi' - [1 - (3kv - 1)(\epsilon x)^k]\phi = 0,$$
(4.85)

where ϵ, k and v are constants, x is proportional to the distance from the center of the plate and ϕ is the angle between the normal to the deformed surface of the plate and the normal to that surface at the center of the plate. Show that the successive substitutions $z = \epsilon x$, $\phi = z\psi$, $z^k = \sigma$ transform (4.85) to the hypergeometric equation

$$\sigma(1-\sigma)\frac{d^2\psi}{d\sigma^2} + \left[\frac{k+2}{k} - \left(1 + \frac{2+3k}{k}\right)\sigma\right]\frac{d\psi}{d\sigma} - \frac{3(v+1)}{k}\psi = 0$$

with $c = (k+2)/k$, $a+b = (2+3k)/k$, $ab = 3(v+1)/k$. Hence, if α and β are the roots of $k\lambda^2 - (2+3k)\lambda + 3(v+1) = 0$, the solution of (4.85) can be written as $\phi(x) = \epsilon x F(\alpha, \beta, (k+2)/k, \epsilon^k x^k)$.

4.35 In the van der Pol's equation (4.73) consider μ as the perturbing parameter, and seek its solution in the form $y(x) = \sum_{n=0}^{\infty} \mu^n y_n(x)$. Show that

$$y(x) = a\cos(x+b) - \mu\left(\frac{a^3 - 4a}{8}x\cos(x+b) + \frac{1}{32}a^3\sin(3(x+b))\right) + O(\mu^2).$$

4.36 Consider Duffing's equation (4.75) with $\delta = 0$, $\alpha = 1$, and $\gamma = 0$, i.e.,

$$y'' + y + \beta y^3 = 0$$
(4.86)

together with the initial conditions

$$y(0) = y_0, \quad y'(0) = 0.$$
(4.87)

In (4.86) let β be the perturbing parameter. We seek the solution of (4.86), (4.87) in the form $y(x) = \sum_{n=0}^{\infty} \beta^n y_n(x)$. Show that

$$y(x) = y_0\cos x + \beta y_0^3\left(-\frac{3}{8}x\sin x + \frac{1}{32}[\cos 3x - \cos(x)]\right) + O(\beta^2).$$

References

1. V.G. Jenson, G.V. Jefferys, *Mathematical Methods in Chemical Engineering* (Academic Press, London, 1977)
2. R.P. Agarwal, D. O'Regan, Appl. Math. Lett. **20**, 1198–1203 (2007)
3. H.T. Davis, *Introduction to Nonlinear Differential and Integral Equations* (Dover, New York, 1962)
4. R.P. Agarwal, D. O'Regan, *Ordinary and Partial Differential Equations with Special Functions*, Fourier Series and Boundary Value Problems (Springer, New York, 2009)
5. E. Cambi, Proc. I.R.E. **36**, 42–49 (1948)
6. E.L. Ince, *Ordinary Differential Equations* (Dover, New York, 1956)

Chapter 5
Systems of First-Order Differential Equations

In the current chapter we will mention about partial derivatives in the context of analyzing systems of differential equations [1]. Partial derivatives are derivatives of a function with multiple variables, with respect to one of those variables. Mathematical models of physical phenomena involving more than one variable are solved with partial differential equations, but we will not cover these applications in this book, due to the broad array of mathematical tools needed to present different techniques of solving partial differential equations.

Consider the linear system of first-order differential equations with variable coefficients

$$
\begin{aligned}
u_1' &= a_{11}(t)u_1 + a_{12}(t)u_2 + \cdots + a_{1n}(t)u_n + b_1(t) \\
u_2' &= a_{21}(t)u_1 + a_{22}(t)u_2 + \cdots + a_{2n}(t)u_n + b_2(t) \\
&\cdots \quad \cdots \quad \cdots \\
u_n' &= a_{n1}(t)u_1 + a_{n2}(t)u_2 + \cdots + a_{nn}(t)u_n + b_n(t).
\end{aligned}
\tag{5.1}
$$

In the matrix form this system can be conveniently written as

$$
\mathbf{u}' = \mathbf{A}(t)\mathbf{u} + \mathbf{B}(t),
\tag{5.2}
$$

where $\mathbf{u} = \mathbf{u}(t) = (u_1(t), \cdots, u_n(t))^T$ and $\mathbf{B}(t) = (b_1(t), \cdots, b_n(t))^T$ are n-dimensional vector-valued functions, and $\mathbf{A}(t) = (a_{ij}(t))$, $1 \leq i, \ j \leq n$ is an $n \times n$ matrix-valued function. In what follows, we shall assume that $\mathbf{A}(t)$ and $\mathbf{B}(t)$ are continuous in the interval $I = (\alpha, \beta)$.

The original version of this chapter was revised: The title of chapter 5 has been updated. The correction to this chapter is available at https://doi.org/10.1007/978-3-030-26384-3_10.

© Springer Nature Switzerland AG 2019
R. P. Agarwal et al., *500 Examples and Problems of Applied Differential Equations*, Problem Books in Mathematics, https://doi.org/10.1007/978-3-030-26384-3_5

A linear differential equation of nth order with variable coefficients of the form

$$y^{(n)} + p_1(t)y^{(n-1)} + \cdots + p_n(t)y = r(t)$$

can be written as (5.2) with $\mathbf{u}(t) = (y(t), y'(t), \cdots, y^{(n-1)}(t))^T$, $\mathbf{B}(t) = (r(t), 0, \cdots, 0)^T$, and

$$\mathbf{A}(t) = \begin{pmatrix} 0 & 1 & 0 & \cdots & 0 \\ 0 & 0 & 1 & \cdots & 0 \\ \vdots & \vdots & \vdots & \cdots & \vdots \\ 0 & 0 & 0 & \cdots & 1 \\ -p_n(t) & -p_{n-1}(t) & -p_{n-2}(t) & \cdots & -p_1(t) \end{pmatrix}.$$

For (5.2) the corresponding homogeneous system

$$\mathbf{u}' = \mathbf{A}(t)\mathbf{u} \tag{5.3}$$

plays an important role.

We recall that the vector-valued functions $\mathbf{u}^1(t), \cdots, \mathbf{u}^m(t)$ defined in an interval I are said to be *linearly independent* in I, if the relation $c_1\mathbf{u}^1(t) + \cdots + c_m\mathbf{u}^m(t) = 0$ for all $t \in I$ implies that $c_1 = \cdots = c_m = 0$. Conversely, these functions are said to be *linearly dependent* if there exist constants $c_1, \cdots, c_m$ not all zero such that $c_1\mathbf{u}^1(t) + \cdots + c_m\mathbf{u}^m(t) = 0$ for all $t \in I$.

Although, the linear system (5.3) cannot be solved in terms of the known functions, the following results are well known.

Theorem 5.1 *There exist exactly n solutions* $\mathbf{u}^1, \cdots, \mathbf{u}^n$ *of (5.3) which are linearly independent in I.*

Theorem 5.2 *The solutions* $\mathbf{u}^1, \cdots, \mathbf{u}^n$ *of (5.3) are linearly independent in I if and only if the Wronskian* $\mathbf{W}(t)$ *defined by*

$$\mathbf{W}(t) = \mathbf{W}(\mathbf{u}^1, \cdots, \mathbf{u}^n)(t) = \begin{vmatrix} u_1^1(t) & \cdots & u_1^n(t) \\ u_2^1(t) & \cdots & u_2^n(t) \\ \vdots & \cdots & \vdots \\ u_n^1(t) & \cdots & u_n^n(t) \end{vmatrix} \tag{5.4}$$

is different from zero for some t in I.

The set $\{\mathbf{u}^1, \cdots, \mathbf{u}^n\}$ of linearly independent solutions of (5.3) is called a *fundamental system of solutions*. Further, the $n \times n$ matrix $\Psi(t) = [\mathbf{u}^1(t), \cdots, \mathbf{u}^n(t)]$ is called the *fundamental matrix* of (5.3).

Theorem 5.3 *For the Wronskian defined in (5.4) the following Abel's identity holds*

$$\mathbf{W}(t) \;=\; \mathbf{W}(t_0) \exp\left(\int_{t_0}^{t} \operatorname{Tr}\mathbf{A}(s)ds\right), \quad t_0 \in I, \tag{5.5}$$

where

$$\operatorname{Tr}\mathbf{A}(t) \;=\; \sum_{i=1}^{n} a_{ii}(t).$$

Thus, if the Wronskian is zero at some $t_0 \in I$, then it is zero for all $t \in I$.

Theorem 5.4 *If $\mathbf{u}^1, \cdots, \mathbf{u}^n$ are solutions of (5.3) and $c_1, \cdots, c_n$ are arbitrary constants, then $c_1\mathbf{u}^1 + \cdots + c_n\mathbf{u}^n$ is also a solution of (5.3). Further, if $\Psi(t)$ is a fundamental matrix of the d.s. (5.3), then its general solution can be written as*

$$\mathbf{u}(t) \;=\; \Psi(t)\mathbf{c}, \tag{5.6}$$

where $\mathbf{c}$ is an arbitrary constant vector.

As a consequence of this theorem, we find that the solution of the initial value problem: (5.3) together with the initial condition

$$\mathbf{u}(t_0) \;=\; \mathbf{u}^0 \tag{5.7}$$

can be written as

$$\mathbf{u}(t) \;=\; \Psi(t)\Psi^{-1}(t_0)\mathbf{u}^0. \tag{5.8}$$

The matrix $\Phi(t, t_0)$ defined by

$$\Phi(t, t_0) \;=\; \Psi(t)\Psi^{-1}(t_0) \tag{5.9}$$

is called *principal fundamental matrix*. Note that $\Phi(t_0, t_0) = \mathbf{I}_n$, where $\mathbf{I}_n$ is the $n \times n$ identity matrix. In the literature $\Phi(t, t_0)$ is also known as the *evolution* or *transition* *matrix*.

In terms of a fundamental matrix $\Psi(t)$ the general solution of the nonhomogeneous system (5.2) can be written as

$$\mathbf{u}(t) \;=\; \Psi(t)\mathbf{c} + \int^{t} \Psi(t)\Psi^{-1}(s)\mathbf{B}(s)ds. \tag{5.10}$$

Also, the solution of the initial value problem (5.2), (5.7) in terms of the principal fundamental matrix $\Phi(t, t_0)$ appears as

$$\mathbf{u}(t) \;=\; \Phi(t, t_0)\mathbf{u}^0 + \int_{t_0}^{t} \Phi(t, s)\mathbf{B}(s)ds. \tag{5.11}$$

Now we shall consider the homogeneous system (5.3) with $\mathbf{A}(t) \equiv \mathbf{A}$, i.e.,

$$\mathbf{u}' = \mathbf{A}\mathbf{u}, \tag{5.12}$$

where $\mathbf{A}$ is a constant matrix. The general solution of (5.12) can be written as

$$\mathbf{u}(t) = e^{\mathbf{A}t}\mathbf{c}, \tag{5.13}$$

where $\mathbf{c}$ is an arbitrary constant vector. Thus, to find the solution of (5.12) all we need is the computation of $e^{\mathbf{A}t}$. For this, several algorithms are now known, however, the following which we state as a theorem is very simple to apply.

Theorem 5.5 *Let $\lambda_1, \cdots, \lambda_k$, $k \leq n$ be distinct eigenvalues of the matrix $\mathbf{A}$ with multiplicities $r_1, \cdots, r_k$, respectively. Define*

$$p(\lambda) = (\lambda - \lambda_1)^{r_1} \cdots (\lambda - \lambda_k)^{r_k}. \tag{5.14}$$

Then,

$$e^{\mathbf{A}t} = \sum_{i=1}^{k} \left[e^{\lambda_i t} a_i(\mathbf{A}) q_i(\mathbf{A}) \sum_{j=0}^{r_i-1} \left\{ \frac{1}{j!}(\mathbf{A} - \lambda_i \mathbf{I})^j t^j \right\} \right], \tag{5.15}$$

where

$$q_i(\lambda) = p(\lambda)(\lambda - \lambda_i)^{-r_i}, \quad 1 \leq i \leq k \tag{5.16}$$

and $a_i(\lambda)$, $1 \leq i \leq k$ are the polynomials of degree less than r_i in the expansion

$$\frac{1}{p(\lambda)} = \frac{a_1(\lambda)}{(\lambda - \lambda_1)^{r_1}} + \cdots + \frac{a_k(\lambda)}{(\lambda - \lambda_k)^{r_k}}. \tag{5.17}$$

Corollary 5.1 *If $k = n$, i.e., $\mathbf{A}$ has n distinct eigenvalues, then $a_i(\mathbf{A}) = (1/q_i(\lambda_i))\mathbf{I}$, and hence (5.15) reduces to*

$$
\begin{aligned}
e^{\mathbf{A}t} &= \sum_{i=1}^{n} \frac{q_i(\mathbf{A})}{q_i(\lambda_i)} e^{\lambda_i t} \\
&= \sum_{i=1}^{n} \frac{(\mathbf{A} - \lambda_1\mathbf{I}) \cdots (\mathbf{A} - \lambda_{i-1}\mathbf{I})(\mathbf{A} - \lambda_{i+1}\mathbf{I}) \cdots (\mathbf{A} - \lambda_n\mathbf{I})}{(\lambda_i - \lambda_1) \cdots (\lambda_i - \lambda_{i-1})(\lambda_i - \lambda_{i+1}) \cdots (\lambda_i - \lambda_n)} e^{\lambda_i t}.
\end{aligned} \tag{5.18}
$$

If for the matrix $\mathbf{A}$, corresponding to distinct eigenvalues $\lambda_1, \cdots, \lambda_n$ the linearly independent eigenvectors are $\mathbf{v}^1, \cdots, \mathbf{v}^n$, then it follows that

$$\frac{(\mathbf{A} - \lambda_1\mathbf{I}) \cdots (\mathbf{A} - \lambda_{i-1}\mathbf{I})(\mathbf{A} - \lambda_{i+1}\mathbf{I}) \cdots (\mathbf{A} - \lambda_n\mathbf{I})}{(\lambda_i - \lambda_1) \cdots (\lambda_i - \lambda_{i-1})(\lambda_i - \lambda_{i+1}) \cdots (\lambda_i - \lambda_n)} \mathbf{v}^k = \begin{cases} 0, & k \neq i \\ v^i, & k = i. \end{cases}$$

Thus, from (5.18) we have

$$e^{\mathbf{A}t}\mathbf{v}^i = \mathbf{v}^i e^{\lambda_i t},$$

which in turn gives

$$e^{\mathbf{A}t}[\mathbf{v}^1 \ \mathbf{v}^2 \ \cdots \ \mathbf{v}^n] = [\mathbf{v}^1 e^{\lambda_1 t} \ \mathbf{v}^2 e^{\lambda_2 t} \ \cdots \ \mathbf{v}^n e^{\lambda_n t}],$$

and hence

$$e^{\mathbf{A}t}\mathbf{d} = [\mathbf{v}^1 e^{\lambda_1 t} \ \mathbf{v}^2 e^{\lambda_2 t} \ \cdots \ \mathbf{v}^n e^{\lambda_n t}][\mathbf{v}^1 \ \mathbf{v}^2 \ \cdots \ \mathbf{v}^n]^{-1}\mathbf{d}, \qquad (5.19)$$

where $\mathbf{d}$ is any constant vector. In (5.19), we let $[\mathbf{v}^1 \ \mathbf{v}^2 \ \cdots \ \mathbf{v}^n]^{-1}\mathbf{d} = \mathbf{c}$, clearly $\mathbf{c}$ is an arbitrary vector. Thus, in view of (5.13) the general solution of (5.12) can be written as

$$\mathbf{u}(t) = \sum_{i=1}^{n} c_i \mathbf{v}^i e^{\lambda_i t}. \qquad (5.20)$$

Corollary 5.2 *If $k = 1$, i.e., A has all the eigenvalues equal to λ_1, then $a_i(\mathbf{A}) = q_i(\mathbf{A}) = \mathbf{I}$, and hence (5.15) reduces to*

$$e^{\mathbf{A}t} = e^{\lambda_1 t} \sum_{j=0}^{n-1} \left\{ \frac{1}{j!} (\mathbf{A} - \lambda_1 \mathbf{I})^j t^j \right\}. \qquad (5.21)$$

Thus, in particular, if $n = 2$, then

$$e^{\mathbf{A}t} = \begin{cases} \dfrac{(\mathbf{A} - \lambda_2 \mathbf{I})}{(\lambda_1 - \lambda_2)} e^{\lambda_1 t} + \dfrac{(\mathbf{A} - \lambda_1 \mathbf{I})}{(\lambda_2 - \lambda_1)} e^{\lambda_2 t} & \text{provided } \lambda_2 \neq \lambda_1 \\ [\mathbf{I} + (\mathbf{A} - \lambda_1 \mathbf{I})t] e^{\lambda_1 t} & \text{provided } \lambda_2 = \lambda_1. \end{cases} \qquad (5.22)$$

From (5.10) it is clear that the general solution of the nonhomogeneous system

$$\mathbf{u}' = \mathbf{A}\mathbf{u} + \mathbf{B}(t), \qquad (5.23)$$

can be written as

$$\mathbf{u}(t) = e^{\mathbf{A}t}c + \int^t e^{\mathbf{A}(t-s)}\mathbf{B}(s)ds. \qquad (5.24)$$

Further, from (5.11) the solution of the initial value problem (5.23), (5.7) appears as

$$\mathbf{u}(t) = e^{\mathbf{A}(t-t_0)}u^0 + \int_{t_0}^{t} e^{\mathbf{A}(t-s)}\mathbf{B}(s)ds. \qquad (5.25)$$

Finally, we shall consider the system of nonlinear first-order differential equations

$$u_1' = g_1(t, u_1, \cdots, u_n)$$
$$u_2' = g_2(t, u_1, \cdots, u_n)$$
$$\vdots \qquad\qquad\qquad\qquad (5.26)$$
$$u_n' = g_n(t, u_1, \cdots, u_n).$$

Clearly, the system (5.2) is a particular case of (5.26). Further, each nth-order differential equation

$$y^{(n)} = f(t, y, y', \cdots, y^{(n-1)}) \qquad (5.27)$$

is equivalent to a system of the form (5.26). Indeed, if we take $y^{(i)} = u_{i+1}$, $0 \le i \le n - 1$, then the Eq. (5.27) can be written as

$$u_i' = u_{i+1}, \quad 1 \le i \le n - 1$$
$$u_n' = f(t, u_1, \cdots, u_n), \qquad\qquad (5.28)$$

which is of the type (5.26).

In what follows, we shall assume that the functions $g_1, \cdots, g_n$ are at least continuous in some domain E of $(n + 1)$ dimensional space $\mathbb{R}^{n+1}$. By a solution of (5.26) in an interval I we mean a set of n functions $u_1(t), \cdots, u_n(t)$ such that (i) $u_1'(t), \cdots, u_n'(t)$ exist for all $t \in I$, (ii) for all $t \in I$ the points $(t, u_1(t), \cdots, u_n(t)) \in E$, and (iii) $u_i'(t) = g_i(t, u_1(t), \cdots, u_n(t))$ for all $t \in I$.

By setting

$$\mathbf{u}(t) = (u_1(t), \cdots, u_n(t))^T \quad \text{and} \quad \mathbf{g}(t, \mathbf{u}) = (g_1(t, \mathbf{u}), \cdots, g_n(t, \mathbf{u}))$$

and agreeing that differentiation and integration are to be performed componentwise, i.e., $\mathbf{u}'(t) = (u_1'(t), \cdots, u_n'(t))^T$ and

$$\int_\alpha^\beta \mathbf{u}(t)dt = \left(\int_\alpha^\beta u_1(t)dt, \cdots, \int_\alpha^\beta u_n(t)dt \right)^T$$

the problem (5.26), (5.7) can be written as

$$\mathbf{u}' = \mathbf{g}(t, \mathbf{u}), \quad \mathbf{u}(t_0) = \mathbf{u}^0. \qquad (5.29)$$

It is easy to see that if $\mathbf{g}(t, \mathbf{u})$ is continuous in the domain E, then any solution of (5.29) is also a solution of the integral equation

$$\mathbf{u}(t) = \mathbf{u}^0 + \int_{t_0}^t \mathbf{g}(s, \mathbf{u}(s))ds \qquad (5.30)$$

and conversely.

Each nonlinear system which can be solved explicitly has to be dealt independently with some special techniques. However, easily verifiable conditions which guarantee the existence and uniqueness of solutions of the initial value problem (5.29) are known. We list some of these results in the following theorems.

Theorem 5.6 (Local Existence Theorem) *Let the following conditions hold*
(i) $\mathbf{g}(t, \mathbf{u})$ *is continuous in* $\Omega : |t - t_0| \leq a$, $\|\mathbf{u} - \mathbf{u}^0\| \leq b$ *and hence there exists a* $M > 0$ *such that* $\|\mathbf{g}(t, \mathbf{u})\| \leq M$ *for all* $(t, \mathbf{u}) \in \Omega$; *here* $\| \cdot \|$ *is any convenient norm,*
(ii) $\mathbf{g}(t, \mathbf{u})$ *satisfies a uniform Rudolf Otto Sigismund Lipschitz (1832–1903) condition, i.e., for all* $(t, \mathbf{u}), (t, \mathbf{v}) \in \Omega$,

$$\|\mathbf{g}(t, \mathbf{u}) - \mathbf{g}(t, \mathbf{v})\| \leq L \|\mathbf{u} - \mathbf{v}\|. \tag{5.31}$$

(This condition is always satisfied if the partial derivatives $\partial \mathbf{g}/\partial u_k$, $k = 1, \cdots, n$ *exist and* $\|\partial \mathbf{g}/\partial \mathbf{u}\| \leq L$.)
Then, (5.29) has a unique solution in the interval $J_h : |t - t_0| \leq h = \min\{a, b/M\}$.

Theorem 5.7 (Global Existence Theorem) *Let the following conditions hold*

(i) $\mathbf{g}(t, \mathbf{u})$ *is continuous in* $\Delta : |t - t_0| \leq a$, $\|\mathbf{u}\| < \infty$,
(ii) $\mathbf{g}(t, \mathbf{u})$ *satisfies a uniform Lipschitz condition (5.31) in* Δ.

Then, (5.29) has a unique solution in the entire interval $|t - t_0| \leq a$.

Example 5.1 Let $x(t)$ and $y(t)$ represent the numerical strengths of two armies F_1 and F_2 engaged in combat. We assume that the two sides are engaged in open combat, so that each side is wholly exposed to the fire of the other side. Lanchester [17, 18] (also see [12, 14]) postulated that under this assumption the combat loss rate of an army will vary linearly with the size of the opposing force with proportionality constants, say, a and b indicating the efficiency of the opposite force. In addition both armies will suffer noncombat losses which for simplicity we assume to be constants c and d, respectively. We also assume that $f(t)$ and $g(t)$, respectively, are the reinforcement rates for the two armies during the battle. This battle can be modeled by the following system of first-order differential equations:

$$\begin{aligned}
\frac{dx}{dt} &= -ay - c + f(t) \\
\frac{dy}{dt} &= -bx - d + g(t),
\end{aligned} \tag{5.32}$$

which in system form is the same as (5.23) with

$$\mathbf{u} = (x, y)^T, \quad \mathbf{A} = \begin{pmatrix} 0 & -a \\ -b & 0 \end{pmatrix}, \quad \mathbf{B}(t) = (-c + f(t), -d + g(t))^T. \tag{5.33}$$

In what follows, for simplicity, we shall assume that $\mathbf{B}(t) \equiv 0$. This can happen if $f(t) \equiv c$, $g(t) \equiv d$ or $f(t) = c = 0$, $g(t) = d = 0$. The meaning in both the cases is clear. Thus, we shall consider the system

$$\begin{pmatrix} x \\ y \end{pmatrix}' = \begin{pmatrix} 0 & -a \\ -b & 0 \end{pmatrix}\begin{pmatrix} x \\ y \end{pmatrix}. \tag{5.34}$$

Since for this matrix $\mathbf{A}$ the eigenvalues are $\lambda_1 = \sqrt{ab}$ and $\lambda_2 = -\sqrt{ab}$, from (5.22) it follows that

$$\begin{aligned} e^{\mathbf{A}t} &= \frac{1}{2\sqrt{ab}}\begin{pmatrix} \sqrt{ab} & -a \\ -b & \sqrt{ab} \end{pmatrix}e^{\sqrt{ab}t} - \frac{1}{2\sqrt{ab}}\begin{pmatrix} -\sqrt{ab} & -a \\ -b & -\sqrt{ab} \end{pmatrix}e^{-\sqrt{ab}t} \\ &= \frac{1}{\sqrt{ab}}\begin{pmatrix} \sqrt{ab}\cosh\sqrt{ab}t & -a\sinh\sqrt{ab}t \\ -b\sinh\sqrt{ab}t & \sqrt{ab}\cosh\sqrt{ab}t \end{pmatrix}. \end{aligned}$$

Hence, the solution of (5.34) satisfying the initial conditions

$$x(0) = \alpha, \quad y(0) = \beta \tag{5.35}$$

can be written as

$$\begin{aligned} \begin{pmatrix} x(t) \\ y(t) \end{pmatrix} &= \frac{1}{\sqrt{ab}}\begin{pmatrix} \sqrt{ab}\cosh\sqrt{ab}t & -a\sinh\sqrt{ab}t \\ -b\sinh\sqrt{ab}t & \sqrt{ab}\cosh\sqrt{ab}t \end{pmatrix}\begin{pmatrix} \alpha \\ \beta \end{pmatrix} \\ &= \begin{pmatrix} \alpha\cosh\sqrt{ab}t - \beta\sqrt{\frac{a}{b}}\sinh\sqrt{ab}t \\ -\alpha\sqrt{\frac{b}{a}}\sinh\sqrt{ab}t + \beta\cosh\sqrt{ab}t \end{pmatrix} \tag{5.36} \\ &= \cosh\sqrt{ab}t\begin{pmatrix} \alpha - \beta\sqrt{\frac{a}{b}}\tanh\sqrt{ab}t \\ -\alpha\sqrt{\frac{b}{a}}\tanh\sqrt{ab}t + \beta \end{pmatrix}. \tag{5.37} \end{aligned}$$

Now, since $\tanh(\theta)$ is an increasing function and $\tanh(\theta) \to 1$ as $\theta \to \infty$, from (5.37) it is clear that the army F_1 survives all the time if $\alpha - \beta\sqrt{a/b} > 0$, whereas the army F_2 survives if $-\alpha\sqrt{b/a} + \beta = -\sqrt{b/a}\left(\alpha - \beta\sqrt{a/b}\right) > 0$. From this the conclusion that both the armies cannot survive forever is immediate. In fact, we have:

(i) The army F_1 is annihilated if $\alpha - \beta\sqrt{a/b} < 0$, and the annihilation time is

$$t_1 = \frac{1}{\sqrt{ab}}\tanh^{-1}\left(\frac{\alpha}{\beta}\sqrt{\frac{b}{a}}\right).$$

(ii) The army F_2 is annihilated if $\alpha - \beta\sqrt{a/b} > 0$, and the annihilation time is

$$t_2 = \frac{1}{\sqrt{ab}}\tanh^{-1}\left(\frac{\beta}{\alpha}\sqrt{\frac{a}{b}}\right).$$

Finally, we note that if $\alpha = \beta\sqrt{a/b}$ the solution given in (5.36) reduces to

$$\begin{pmatrix} x(t) \\ y(t) \end{pmatrix} = \begin{pmatrix} \alpha[\cosh(\sqrt{abt}) - \sinh(\sqrt{abt})] \\ \beta[\cosh(\sqrt{abt}) - \sinh(\sqrt{abt})] \end{pmatrix},$$

which suggest that neither army is annihilated, however, since $[\cosh(\theta) - \sinh(\theta)] \to 0$ exponentially as $\theta \to \infty$, both the armies decrease exponentially to zero as $t \to \infty$.

Example 5.2 Suppose Tom is in love with Linda, but Linda is a fickle lover. The more Tom loves Linda, the more she dislikes him; but when he loses interest in her, her feelings for him warm up. On the other hand, when she loves him, his love for her grows; and when she loses interest, he also loses interest. Let $x(t)$ and $y(t)$ represent Tom and Linda's feelings at time t, then their love affair is governed by the system of differential equations (see [28])

$$\begin{aligned} \frac{dx}{dt} &= ay, \quad x(0) = \alpha \\ \frac{dy}{dt} &= -bx, \quad y(0) = \beta, \end{aligned} \tag{5.38}$$

where a and b are positive constants, and α and β denote the initial feelings of Tom and Linda, respectively.

The solution of (5.38) in view of (5.36) (replacing a by $(-a)$) can be written as

$$\begin{aligned} \begin{pmatrix} x(t) \\ y(t) \end{pmatrix} &= \begin{pmatrix} \alpha \cos(\sqrt{abt}) + \beta\sqrt{\frac{a}{b}} \sin(\sqrt{abt}) \\ -\alpha\sqrt{\frac{b}{a}} \sin(\sqrt{abt}) + \beta \cos(\sqrt{abt}) \end{pmatrix} \\ &= \sqrt{\alpha^2 + \frac{a}{b}\beta^2} \begin{pmatrix} \cos(\sqrt{abt} - \theta) \\ -\sqrt{\frac{b}{a}} \sin(\sqrt{abt} - \theta) \end{pmatrix}, \end{aligned} \tag{5.39}$$

where

$$\theta = \tan^{-1}\left(\sqrt{\frac{a}{b}}\frac{\beta}{\alpha}\right).$$

As positive and negative values of the variables x and y represent like and dislike, Tom's love is maximum at time $t = t_0 = \theta/\sqrt{ab}$ and is given by $x(t_0) = \sqrt{\alpha^2 + a\beta^2/b}$. However, since $y(t_0) = 0$, Linda's attitude at time $t = t_0$ is neutral. As time passes by Tom's love starts diminishing until it is zero at time $t = t_1 = (2\theta + \pi)/(2\sqrt{ab})$ $(x(t_1) = 0)$, and since at this time $y(t_1) = -\sqrt{\beta^2 + b\alpha^2/a}$, Linda's dislike for Tom is at its worst. Now, for $t > t_1$ until $t = t_2 = (\theta + \pi)/\sqrt{ab}$ Tom dislikes her and at $t = t_2$ while Tom's dislike is at its worst $\left(x(t_2) = -\sqrt{\alpha^2 + a\beta^2/b}\right)$, Linda losses her dislike for Tom $(y(t_2) = 0)$. For $t > t_2$ until $t = t_3 = (2\theta + 3\pi)/2\sqrt{ab}$ Linda is in love with Tom, and Tom also losses his dislike for her. In fact, at

$t = t_3$, $x(t_3) = 0$, i.e., Tom's dislike is over, whereas $y(t_3) = \sqrt{\beta^2 + b\alpha^2/a}$ shows that Linda's love is maximum. For $t > t_3$ until $t = t_4 = (\theta + 2\pi)/2\sqrt{ab}$ Tom loves her, whereas Linda's love cools down $\left(x(t_4) = \sqrt{\alpha^2 + abeta^2/b},\ y(t_4) = 0\right)$. Clearly, at $t = t_4$ the situation is the same as at $t = t_0$, and hence the whole process repeats itself. Thus, Tom and Linda are in a continuous cycle of love and hate. They both are in love only during the time $t_3 < t < t_4$.

From (5.39) it follows that

$$\frac{x^2}{\alpha^2 + \frac{a}{b}\beta^2} + \frac{y^2}{\frac{b}{a}\alpha^2 + \beta^2} = 1, \tag{5.40}$$

and hence if $a > b$ their love affair is best depicted in the following Fig. 5.1.

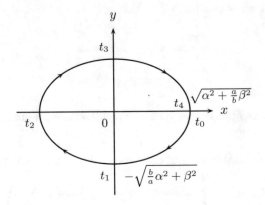

Fig. 5.1

Example 5.3 Let $x(t)$ and $y(t)$ represent the expenditures for armaments by two countries X and Y at time t. Lewis Fry Richardson (1881–1953) model for arms race makes the following assumptions:

1. The expenditure of each country increases at a rate that is proportional to the other country's expenditure.
2. The expenditure of each country decreases at a rate proportional to its own expenditure.
3. The rate of change of arms expenditure for each country has a constant component that depends on the level of antagonism of one country toward the other.
4. The effects of the assumptions 1–3 are additive.

These assumptions lead to the system

$$
\begin{aligned}
\frac{dx}{dt} &= ay - px + r \\
\frac{dy}{dt} &= bx - qy + s,
\end{aligned}
\tag{5.41}
$$

where the constants a, b, p, q are positive, whereas r and s may have any value, positive in the case of mutual suspicion and negative in the case of mutual goodwill. Equations (5.41) in the system form are the same as (5.23) with

$$
\mathbf{u} = (x, y)^T, \quad \mathbf{A} = \begin{pmatrix} -p & a \\ b & -q \end{pmatrix}, \quad \mathbf{B}(t) = \mathbf{B} = (r, s)^T.
\tag{5.42}
$$

The eigenvalues of the matrix A are the roots of the equation

$$
\begin{vmatrix} -p - \lambda & a \\ b & -q - \lambda \end{vmatrix} = \lambda^2 + (p+q)\lambda + (pq - ab) = 0,
$$

i.e.,

$$
\frac{-(p+q) \pm \sqrt{(p+q)^2 - 4(pq - ab)}}{2} = \frac{-(p+q) \pm \sqrt{(p-q)^2 + 4ab}}{2}
$$

and hence both the eigenvalues λ_1 and λ_2 are real and distinct. Thus, the solution of (5.41) satisfying the initial condition (5.35) in view of (5.25) can be written as

$$
\begin{aligned}
\begin{pmatrix} x(t) \\ y(t) \end{pmatrix} &= e^{At} \begin{pmatrix} \alpha \\ \beta \end{pmatrix} + \int_0^t e^{A(t-s)} \mathbf{B} \, ds \\
&= e^{At} \left[\begin{pmatrix} \alpha \\ \beta \end{pmatrix} + A^{-1}\mathbf{B} \right] - A^{-1}\mathbf{B},
\end{aligned}
\tag{5.43}
$$

where $A^{-1} = \frac{1}{ab - pq} \begin{pmatrix} q & a \\ b & p \end{pmatrix}$ exists provided $ab - pq \neq 0$, and hence

$$
A^{-1}\mathbf{B} = \frac{1}{ab - pq} \begin{pmatrix} qr + as \\ br + ps \end{pmatrix}.
\tag{5.44}
$$

From (5.22) it is also clear that

$$
e^{At} = \frac{(A - \lambda_2 I)}{(\lambda_1 - \lambda_2)} e^{\lambda_1 t} + \frac{(A - \lambda_1 I)}{(\lambda_2 - \lambda_1)} e^{\lambda_2 t}.
$$

Thus, the nature of the solution (5.43) depends on the eigenvalues of $\mathbf{A}$. Clearly, (i) if $pq - ab > 0$, both the eigenvalues λ_1 and λ_2 are negative and the solution tends to $-A^{-1}\mathbf{B}$, i.e., countries X and Y are not racing for the arms;

(ii) if $pq - ab < 0$, then the eigenvalues λ_1 and λ_2 have opposite signs. The presence of positive eigenvalue causes exponential growth and it becomes unbounded as time increases. This is a situation that may result in a runaway arms race;

(iii) if $pq - ab = 0$, then $\lambda_1 = 0$, $\lambda_2 = -(p + q)$, and hence

$$e^{At} = \frac{1}{p+q}\begin{pmatrix} q & a \\ b & p \end{pmatrix} - \frac{1}{(p+q)}\begin{pmatrix} -p & a \\ b & -q \end{pmatrix}e^{-(p+q)t}.$$

Further, a particular solution of (8.24) can be obtained rather easily and appears as

$$\left(\frac{as+qr}{p+q}t \, , \quad \frac{as+qr}{a(p+q)}(1+pt) - \frac{r}{a} \right)^T.$$

Thus the countries are arm racing or not depends on $(as + qr)$ is positive or negative. For more details see [27].

Example 5.4 Consider the following electrical network with the assumption that when the switch is closed each current is zero, i.e., $I_1(0) = I_2(0) = I_3(0) = 0$.

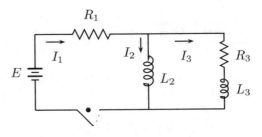

Fig. 5.2

In view of Kirchhoff's laws, we have

$$I_1 = I_2 + I_3 \tag{5.45}$$

$$R_1 I_1 + L_2 I_2' = E \tag{5.46}$$

$$R_1 I_1 + R_3 I_3 + L_3 I_3' = E \tag{5.47}$$

$$R_3 I_3 + L_3 I_3' - L_2 I_2' = 0. \tag{5.48}$$

Note that if we subtract (5.46) from (5.47), we obtain (5.48). Eliminating I_1 from Eqs. (5.45), (5.46) we get

$$R_1(I_2 + I_3) + L_2 I_2' = E. \tag{5.49}$$

In system form Eqs. (5.49), (5.48) involving only I_2 and I_3 can be written as

$$\begin{pmatrix} L_2 & 0 \\ -L_2 & L_3 \end{pmatrix} \begin{pmatrix} I_2' \\ I_3' \end{pmatrix} = -\begin{pmatrix} R_1 & R_2 \\ 0 & R_3 \end{pmatrix} \begin{pmatrix} I_2 \\ I_3 \end{pmatrix} + \begin{pmatrix} E \\ 0 \end{pmatrix},$$

which is the same as

$$\begin{pmatrix} I_2' \\ I_3' \end{pmatrix} = -\begin{pmatrix} L_2 & 0 \\ -L_2 & L_3 \end{pmatrix}^{-1} \begin{pmatrix} R_1 & R_2 \\ 0 & R_3 \end{pmatrix} \begin{pmatrix} I_2 \\ I_3 \end{pmatrix}$$
$$+ \begin{pmatrix} L_2 & 0 \\ -L_2 & L_3 \end{pmatrix}^{-1} \begin{pmatrix} E \\ 0 \end{pmatrix} \qquad (5.50)$$
$$= \mathbf{A} \begin{pmatrix} I_2 \\ I_3 \end{pmatrix} + \mathbf{B},$$

where

$$\mathbf{A} = \begin{pmatrix} -\dfrac{R_1}{L_2} & -\dfrac{R_1}{L_2} \\ -\dfrac{R_1}{L_3} & -\dfrac{R_1 + R_3}{L_3} \end{pmatrix} \quad \text{and} \quad \mathbf{B} = \begin{pmatrix} \dfrac{E}{L_2} \\ \dfrac{E}{L_3} \end{pmatrix},$$

which is exactly of the form (5.23) (Fig. 5.2).

The eigenvalues of the matrix $\mathbf{A}$ are the roots of the equation

$$\begin{vmatrix} -\dfrac{R_1}{L_2} - \lambda & -\dfrac{R_1}{L_2} \\ -\dfrac{R_1}{L_3} & -\dfrac{R_1 + R_3}{L_3} - \lambda \end{vmatrix} = \left(\dfrac{R_1}{L_2} + \lambda \right) \left(\dfrac{R_1 + R_3}{L_3} + \lambda \right) - \dfrac{R_1^2}{L_2 L_3} = 0,$$

or

$$\lambda^2 + \left(\dfrac{R_1}{L_2} + \dfrac{R_1 + R_3}{L_3} \right) \lambda + \dfrac{R_1 R_3}{L_2 L_3} = 0. \qquad (5.51)$$

We claim that both the eigenvalues of the matrix $\mathbf{A}$ are distinct and negative. For this it suffices to show that the discriminant D of (5.51) is positive. Clearly,

$$D = \left(\dfrac{R_1}{L_2} + \dfrac{R_1 + R_3}{L_3} \right)^2 - \dfrac{4 R_1 R_3}{L_2 L_3}$$
$$= \left(\dfrac{R_1}{L_2} - \dfrac{R_3}{L_3} \right)^2 + \dfrac{R_1^2}{L_3^2} + \dfrac{2 R_1}{L_3} \left(\dfrac{R_1}{L_2} + \dfrac{R_3}{L_3} \right) > 0.$$

Now, let the eigenvalues of the matrix $\mathbf{A}$ be $\lambda_1 = -\mu$, $\lambda_2 = -\nu$ where $\mu > 0$, $\nu > 0$. Thus, in view of (5.25) the solution of (5.50) satisfying $I_2(0) = I_3(0) = 0$ can be written as

$$\begin{pmatrix} I_2(t) \\ I_3(t) \end{pmatrix} = \int_0^t e^{A(t-s)} \mathbf{B} ds = \mathbf{A}^{-1}(e^{\mathbf{A}t} - \mathbf{I})\mathbf{B}, \tag{5.52}$$

where

$$\mathbf{A}^{-1} = \frac{L_2 L_3}{R_1 R_3} \begin{pmatrix} -\dfrac{R_1 + R_3}{L_3} & \dfrac{R_1}{L_2} \\ \dfrac{R_1}{L_3} & -\dfrac{R_1}{L_2} \end{pmatrix}$$

so that

$$-\mathbf{A}^{-1}\mathbf{B} = \frac{E}{R_1} \begin{pmatrix} 1 \\ 0 \end{pmatrix}; \tag{5.53}$$

and in view if (5.22),

$$e^{\mathbf{A}t} = \frac{(\mathbf{A} + \nu \mathbf{I})}{(\nu - \mu)} e^{-\mu t} + \frac{(\mathbf{A} + \mu \mathbf{I})}{(\mu - \nu)} e^{-\nu t},$$

which after some computation gives

$$\mathbf{A}^{-1} e^{\mathbf{A}t} \mathbf{B} = \begin{pmatrix} \dfrac{E(R_1 - \nu L_2)}{R_1 L_2 (\nu - \mu)} \\ \dfrac{E(R_1 - \nu L_2)(\mu L_2 - R_1)}{R_1^2 L_2 (\nu - \mu)} \end{pmatrix} e^{-\mu t} + \begin{pmatrix} \dfrac{E(\mu L_2 - R_1)}{R_1 L_2 (\nu - \mu)} \\ \dfrac{E(\mu L_2 - R_1)(\nu L_2 - R_1)}{R_1^2 L_2 (\nu - \mu)} \end{pmatrix} e^{-\nu t}. \tag{5.54}$$

Hence, in view of (5.52)–(5.54), we have

$$I_2(t) = \frac{E(R_1 - \nu L_2)}{R_1 L_2 (\nu - \mu)} e^{-\mu t} + \frac{E(\mu L_2 - R_1)}{R_1 L_2 (\nu - \mu)} e^{-\nu t} + \frac{E}{R_1}, \tag{5.55}$$

$$I_3(t) = \frac{E(R_1 - \nu L_2)(\mu L_2 - R_1)}{R_1^2 L_2 (\nu - \mu)} e^{-\mu t} + \frac{E(\mu L_2 - R_1)(\nu L_2 - R_1)}{R_1^2 L_2 (\nu - \mu)} e^{-\nu t}. \tag{5.56}$$

Finally, from (5.45), we obtain

$$I_1(t) = \frac{E\mu(R_1 - \nu L_2)}{R_1^2 (\nu - \mu)} e^{-\mu t} + \frac{E\nu(\mu L_2 - R_1)}{R_1^2 (\nu - \mu)} e^{-\nu t} + \frac{E}{R_1}. \tag{5.57}$$

From (5.55) to (5.57) it is clear that as $t \to \infty$, $I_1(t) \to E/R_1$, $I_2(t) \to E/R_1$, $I_3(t) \to 0$.

Example 5.5 Every radioactive substance decays by the emission of alpha particles, which are helium nuclei consisting of two protons and two neutrons. The mother substance decays into a daughter substance of lower atomic number, which may or may not be radioactive. Then, the decay substance can decay into another substance.

Once a decay product becomes the element lead, the process stops. Thus the chain of events is a finite number. If we denote $x_1(t)$ as the mother substance at time t, then the law of radioactive decay states that

$$x_1' = -k_1 x_1,$$

where $k_1 > 0$ is the mother disintegration constant. Let $x_2(t)$ denote the daughter substance at time t, then $x_2(t)$ will grow at the same rate k_1 proportional to the quantity of the mother remaining, and if the daughter is radioactive, it will decay at a rate proportional to its remaining quantity of matter. Thus,

$$x_2' = k_1 x_1 - k_2 x_2,$$

where $k_2 > 0$ is the daughter disintegration constant. This process continues until, say, $(n-1)$th substance is reached. This leads to the differential system

$$\begin{aligned}
x_1' &= -k_1 x_1 \\
x_{i+1}' &= k_i x_i - k_{i+1} x_{i+1}, \quad 1 \le i \le n-2 \\
x_n' &= k_{n-1} x_{n-1}, \quad k_j > 0, \quad 1 \le j \le n-1 \quad \text{and distinct.}
\end{aligned} \tag{5.58}$$

If initially the mother substance is α, then we need to solve the system (5.53) satisfying the initial conditions

$$x_1(0) = \alpha, \quad x_i(0) = 0, \quad 2 \le i \le n. \tag{5.59}$$

Adding the equations in the system (5.58), we get

$$x_1' + x_2' + \cdots + x_n' = 0$$

and hence, in view of (5.59),

$$x_1(t) + x_2(t) + \cdots + x_n(t) = x_1(0) + x_2(0) + \cdots + x_n(0) = \alpha, \tag{5.60}$$

which expresses the conservation law.

In matrix form (5.58) is the same as (5.12) with $\mathbf{u} = (x_1, \cdots, x_n)^T$ and

$$A = \begin{pmatrix}
-k_1 & 0 & 0 & 0 & \cdots & 0 & 0 & 0 \\
k_1 & -k_2 & 0 & 0 & \cdots & 0 & 0 & 0 \\
0 & k_2 & -k_3 & 0 & \cdots & 0 & 0 & 0 \\
& \cdots & & \cdots\cdots & & & \cdots & \\
0 & 0 & 0 & 0 & \cdots & k_{n-2} & -k_{n-1} & 0 \\
0 & 0 & 0 & 0 & \cdots & 0 & k_{n-1} & 0
\end{pmatrix}.$$

Clearly, the eigenvalues of A are $\lambda_1 = -k_1, \lambda_2 = -k_2, \cdots, \lambda_{n-1} = -k_{n-1}, \lambda_n = 0$, which are real and distinct. Hence, in view of (5.18) and (5.25) the solution of

(5.58), (5.59) can be written as

$$\begin{pmatrix} x_1(t) \\ x_2(t) \\ \vdots \\ x_n(t) \end{pmatrix} = \sum_{i=1}^{n} \left(\frac{\displaystyle\prod_{j=1, j \neq i}^{n} (A + k_j I)}{\displaystyle\prod_{j=1, j \neq i}^{n} (-k_i + k_j)} \right) e^{-k_i t} \begin{pmatrix} \alpha \\ 0 \\ \vdots \\ 0 \end{pmatrix} \tag{5.61}$$

where $k_n = 0$.

For $n = 3$, (5.61) gives

$$\begin{aligned} x_1(t) &= \alpha e^{-k_1 t}, \\ x_2(t) &= \frac{k_1 \alpha}{k_2 - k_1} \left(e^{-k_1 t} - e^{-k_2 t} \right), \\ x_3(t) &= \alpha \left(1 + \frac{k_1 e^{-k_2 t} - k_2 e^{-k_1 t}}{k_2 - k_1} \right). \end{aligned} \tag{5.62}$$

In (5.62) as $t \to \infty$, we find $x_1(t) \to 0$, $x_2(t) \to 0$ and $x_3(t) \to \alpha$. Finally, we remark that in (5.58) the constants k_i are determined from the known half-lifes.

Example 5.6 Suppose two masses M_1 and M_2 are connected to two springs having spring constants k_1 and k_2, respectively (see Fig. 5.3). Let x_1 and x_2 denote the vertical displacements of the masses from their equilibrium positions. When the system is in motion, the stretching of the lower spring is $(x_2 - x_1)$, exerting a force of $k_2(x_2 - x_1)$ on the upper mass M_1. Also, the upper spring exerts the force $(-k_1 x_1)$ on this mass, and so, in view of Newton's second law of the motion the mass M_1 is governed by

$$M_1 x_1'' = -k_1 x_1 + k_2(x_2 - x_1). \tag{5.63}$$

Similarly, the equation of motion for the lower mass M_2 is

$$M_2 x_2'' = -k_2(x_2 - x_1). \tag{5.64}$$

Thus, the motion of the couple spring–mass system, in the absence of damping effects and external forces, is governed by the system of second-order differential equations (5.63), (5.64).

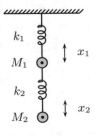

Fig. 5.3

To write (5.63), (5.64) in the system form (5.12), we define $u_1 = x_1$, $u_2 = x_2$, $u_3 = x_1'$, $u_4 = x_2'$ so that

$$\begin{aligned}
u_1' &= u_3 \\
u_2' &= u_4 \\
u_3' &= -\left(\frac{k_1 + k_2}{M_1}\right) u_1 + \frac{k_2}{M_1} u_2 \\
u_4' &= \frac{k_2}{M_2} u_1 - \frac{k_2}{M_2} u_2.
\end{aligned} \tag{5.65}$$

If external forces $F_1(t)$ and $F_2(t)$ are applied to the masses, the system (5.63), (5.64) becomes

$$\begin{aligned}
M_1 x_1'' &= -k_1 x_1 + k_2(x_2 - x_1) + F_1(t) \\
M_2 x_2'' &= -k_2(x_2 - x_1) + F_2(t).
\end{aligned} \tag{5.66}$$

Example 5.7 Consider a four compartmental system sketched in Fig. 5.4. The substance is transferred between compartments as indicated by the arrows. The rates of transfer are indicated by the constants $k_i > 0$, $1 \le i \le 4$. If $x_i(t)$ represents the amount of substance at time t in the ith compartment, then it leads to the system

$$\begin{aligned}
x_1' &= -k_1 x_1 - k_2 x_1 \\
x_2' &= k_1 x_1 - k_3 x_2 \\
x_3' &= k_2 x_1 - k_4 x_3 + k_5 x_4 \\
x_4' &= k_3 x_2 + k_4 x_3 - k_5 x_4.
\end{aligned} \tag{5.67}$$

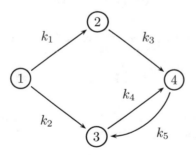

Fig. 5.4

Example 5.8 Let P_i, D_i and S_i $(i = 1, 2)$ denote the price, demand and supply of the ith commodity in a interrelated market. Allen's model [4] assumes that the rate of change of P_i is proportional to $(D_i - S_i)$, i.e.,

$$\frac{dP_i}{dt} = \alpha_i(D_i - S_i), \quad i = 1, 2.$$

If D_i and S_i are linear functions of P_1 and P_2, then this system takes the form

$$\frac{dP_1}{dt} = a_{11}P_1 + a_{12}P_2 + b_1$$
$$\frac{dP_2}{dt} = a_{21}P_1 + a_{22}P_2 + b_2,$$

(5.68)

which is exactly the same as (5.23) with $b_1(t) = b_1$, $b_2(t) = b_2$.

Example 5.9 In certain bacterial colonies toxic substances produced by the bacteria become a limiting factor to the further growth of the bacteria. The number of organisms will increase at a rate proportional to the number present at any time and decrease at a rate (per organism) proportional to the concentration of the toxic substance. The differential equation

$$\frac{dn}{dt} = kn(1 - ac)$$

(5.69)

describes the relation between the number of organisms n in the culture and the concentration c of the toxic metabolic product. Here, k and a are positive constants. If the toxic substance is formed at a constant rate r per organism we have the equation

$$\frac{dc}{dt} = rn.$$

(5.70)

Clearly, (5.69), (5.70) is a system of the form (5.26). We note that Eq. (5.70) is the same as

$$c(t) = r \int_0^t n(s)ds$$

and hence (5.69) can be written as an integro-differential equation

$$\frac{dn}{dt} = kn\left(1 - ar \int_0^t n(s)ds\right).$$

(5.71)

To find the solution of (5.71) we shall follow a clever technique proposed by Reid [26]. We divide both sides of (5.71) by n and make the substitution $n = e^y$, to obtain

$$\frac{dy}{dt} = k - ark \int_0^t e^{y(s)}ds.$$

A differentiation of this equation with respect to t gives the second-order differential equation

$$\frac{d^2y}{dt^2} = -arke^y.$$

(5.72)

We multiply (5.72) by $2(dy/dt)$ and integrate, to get

$$\left(\frac{dy}{dt}\right)^2 = -2arke^y + c.$$

However, since $n(0) = n_0$, $(dy/dt)(0) = k$ it follows that $c = k^2 + 2arkn_0$, and hence

$$\left(\frac{dy}{dt}\right)^2 = k^2 + 2ark(n_0 - e^y),$$

which gives

$$\frac{1}{n}\frac{dn}{dt} = \pm\sqrt{k^2 + 2ark(n_0 - n)}.$$

Thus, over the range where n' is positive, we have

$$\int_0^t ds = \int_{n_0}^n \frac{du}{u\sqrt{k^2 + 2ark(n_0 - u)}},$$

or

$$t = \frac{1}{C}\ln\left(\frac{(\sqrt{C^2 - 2arkn} - C)(k + C)}{(\sqrt{C^2 - 2arkn} + C)(k - C)}\right),$$

where $C = \sqrt{k^2 + 2arkn_0}$. Let $B = (k - C)/(k + C)$, then we can solve for $n(t)$, to obtain

$$n(t) = \frac{Ae^{Ct}}{(1 - Be^{Ct})^2},$$

where $A = -2C^2B/ark$.

Now we differentiate the solution and set the derivative equal to zero, i.e.,

$$\frac{dn}{dt} = \frac{ACe^{Ct}}{(1 - Be^{Ct})^2} + \frac{2ABCe^{2Ct}}{(1 - Be^{Ct})^3} = 0,$$

which immediately gives the critical point

$$t^* = \frac{1}{C}\ln\left(\frac{C + k}{C - k}\right).$$

It is easy to verify that $\left(d^2n/dt^2\right)\big|_{t=t^*} < 0$, and hence $n(t)$ at t^* has its maximum value $n_{max} = n_0 + (k/2ar)$. Clearly, the maximum number of organisms n_{max} depends only on the initial number and parameters of the species involved. For $t > t^*$ it is clear that dn/dt is negative. Further, analysis shows that the function $n(t)$ is symmetric about the line $t = t^*$. We illustrate the growth curve in Fig. 5.5.

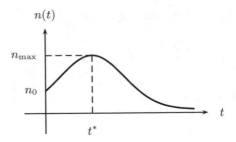

Fig. 5.5

Example 5.10 Consider two populations sharing the same habitat. This may result in *competition*, i.e., both populations use the same limited resources, such as food supply; *predation*, i.e., one population feeds on the other; or *symbiosis*, i.e., the presence of each population is beneficial to the other. Here we shall develop a model in the case of predation. We shall assume that the species eaten (the *prey*) makes up an essential part of the diet of the eating species (the *predators*). For example, the habitat may be a cornfield, the prey species may be field mice, and the predator species can be wild cats. As another example, habitant may be vegetation in a forest, prey species may be rabbits, and the predator species can be wolves. The model we shall develop is due to a mathematician Vito Volterra (1860–1940), and a biophysicist Alfred James Lotka (1880–1949), and now in the literature it is known as Lotka–Volterra system.

Let $x(t)$, $y(t)$ respectively represent the size of predator and prey at time t. We have seen in Chap. 1, if each of the species existed alone, they would obey simple growth and/or decay according as $x' = \alpha x$, $y' = \beta y$ where the parameters α and β are the respective differences in birth and death rates for the species. If, however, predators eat the prey, we expect that the excess birth rate of the prey will depend on the number of predators: The more predators are, the more prey (die) are eaten, so β will be a decreasing function of x. The simplest possible choice of such a function is $\beta = K - Lx$. Next, since an essential part of the predators' diet is prey, if there are no prey, the predators would not survive. Thus, α should be negative for $y = 0$ and must be an increasing function of y. Again the simplest possible form of such a function is $\alpha = -M + Py$. This leads to the Lotka–Volterra system

$$\frac{dx}{dt} = x(-M + Py)$$
$$\frac{dy}{dt} = y(K - Lx), \tag{5.73}$$

where K, L, M and P are positive constants.

Clearly, (5.73) is a coupled pair of nonlinear differential equations. We cannot solve this system analytically. In fact, if $x \neq 0$, it follows that (5.73) is equivalent to the system

$$y = \frac{1}{Px}\frac{dx}{dt} + \frac{M}{P}$$
$$\frac{d^2x}{dt^2} - \frac{1}{x}\left(\frac{dx}{dt}\right)^2 + (Lx - K)\frac{dx}{dt} + (LMx^2 - KMx) = 0.$$

Thus, although y has been eliminated, the resulting equation for x is highly nonlinear. However, we can find a relation between x and y as follows: Clearly, the number of prey depends on the number of predator, we can assume that y is a function of x, i.e., $y = y(x)$, then from (5.73) it follows that

$$\frac{dy}{dx} = \frac{dy}{dt} \Big/ \frac{dx}{dt} = \frac{y(K - Lx)}{x(-M + Py)},$$

which can be written as

$$\left(P - \frac{M}{y}\right) dy = \left(\frac{K}{x} - L\right) dx. \tag{5.74}$$

An integration of (5.74) gives

$$Py - M\ln(y) = K\ln(x) - Lx + \ln(C), \tag{5.75}$$

where C is a constant. Thus, the required relation is

$$\frac{e^{Py}}{y^M} = C\frac{x^K}{e^{Lx}}. \tag{5.76}$$

If the initial population of predator and prey is $x(0) = x_0$ and $y(0) = y_0$, then from (5.76) it follows that

$$e^{P(y-y_0)}(y_0/y)^M = e^{L(x_0-x)}(x/x_0)^K. \tag{5.77}$$

For fixed values of K, L, M, P, x_0 and y_0, the graph of (5.77) is a closed curve in the xy-plane, where (x_0, y_0) is located somewhere on the curve. From this graph it can easily be concluded that

(i) As the number of prey increases, the prey are easier to catch; this leads to an increase in the predator population, until

(ii) the number of predators is so large that they severely deplete the quantity of prey, causing

(iii) a large number of deaths by starvation in the predator population, and allowing

(iv) a reestablishment of the growth in the prey population.

The system (5.73) in view of Verhulst model can be generalized to

$$\frac{dx}{dt} = x(-M + Py - Qx)$$
$$\frac{dy}{dt} = y(K - Lx - Ry),$$
(5.78)

where Q and R are also positive constants.

Example 5.11 Kermack and McKendrick [19] in 1927 proposed a model which explains the sharp rise and fall in the number of infected individuals during epidemic outbreaks. The model was based on data sets available from plague in London (1664–1666) and cholera in London (1865). For decades the model went largely ignored until it was brought back to prominence by Sir Roy Malcolm Anderson (born 1947) and Robert McCredie May (born 1936) in their publication *Population Biology of Infectious Diseases: Part I* in 1979. Here we shall discuss a simple case of this model, which is due to Bailey [6], also see [20, 30]. Assume that a population of fixed size N can be separated into three distinct classes: $S(t)$ denotes the number of individuals in the population at time t who are *susceptible* to a given contagious disease; $I(t)$ is the number of individuals at time t in the population who are *infective*, that is, they have the disease and are contagious; and, $R(t)$ denotes the number of individuals at time t who have *recovered* from the disease and are now immune to further infection. This model is usually called an *SIR epidemic*. Note that this classification leads to the equation

$$S(t) + I(t) + R(t) = N. \tag{5.79}$$

We now assume that the spread of the disease is governed by the following rules:

(i)   The rate of change of the susceptible population is proportional to the size of the susceptible population when the number of infectives exceeds a given cutoff value and

(ii)   The rate at which infectives recover from the disease is proportional to the size of the infective population.

These rules are for simplicity, but have some connection with reality. We further assume that when the number of infectives $I(t)$ exceeds the cutoff value I_*, they are able to infect susceptible individuals in the population. This takes into account the fact that often infectives are quarantined or kept apart from the susceptible population. Note that this can be written as a differential equation

$$\frac{dS}{dt} = \begin{cases} -\alpha S & \text{if } I(t) > I_* \\ 0 & \text{if } I(t) \leq I_*. \end{cases} \tag{5.80}$$

Since each susceptible who catches the disease becomes infected, the rate of change of the infective population is the difference between the newly infected individuals and those who are recovering from the disease. Thus, it follows that

$$\frac{dI}{dt} = \begin{cases} \alpha S - \beta I & \text{if } I(t) > I_* \\ -\beta I & \text{if } I(t) \leq I_*. \end{cases} \tag{5.81}$$

The constants of proportionality α and β are called the *infection rate* and *recovery rate*, respectively. Finally, the rate of change of the recovered individuals is

$$\frac{dR}{dt} = \beta I. \tag{5.82}$$

To determine the solution uniquely we need initial conditions. For simplicity, let us assume that no individuals in the population are immune to the disease, that is, $R(0) = 0$, and that initially $I(0)$ individuals are infective. Further, we assume that the infection and recovery rates are equal, hence $\alpha = \beta$ We then have two cases:
Case 1. If $I(0) \leq I_*$, no further individuals will be infected since $dS/dt = 0$, imply that

$$S(t) = S(0) = N - I(0),$$

for all t, by (5.79) since $R(0) = 0$. This represents the situation in which a very few infectives are immediately quarantined. Then, from (5.81) and in view of $\alpha = \beta$, we have

$$\frac{dI}{dt} = -\alpha I,$$

so that $I(t) = I(0)e^{-\alpha t}$ and

$$R(t) = N - S(t) - I(t) = I(0)(1 - e^{-\alpha t}).$$

These expressions indicate that for the three populations the disease ran its course without causing any further infections.
Case 2. If $I(0) > I_*$, there must be an interval $0 \leq t < T$ on which $I(t) > I_*$, since I must be continuous. For all t in $[0, T)$ the disease will spread to the susceptible population. Thus, from (5.80), we have

$$S(t) = S(0)e^{-\alpha t} \quad \text{for } 0 \leq t < T. \tag{5.83}$$

Substituting this function into (5.81), we find

$$\frac{dI}{dt} + \alpha I = \alpha S(0)e^{-\alpha t},$$

which gives

$$I(t) = ce^{-\alpha t} + \alpha S(0)te^{-\alpha t}. \tag{5.84}$$

Setting $t = 0$, we see that $c = I(0)$ so (5.84) becomes

$$I(t) = [I(0) + \alpha S(0)t]e^{-\alpha t} \quad \text{for} \ \ 0 \le t < T. \tag{5.85}$$

Now two questions are of particular importance:
1. What is the value of T?
2. When will the number of infectives be at a maximum?
    The first question is important since at time T no further susceptibles become infective, that is, the epidemic has run its course. We know that at $t = T$, the right-hand side of (5.85) will be equal to I_*, i.e.,

$$I_* = [I(0) + \alpha S(0)T]e^{-\alpha T}. \tag{5.86}$$

But $S(T) = \lim_{t \to \infty} S(t) = S(\infty)$ is the susceptible population that escapes infection and

$$S(\infty) = S(T) = S(0)e^{-\alpha T},$$

so that

$$T = \frac{1}{\alpha} \ln\left(\frac{S(0)}{S(\infty)} \right). \tag{5.87}$$

Thus, if we can find an expression for $S(\infty)$, we can use (5.87) to predict the end of the epidemic. Substituting (5.87) for T in (5.86), we get

$$I_* = \left[I(0) + S(0) \ln\left(\frac{S(0)}{S(\infty)} \right) \right] \frac{S(\infty)}{S(0)}$$

from which we obtain the expression

$$\frac{I_*}{S(\infty)} + \ln(S(\infty)) = \frac{I(0)}{S(0)} + \ln(S(0)). \tag{5.88}$$

Since I_* and all the terms on the right-hand side of (5.88) are known, we can use this equation to determine $S(\infty)$. The quantity $(N - S(\infty))$ is called the size of the epidemic, since it consists of all the individuals who contracted the disease.
    To answer the second question, we maximize $I(t)$ in (5.85). Letting

$$I'(t) = [\alpha S(0) - \alpha I(0) - \alpha^2 S(0)t]e^{-\alpha t} = 0,$$

we obtain the time at which I attains its maximum value

$$t_{\max} = \frac{1}{\alpha}\left(1 - \frac{I(0)}{S(0)} \right).$$

Hence, substituting $t_{\max}$ in (5.85), we obtain

$$I_{\max} = S(0)e^{-[1-I(0)/S(0)]} = S(t_{\max}),\tag{5.89}$$

which indicates that at time $t_{\max}$ there are just as many susceptibles as infectives.

When $t > T$, no further susceptibles become infected, so the number of infectives is given by

$$I(t) = I_* e^{-\alpha(t-T)}.$$

Example 5.12 Continuing with Examples 5.11, if we assume that a person becomes immune to a disease after recovering from it, and births and deaths in the population are not taken into account, then the percent of persons susceptible to becoming infected with the disease, $S(t)$, the percent of people in the population infected with the disease, $I(t)$, and the percent of the population recovered and immune to the disease, $R(t)$, can be modeled by the system

$$\begin{aligned} S' &= -\lambda S I \\ I' &= \lambda S I - \gamma I \\ R' &= \gamma I \end{aligned}\tag{5.90}$$

$$S(0) = S_0, \quad I(0) = I_0, \quad R(0) = 0,\tag{5.91}$$

where λ and γ are positive numbers. Because $S(t) + I(t) + R(t) = 1$, once we know S and I, we can compute R with $R(t) = 1 - S(t) - I(t)$. This model is called an *SIR model without vital dynamics* because once a person has had the disease he becomes immune to it, and because births and deaths are not taken into consideration. This model might also be used to model diseases that are *epidemic* to a population: those diseases that persist in a population for short periods of time (less than one year). Such diseases typically include influenza, measles, rubella, and chicken pox. For the system (5.90) the following properties can be established rather easily.

(1) $S(t) = S_0 \exp\left(-\lambda \int_0^t I(\tau)dt\right) > 0$ is decreasing,

$I(t) = I_0 \exp\left(-\gamma t + \lambda \int_0^t S(\tau)d\tau\right)$, and $R(t)$ is increasing.

(2) $\lim_{t\to\infty} S(t) = S_\infty$ and $\lim_{t\to\infty} R(t) = R_\infty$ exist.

(3) If $S_0 < \rho$, the disease dies out, while an epidemic results if $S_0 > \rho$, where $\rho = \gamma/\lambda$.

(4) The solution of $\dfrac{dI}{dS} = -1 + \dfrac{\rho}{S}$ can be written as

$$I + S - \rho \ln S = I_0 + S_0 - \rho \ln S_0.$$

(5) If $S_0 > \rho$, then $I(t)$ increases to a maximum value

$$I_{\max} = 1 - \rho - \rho \ln(S_0/\rho)$$

and then decreases to zero, and $S(t)$ decreases to a fixed value that is the unique root of

$$x = 1 + \rho \ln(x/S_0)$$

in the interval $(0, \rho)$.

(6) The following relation holds

$$\frac{\lambda}{\gamma}(1 - S_\infty) = -\ln\left(\frac{S_\infty}{S_0}\right),$$

which shows that $S_\infty > 0$.

In recent years there have been many complications introduced to the above model (5.90). These complications generate other SIR-type models that can be used to study different types of diseases which don't conform to the guidelines of the simple SIR model. Some of these complications are:

Recurrent diseases: Those recovered can again become susceptible.

Births and Deaths (from other causes) can be added to the model.

Latent class individuals may be infected, but not yet infectious.

Interacting classes: STDs, malaria, AIDS.

Variable incubation periods: AIDS.

Age-dependent classes: Some diseases affect primarily children.

Seasonal changes, for example, children share diseases more during school terms.

Interacting SIR models (humans and animals): Sometimes a disease is spread through an animal such as a rat or mosquito. The models must account for these other organisms as well.

Example 5.13 Consider the reaction (see [5])

$$A_1 + A_2 \xrightarrow{k_1} A_3$$
$$A_3 + A_2 \xrightarrow{k_2} A_4$$
$$A_5 + A_2 \xrightarrow{k_3} A_6$$
$$A_6 + A_2 \xrightarrow{k_4} A_7,$$

where reactant A_1 has an effective constant concentration C because it is in excess, goes into solution as the reaction progresses and therefore C moles per unit volume are always present.

The equations for the constant volume reaction are

$$y_2' = -y_2[k_1 C + k_2 y_3 + k_3 y_5 + k_4 y_6] \tag{5.92}$$

$$y_3' = k_1 C y_2 - k_2 y_2 y_3 \tag{5.93}$$

$$y_4' = k_2 y_2 y_3 \tag{5.94}$$

$$y_5' = -k_3 y_2 y_5 \tag{5.95}$$

$$y_6' = k_3 y_2 y_5 - k_4 y_2 y_6 \tag{5.96}$$

$$y_7' = k_4 y_2 y_6, \tag{5.97}$$

where y_i is the concentration of A_i in moles per unit volume and k_i are the rate constants in liters per gram mole second. From material balances we have the additional relations

$$y_5 = y_5(0) - y_6 - y_7 \tag{5.98}$$

$$y_2 = y_2(0) - y_3 - 2y_4 - y_6 - 2y_7. \tag{5.99}$$

These relations also follow from the above Eqs. (5.92)–(5.97). For example an addition of (5.95)–(5.97) and then integration gives (5.98).

We cannot solve the system (5.92)–(5.97) analytically, i.e., there is no way to find the rate constants. However, it is possible to obtain some ratios between rate constants. For example, from (5.93) and (5.94) we have

$$\frac{dy_3}{dy_4} = \frac{a}{y_3} - 1, \quad a = \frac{k_1 C}{k_2}$$

and hence

$$a\left(1 - e^{-(y_3+y_4)/a}\right) - y_3 = 0,$$

which is an implicit relation for the rate constant ratio in terms of y_3 and y_4.

Next, from the Eq. (5.95) we have

$$\frac{1}{k_3} \ln\left(\frac{y_5(0)}{y_5(t)}\right) = \int_0^t y_2(s)\,ds$$

and the sum of (5.93) and (5.94) gives

$$\frac{1}{Ck_1}[y_3(t) + y_4(t)] = \int_0^t y_2(s)\,ds.$$

Equating these two equations we obtain an explicit relation

$$\frac{k_3}{k_1} = \frac{C}{y_3 + y_4} \ln\left(\frac{y_5(0)}{y_5}\right).$$

To obtain the final ratio we note that Eqs. (5.95) and (5.97) can be written as

$$y_6^{-1} y_7' = k_4 y_2 = -\frac{k_4}{k_3} y_5^{-1} y_5'.$$

However, since from (5.98), $y_6 = y_5(0) - y_5 - y_7$ it follows that

$$y_5 dy_7 + b[y_5(0) - y_5 - y_7] dy_5 = 0, \quad b = k_4/k_3.$$

This equation can be solved rather easily, to obtain the implicit relation

$$b - 1 + y_6^{-1} \left\{ y_5(0)[y_5/y_5(0)]^b - y_5 \right\} = 0.$$

Example 5.14 The firing of the axon of a nerve cell is the result of the change in concentration of potassium ions within the cell and sodium ions outside the cell. Although the firing process is complex, basically it begins when this difference in concentration is large enough to cause sodium ions to flow through the cell walls into interior of the axon. As sodium ions are introduced into the axon, the cell walls become more permeable, resulting in even more sodium ions flowing into the axon. Once a critical level of membrane permeability is reached, a rapid rise in the potential difference occurs and the axon fires. Following the firing, the sodium ions gradually flow outward and the potassium ions rapidly flow into the axon, and the system returns to its initial state of rest.

Many mathematical models have been developed for this phenomenon, including one proposed in 1952 by physiologists Sir Alan Lloyd Hodgkin (1914–1998) and Sir Andrew Fielding Huxley (1917–2012), who were awarded the Nobel prize for this work. A simplified model of much interest at present was proposed by Richard FitzHugh (1922–2007) in 1961 and developed by Jin-ichi Nagumo (1926–1999) in 1962 and takes the form

$$\frac{\partial u}{\partial t} = \frac{\partial^2 u}{\partial x^2} + u(1 - u)(u - a) - w$$
$$\frac{\partial w}{\partial t} = bu - \gamma w,$$

where a, b, and γ are positive constants with $0 < a < 1$. In this model, u represents membrane potential, with x measuring the distance along the axon and t referring to time. The cubic term $u(1 - u)(u - a)$ represents the flow inward of sodium ions, and w is a recovery variable analogous to the flow of potassium ions following firing. The solutions depend on $(x + ct)$ and are of the form

$$u(x, t) = \phi(x + ct) = \phi(s)$$
$$w(x, t) = \psi(x + ct) = \psi(s).$$

By substituting the solution forms into the system, a coupled system of ordinary differential equations is obtained

$$\phi'' = c\phi' - \phi(1 - \phi)(\phi - a) + \psi$$
$$c\psi' = b\phi - \gamma\psi,$$

where prime denotes differentiation with respect to s. By setting $\theta = \phi'$, the equations can be written as

$$\theta' = c\theta - \phi(1 - \phi)(\phi - a) + \psi$$
$$\phi' = \theta$$
$$\psi' = \frac{b}{c}\phi - \frac{\gamma}{c}\psi.$$

Mathematical investigation and analysis of the differential equations reveal that $\psi \equiv 0$. Thus the system simplifies to

$$\theta' = c\theta - \phi(1 - \phi)(\phi - a)$$
$$\phi' = \theta.$$

Experimenatl evidence shows that when $c = \sqrt{2}(1/2 - a)$ and $0 < a < 1/2$, then a solution curve exists in the (ϕ, θ) plane that leaves $(0, 0)$ and approaches $(1, 0)$. The analytical solution of this system is

$$\phi(s) = \left(1 + e^{-s/\sqrt{2}}\right)^{-1}$$
$$\theta(s) = \frac{1}{\sqrt{2}}e^{-s/\sqrt{2}}\left(1 + e^{-s/\sqrt{2}}\right)^{-2}.$$

Problems

5.1 In a study related to automobile driving, Rashevsky [25] found that a driver's decision to make a correction in steering is correlated with his perception of nearness of the edge of the road and the rate at which he is approaching that edge. If y represents the distance of the center of the car from the center of the driving lane, then

$$\frac{d^3y}{dt^3} + k\frac{d^2y}{dt^2} + \gamma B\frac{dy}{dt} + By = 0,$$

where k, γ and B are positive constants, and t is the time. Write this equation in a system of the form (5.12).

5.2 In the development of a mathematical theory of thyroid–pituitary interactions the following third-order differential equations occur (see [10]):

$$\alpha_3\theta''' + \alpha_2\theta'' + \alpha_1\theta' + (1+K)\theta = KC, \quad \theta < C$$

and

$$\alpha_3\theta''' + \alpha_2\theta'' + \alpha_1\theta' + \theta = 0, \quad \theta > C.$$

Here $\theta(t)$ is the concentration of thyroid hormone at time t and $\alpha_1, \alpha_2, \alpha_3$, K and C are constants. Write these equations in systems of the form (5.12).

5.3 A double pendulum oscillates in a vertical plane under the influence of gravity alone (see Fig. 5.6). For small displacements it can be shown that the equations of motion are

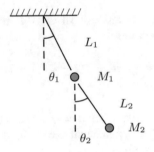

Fig. 5.6

$$(M_1 + M_2)L_1^2\theta_1'' + M_2L_1L_2\theta_2'' + (M_1 + M_2)L_1g\theta_1 = 0 \qquad (5.100)$$

$$M_2L_2^2\theta_2'' + M_2L_1L_2\theta_1'' + M_2L_2g\theta_2 = 0. \qquad (5.101)$$

Show that both θ_1 and θ_2 are solutions of the fourth-order differential equation

$$M_1M_2L_1^2L_2^2\frac{d^4\theta}{dt^4} + M_2L_1L_2g(M_1 + M_2)(L_1 + L_2)\frac{d^2\theta}{dt^2}$$
$$+ M_2L_1L_2g^2(M_1 + M_2)\theta = 0.$$

Further, write (5.100), (5.101) in a system of the form (5.12).

5.4 In the study of a cat's spinal cord motoneuron, the following first-order system appears

$$\frac{dx_1}{dt} = x_2$$
$$\frac{dx_2}{dt} = -x_2\left[\left(\frac{T_2 + T_3}{T_2T_3}\right) - \left(\frac{KT_1}{T_2T_3}\right)f'(x)\right] - \frac{1}{T_2T_3}x_1 + \frac{K}{T_2T_3}f(x),$$

where K is a scaling factor, T_1, T_2 and T_3 are time constants, and $f(x)$ is the transmembrane current best described by

$$f(x) = (x_1 + 2x_2)(x_1^2 + 1).$$

Let $K = 2.5$, $T_1 = T_2 = 1$, $T_3 = 0.5$. Write this system as a standardized first-order system, then ascribe the correct terminology to the system.

5.5 Consider the famous n-body problem of classical mechanics: Let n particles with masses m_i be located at points (x_i, y_i, z_i) and assume that they attract one another according to Newton's law of gravitation. If r_{ij} is the distance between m_i and m_j, and if θ is the angle from the positive x axis to the segment joining them, then the x component of the force exerted on m_i by m_j is

$$\frac{Gm_i m_j}{r_{ij}^2} \cos(\theta) = \frac{Gm_i m_j (x_j - x_i)}{r_{ij}^3},$$

where G is the gravitational constant. Since the sum of these components for all $j \neq i$ equals $m_i(d^2 x_i/dt^2)$, we have n second-order differential equations

$$\frac{d^2 x_i}{dt^2} = G \sum_{j \neq i} \frac{m_j (x_j - x_i)}{r_{ij}^3},$$

and similarly

$$\frac{d^2 y_i}{dt^2} = G \sum_{j \neq i} \frac{m_j (y_j - y_i)}{r_{ij}^3},$$

and

$$\frac{d^2 z_i}{dt^2} = G \sum_{j \neq i} \frac{m_j (z_j - z_i)}{r_{ij}^3}.$$

Write these equations as a system of first-order $6n$ nonlinear equations. The existence and uniqueness theorems presented earlier applied to this resulting system conclude that: If the initial positions and initial velocities of the particles at a certain instant $t = t_0$, are known, and if the particles do not collide in the sense that the r_{ij}'s do not vanish, then their subsequent positions and velocities are uniquely determined. This conclusion underlies the once popular philosophy of mechanistic determinism, according to which the universe is nothing more than a gigantic machine whose future is inexorably fixed by its state at any given moment.

5.6 (Domar Macro Model) Let $S(t)$, $I(t)$, $Y(t)$ be the savings, investment and national income at time t. Assume that
(1) savings are proportional to national income, so that $S(t) = \alpha Y(t)$, $\alpha > 0$,
(2) investment is proportional to the rate of increase of national income so that $I(t) = \beta Y'(t)$, $\beta > 0$,

(3) all savings are invested, so that $S(t) = I(t)$.
Show that $Y(t) = Y(0)e^{\alpha t/\beta}$, $I(t) = S(t) = \alpha Y(0)e^{\alpha t/\beta}$.

5.7 (Domar Debt Model) Let $D(t)$, $Y(t)$ be the debt and national income at time t.
Assume that
(1) rate at which national debt changes is proportional to national income, so that
$D'(t) = \alpha Y(t)$, $\alpha > 0$,
(2) rate of increase of national income is proportional to the national income so that
$Y'(t) = \beta Y(t)$, $\beta > 0$.
Show that $Y(t) = Y(0)e^{\beta t}$, $D(t) = D(0) + \frac{\alpha}{\beta}Y(0)(e^{\beta t} - 1)$. Hence, $\lim_{t \to \infty} \frac{D(t)}{Y(t)}$
$= \frac{\alpha}{\beta}$.

5.8 Suppose that a factory has a sufficient stock of raw materials necessary for the production of steel. Denote the maximum capacity (to produce steel) of the factory at time t by $y(t)$ mass per time unit (for example, tons per day). Denote the total production of steel in the interval $[0, t]$ by $x(t)$ and call it the *stockpile* of the factory. If the production capacity is constant, then the stockpile increases by ayT in the interval $[t, t + T]$ where a is a constant between zero and 1. When $a = 1$ the factory produces at its full capacity, and when $a = 0$ there is no production. Obviously, if the production capacity is increased, the stockpile will grow faster. But to increase the capacity, the factory must be enlarged, which, in its turn, requires steel. Steel for this purpose is taken from the stockpile. Let a fixed fraction b of the stockpile be used per time unit at any time t to increase the production capacity. Assume that the increase in the production capacity is directly proportional to the amount of steel used to enlarge the factory. If the initial stockpile is α and the initial production capacity is β. It follows that (see [8])

$$\begin{aligned} x' &= -bx + ay \\ y' &= bcx \\ x(0) &= \alpha, \quad y(0) = \beta, \end{aligned} \qquad (5.102)$$

where c is a constant. Also, show that the solution of (5.102) can be written as

$$\begin{aligned} x(t) &= \frac{\mu\alpha + a\beta}{\mu - \nu}e^{\mu t} - \frac{\nu\alpha + a\beta}{\mu - \nu}e^{\nu t} \\ y(t) &= \frac{\mu\beta + b\beta + bc\alpha}{\mu - \nu}e^{\mu t} - \frac{\nu\beta + b\beta + bc\alpha}{\mu - \nu}e^{\nu t}, \end{aligned}$$

where μ and ν are the zeros of the quadratic equation $\lambda^2 + b\lambda - abc = 0$.

5.9 Consider a political system consists of two parties, say Republicans and Democrats. Suppose pollsters have observed that about 25% of the Democrats reregister as Republicans before each next election. Similarly, Republicans migrate to the Democratic party at the rate of 30%. If $D(t)$, $R(t)$ represent the number of Democrats and Republicans at time t, and Δt is the time between elections, then it follows that

$$D(t + \Delta t) = D(t) - 0.25D \, \Delta t + 0.30R \, \Delta t$$
$$R(t + \Delta t) = R(t) + 0.25D \, \Delta t - 0.30R \, \Delta t.$$

Now taking the limit as $\Delta t \to 0$, it follows that

$$D' = -0.25D + 0.30R$$
$$R' = 0.25D - 0.30R.$$

Find the solution of the above system and analyze the solution as $t \to \infty$.

5.10 Let two solutions of a substance be separated by a membrane. The amount of substance that passes through the membrane at any particular time is proportional to the difference in the concentrations of the two solutions. The constant of proportionality $P(> 0)$ is called the *permeability* of the membrane and describes the ability of substance to permeate the membrane. Let $x(t)$ and $y(t)$ represent the amount of substance at time t on each side of the membrane, and V_1 and V_2 be the respective constant volume of each solution, then $x(t)$ and $y(t)$ are related by the following system of differential equations

$$\frac{dx}{dt} = P\left(\frac{y}{V_2} - \frac{x}{V_1}\right)$$
$$\frac{dy}{dt} = P\left(\frac{x}{V_1} - \frac{y}{V_2}\right).$$

Solve this system with the initial conditions $x(0) = \alpha$, $y(0) = \beta$ and show that $\lim_{t \to \infty} x(t) = V_1(\alpha + \beta)/(V_1 + V_2)$, $\lim_{t \to \infty} y(t) = V_2(\alpha + \beta)/(V_1 + V_2)$.

5.11 In many situations of interest in population dynamics, we need to study the interaction of two species. For example, one species eats the other (predator–prey problem), or both species help each other to survive. As an example of predator–prey problem, we can think of rabbits as the prey and fox as the predator, a small fish as the prey for a larger fish, and so on. Let $x(t)$ and $y(t)$ denote the population of predator and prey and at time t. Let us assume that the rate of change of prey and predator be governed by

$$\frac{dx}{dt} = cy - dx$$
$$\frac{dy}{dt} = ay - bx,$$

where a takes into account the difference between the birth and death rates of the prey, $b > 0$ because an increase in the predator population should cause the rate of change of prey to diminish, $c > 0$ because an increase in the prey population should increase the population of the predator, and d represents the difference between birth and death rates of the predator. Clearly, $d > 0$ accounts the fact that the predators would die out altogether if there were no prey. Find the general solution of this system.

5.12 In the previous problem, let $x(t)$ and $y(t)$ represent the number of fishermen and fish at the lake at time t. If the rate of stocking the lake is denoted by $S(t)$, the system becomes

$$\frac{dx}{dt} = cy - dx$$
$$\frac{dy}{dt} = ay - bx + S(t).$$

Find the general solution of this system for $S(t) = S_0[1 - \cos(t)]$.

5.13 The cooling of a bare homogeneous metal ingot with a tubular shell is governed by the system (see [29])

$$\frac{dT_1}{dt} = -a_1 T_1 + a_1 T_2 + A_1(T_1^4 - T_0^4)$$
$$\frac{dT_2}{dt} = a_1 T_1 - (a_1 + a_2)T_2 + A_2(T_2^4 - T_0^4) + c,$$

where T_1 is the temperature of the ingot, T_2 is the temperature of a tubular shell that surrounds the ingot, T_0 is the temperature of the surrounding atmosphere (assumed constant), and the constants a_1, a_2 are positive. Find the general solution of this system when the radiation terms are negligible, i.e., $A_1 = A_2 = 0$. Further, show that $\lim_{t \to \infty} T_1(t) = \lim_{t \to \infty} T_2(t) = c/a_2$. What is the significance of these limits.

5.14 Suppose a substance S_1 is produced in a cell at a rate which is proportional to its own concentration c_1. A second substance S_2 catalyze the decomposition of S_1, so that S_1 is decomposed at a rate which is proportional to the concentration c_2 of S_2. Assume that the production of S_2 is catalyzed by S_1 at a rate which is proportional to c_1, while S_2 decomposes at a rate which is proportional to its own concentration c_2. This leads to the system (see [24]),

$$\frac{dc_1}{dt} = \alpha_1 c_1 + \beta_1 c_2 + \gamma_1$$
$$\frac{dc_2}{dt} = \alpha_2 c_1 + \beta_2 c_2 + \gamma_2.$$

Solve this system with the actual experimental data

$$\alpha_1 = 0.1, \quad \beta_1 = -0.1, \quad \gamma_1 = 2 \times 10^{-5}$$
$$\alpha_2 = 0.3, \quad \beta_2 = -0.1, \quad \gamma_2 = 10^{-5}.$$

5.15 For the detection of diabetes an oral glucose tolerance test leads to the system (see [2])

$$\frac{dg}{dt} = -m_1 g - m_2 h + J(t)$$
$$\frac{dh}{dt} = m_3 g - m_4 h + k(t), \tag{5.103}$$

where all the constants m_1, m_2, m_3, m_4 are positive, g represents the concentration of glucose above a fasting state level, h is the effective hormonal concentration above fasting state level, and J and k denote exogenous inputs into the system via intestines, or intravenously. Find the general solution of (5.103) in the special case $J(t) \equiv J > 0$, $k(t) \equiv k > 0$, and $\omega^2 = m_2 m_3 - \frac{1}{4}(m_1 - m_4)^2 > 0$ so that glucose levels are periodic (research has shown that lab results of $\pi > 2\omega$ indicate a mild case of diabetes). Further, compute $\lim_{t \to \infty} g(t)$ and $\lim_{t \to \infty} h(t)$ and interpret these values in relation to the model.

The system (5.54) with $J(t) \equiv J > 0$ and $k(t) \equiv 0$ has also been used for the reactions of the pituitary and thyroid glands in the regulation of the metabolic rate and reactions of the pituitary and ovary glands in the control of the menstrual cycle (see [11, 23]).

5.16 In a study of chemical transmitter in cellular control, Grossberg [13] obtained the following initial value problem

$$\frac{dx}{dt} = a(b - x) - cfy$$

$$\frac{dy}{dt} = d(x - y) - cfy - ay$$

$$x(0) = b, \quad y(0) = \frac{db}{a + d},$$

where a, b, c, d, and f are positive constants. Show that

(1) $z(t) = x(t) - y(t) = \dfrac{ab}{a + d}$.

(2) $y(t) = \dfrac{bd}{a + d} \left[e^{-(a+cf)t} + \dfrac{a}{a + cf} \left(1 - e^{-(a+cf)t} \right) \right]$.

(3) $y(t)$ is a monotone decreasing function, and $y(\infty) = \dfrac{abd}{(a + d)(a + cf)}$.

5.17 Consider the two compartment system depicted in Fig. 5.7, where I_i is the rate at which material enters the compartment i, f_{ij} is the rate at which material passes from compartment j to compartment i, f_{0i} is the rate at which material is removed from compartment i, q_i is the concentration of material in compartment i, and f_{01}, f_{02}, f_{12}, f_{21} are nonnegative. Then, it follows that (see [15])

$$\frac{dq_1}{dt} = -(f_{01} + f_{21})q_1 + f_{12}q_2 + I_1(t)$$

$$\frac{dq_2}{dt} = f_{21}q_1 - (f_{02} + f_{12})q_2 + I_2(t)$$

$$q_1(0) = \alpha, \quad q_2(0) = \beta,$$

where α and β are initial concentrations.

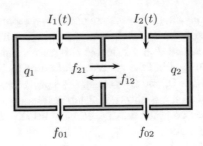

Fig. 5.7

(1) The system is called closed if $I_1(t) = I_2(t) = f_{01} = f_{02} = 0$. Show that the general solution in this case is

$$c_1 \begin{bmatrix} f_{12} \\ f_{21} \end{bmatrix} + c_2 e^{-(f_{12}+f_{21})t} \begin{bmatrix} 1 \\ -1 \end{bmatrix}.$$

(2) The system is called open without intakes if $I_1 = I_2 = 0$. Show that the general solution for the special case $f_{01} = f_{02} = f_{12} = f_{21} = a$ is

$$c_1 e^{-at} \begin{bmatrix} 1 \\ 1 \end{bmatrix} + c_2 e^{-3at} \begin{bmatrix} 1 \\ -1 \end{bmatrix}.$$

(3) The system is called open with intakes if $I_1(t) = k_1$, $I_2(t) = k_2$ and $f_{01} \neq 0 \neq f_{02}$. Find the general solution for the special case $f_{12} = f_{21} = f_{02} = f_{01} = a$.
(4) The system is called a mammillary system with intake if $I_1(t) = p_1$, $I_2(t) = 0$, $f_{01} = 0$, but $f_{02} \neq 0$. Find the general solution for the special case $f_{12} = f_{21} = f_{02} = a$.

5.18 Mathematical modeling has proven to be valuable in understanding the dynamics of HIV-1 infection. In 1996, Perelson et al. [21] considered the following system

$$\frac{dT^*}{dt} = kV_I T - \delta T^*$$
$$\frac{dV_I}{dt} = -cV_I$$
$$\frac{dV_{NI}}{dt} = N\delta T^* - cV_{NI},$$

where T represents the concentrations of uninfected target cells and T^* infected cells that are producing virus with a constant rate k. After protease inhibitors are given, virus V is classified as either infectious V_I, i.e., not influenced by the protease inhibitor, or as noninfectious V_{NI} due to the action of the protease inhibitor which prevents virion maturation into infectious particles ($V = V_I + V_{NI}$). δ is the rate of loss of virus producing cells, N is the number of new virions produced per infected cell during its lifetime, and c is the rate constant for virion clearance. Initially it is assumed that $V_I(0) = V_0$ and $V_{NI}(0) = 0$. Show that

$$V_I(t) = V_I(0)e^{-ct}$$
$$T^*(t) = e^{-\delta t}T^*(0) + \frac{kV_0T}{c - \delta}\left(e^{-\delta t} - e^{-ct}\right)$$
$$V_{NI}(t) = \frac{N\delta T^*(0)}{c - \delta}\left(e^{-\delta t} - e^{-ct}\right) + \frac{N\delta kV_0T}{c - \delta}\left(\left\{\frac{e^{-\delta t} - e^{-ct}}{c - \delta}\right\} - te^{-ct}\right).$$

Further, show that if initially the system is at quasi-steady state, i.e., at $t = 0$, $\dfrac{dT^*}{dt} = \dfrac{dV}{dt} = 0$, which give the relations $kV_0T = \delta T^*(0)$, $N\delta T^*(0) = cV_0$, and $NkT = c$, it follows that the total concentration of plasma virions is

$$V(t) = V_0e^{-ct} + \frac{cV_0}{c - \delta}\left[\frac{c}{c - \delta}\left(e^{-\delta t} - e^{-ct}\right) - \delta te^{-ct}\right].$$

5.19 The tumor growth occurs in the colon has 10^7 crypts (or investigations in the lining of the colon) to help prevent uncontrolled growth of cancer cells (see [1]). Each crypt has stem cells at its base, which produce a variety of cell types needed for tissue renewal and regeneration after injury. Stem cells produce semi-differentiated cells, which migrate up the crypt wall to the surface; along the way, these cells differentiate into several types of cells. Once they reach the surface, these cells either die or are transported away. To build a basic model, we divide the cells into three groups, stem cells, semi-differentiated cells, and fully differentiated cells, denoted by N_0, N_1 and N_2, respectively.

We assume that per-capita rate of stem (respectively semidifferentiated) cell proliferation by α_3 (respectively, β_3), diffcrentiation by α_2 (respectively, β_2), and death by α_1 (respectively, β_1), and the per-capita removal rate of fully differentiated cells by γ, this leads to the system (see [16])

$$\frac{dN_0}{dt} = \alpha N_0, \quad \text{where} \quad \alpha = \alpha_3 - \alpha_1 - \alpha_2$$
$$\frac{dN_1}{dt} = \beta N_1 + \alpha_2 N_0, \quad \text{where} \quad \beta = \beta_3 - \beta_1 - \beta_2 \qquad (5.104)$$
$$\frac{dN_2}{dt} = \beta_2 N_1 - \gamma N_2.$$

Show that
$$N_0(t) = N_{00}e^{\alpha t}$$
$$N_1(t) = Ae^{\alpha t} + (N_{10} - A)e^{\beta t}$$
$$N_2(t) = Be^{\alpha t} + Ce^{\beta t} + (N_{20} - B - C)e^{-\gamma t},$$

where
$$A = \frac{\alpha_2 N_{00}}{\alpha - \beta}, \quad B = \frac{\beta_2 A}{\gamma + \alpha}, \quad C = \frac{\beta_2 (N_{10} - A)}{\gamma + \beta}.$$

Thus, the stem cell population $N_0(t)$ grows exponentially if $\alpha > 0$, decays exponentially if $\alpha < 0$, or remains constant if $\alpha = 0$. Use the fact that $\alpha_1 + \alpha_2 + \alpha_3 = 1$ to show that the stem cell population size remains constant if $\alpha_3 = 1/2$. Further, show that if the stem cell population size is constant, i.e., $\alpha = 0$, then the limiting population sizes for N_1 and N_2 when $\beta < 0$ and $\gamma > 0$ are $\alpha_2 N_{00}/(\alpha - \beta)$ and $\beta_2 \alpha_2 N_{00}/(\alpha - \beta)(\gamma + \alpha)$, respectively.

5.20 The following Fig. 5.8 represents a model for the movement of iodine through a dog. If f_i is the amount of iodine in compartment i, then the differential equations corresponding to this model are (see [9])

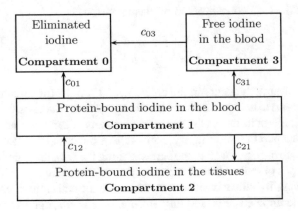

Fig. 5.8

$$f_1' = -(c_{21} + c_{01} + c_{31})f_1 + c_{12}f_2$$
$$f_2' = c_{21}f_1 - c_{12}f_2$$
$$f_3' = c_{31}f_1 - c_{03}f_3,$$

where all c_{ij} are positive. Show that the eigenvalues of the corresponding matrix A are real, distinct, and negative. Then, use (5.18) to show that all solutions of this system tend to zero as $t \to \infty$.

5.21 Consider the following electrical network
where $I_1(0) = I_2(0) = I_3(0) = 0$. Find $I_1(t)$, $I_2(t)$, $I_3(t)$ for all time t (Fig. 5.9).

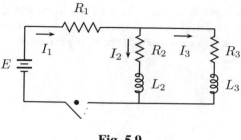

Fig. 5.9

5.22 Consider the following electrical network

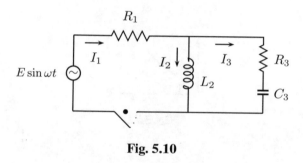

Fig. 5.10

where $I_1(0) = I_2(0) = I_3(0) = Q_3(0) = 0$. Find $I_1(t)$, $I_2(t)$, $I_3(t)$, $Q_3(t)$ for all time t (Fig. 5.10).

5.23 Consider the following electrical network

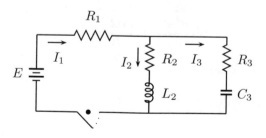

Fig. 5.11

where $I_1(0) = I_2(0) = I_3(0) = Q_3(0) = 0$. Find $I_1(t)$, $I_2(t)$, $I_3(t)$, $Q_3(t)$ for all time t (Fig. 5.11).

5.24 A projectile of mass M is fired into the air with the initial velocity u_0 and an angle α from the horizontal. If xy-coordinate system is placed as in Fig. 5.12 and if air resists the motion with a force per unit mass of k times the velocity, the resulting equations of motion are

$$M\frac{d^2x}{dt^2} = -k\frac{dx}{dt}$$

$$M\frac{d^2y}{dt^2} = -k\frac{dy}{dt} - Mg$$

$$x(0) = y(0) = 0, \quad x'(0) = u_0\cos(\alpha), \quad y'(0) = u_0\sin(\alpha).$$

Find its solution.

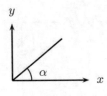

Fig. 5.12

5.25 Let M and e represent the mass and charge of an electron. If electrons are subjected to an electric field of intensity E and a magnetic field of intensity H, then Sir Joseph John Thomson's (1856–1940) model to determine the path of the electron is governed by the system

$$M\frac{d^2x}{dt^2} + H\,e\frac{dy}{dt} = E\,e$$

$$M\frac{d^2y}{dt^2} - H\,e\frac{dx}{dt} = 0$$

subject to the initial conditions

$$x(0) = x'(0) = y(0) = y'(0) = 0.$$

Show that

$$x(t) = \frac{EM}{H^2e}\left[1 - \cos\left(\frac{H\,e}{M}\right)t\right]$$

$$y(t) = \frac{EM}{H^2e}\left[\left(\frac{H\,e}{M}\right)t - \sin\left(\frac{H\,e}{M}\right)t\right],$$

which is a parametric representation of a cycloid.

5.26 Consider two pendulums of equal length L and equal mass M coupled by a spring (see Fig. 5.13). Let x_1 and x_2 denote the vertical displacements of the masses from their equilibrium positions. The action of the spring is to introduce a force proportional to the difference of the displacements, i.e., a force $k(x_1 - x_2)$, where k is the spring constant. The motion then is described by the solutions of

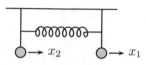

Fig. 5.13

$$Mx_1'' + M\omega^2 x_1 = -k(x_1 - x_2)$$
$$Mx_2'' + M\omega^2 x_2 = -k(x_2 - x_1), \quad \omega^2 = g/L. \tag{5.105}$$

(1) Show that (5.105) is equivalent to the differential system (5.12) with

$$A = \begin{bmatrix} 0 & 1 & 0 & 0 \\ -\omega^2 - k_1 & 0 & k_1 & 0 \\ 0 & 0 & 0 & 1 \\ k_1 & 0 & -\omega^2 - k_1 & 0 \end{bmatrix},$$

where $k_1 = k/M$.

(2) The general solution of this first-order differential system can be written as

$$u(t) = \begin{bmatrix} \cos(\omega t) & \sin(\omega t) & \cos(\omega_1 t) & \sin(\omega_1 t) \\ -\omega \sin(\omega t) & \omega \cos(\omega t) & -\omega_1 \sin(\omega_1 t) & \omega_1 \cos(\omega_1 t) \\ \cos(\omega t) & \sin(\omega t) & -\cos(\omega_1 t) & -\sin(\omega_1 t) \\ -\omega \sin(\omega t) & \omega \cos(\omega t) & \omega_1 \sin(\omega_1 t) & -\omega_1 \cos(\omega_1 t) \end{bmatrix} \begin{bmatrix} c_1 \\ c_2 \\ c_3 \\ c_4 \end{bmatrix},$$

where $\omega_1 = \sqrt{\omega^2 + 2k_1}$.

(3) Deduce the solutions of this first-order differential system satisfying the initial conditions $\mathbf{u}(0) = (1, 0, 1, 0)^T$ and $\mathbf{u}(0) = (1, 0, -1, 0)^T$, and interpret these solutions in relation to the original system (5.105).

5.27 Consider the mechanical system consisting of two masses M_1 and M_2 and three springs with spring constants k_1, k_2 and k_3 (see Fig. 5.14), and on each mass a frictional force proportional to the velocity is acting. Let x_1 and x_2 denote the displacement from equilibrium of the masses M_1 and M_2, respectively. Then, Newton's second law provides

$$M_1 \frac{d^2 x_1}{dt^2} = \mu_1 \frac{dx_1}{dt} - (k_1 + k_2)x_1 + k_2 x_2$$

$$M_2 \frac{d^2 x_2}{dt^2} = \mu_2 \frac{dx_2}{dt} + k_2 x_1 - (k_2 + k_3)x_2.$$

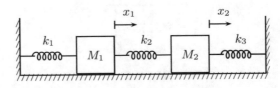

Fig. 5.14

Show that these equations can be written as (5.12) with $\mathbf{u}(t) = (x_1(t), \frac{dx_1}{dt},$
$x_2(t), \frac{dx_2}{dt})^T$ and

$$\mathbf{A} = \begin{pmatrix} 0 & 1 & 0 & 0 \\ -\dfrac{k_1 + k_2}{M_1} & \dfrac{\mu_1}{M_1} & \dfrac{k_2}{M_1} & 0 \\ 0 & 0 & 0 & 1 \\ \dfrac{k_2}{M_2} & 0 & -\dfrac{k_2 + k_3}{M_2} & \dfrac{\mu_2}{M_2} \end{pmatrix}.$$

Take some reasonable values of the constants $k_1, k_2, k_3, \mu_1, \mu_2, M_1$ and M_2 to find
the solution of this system.

5.28 Consider a uniform shaft with three identical disks equally spaced at $x = L$, $2L$
and $3L$ along it. At the left end $x = 0$ the shaft is fixed while at the right end $x = 3L$
it is free. If w_1, w_2 and w_3 denote the angular deflections of the three disks, then the
torsional oscillations of the disks are described by the differential equations

$$\frac{d^2 w_1}{dt^2} = k(-2w_1 + w_2)$$

$$\frac{d^2 w_2}{dt^2} = k(w_1 - 2w_2 + w_3) \qquad\qquad (5.106)$$

$$\frac{d^2 w_3}{dt^2} = k(w_2 - w_3),$$

where k is a constant that depends on the torsional stiffness of the shaft, the moment
of inertia of the disks, and the length of the shaft. In what follows we assume that
$k = 1$.
(1) Write (5.106) in a system of the form (5.12) and show that the characteristic
equation of the resulting matrix $\mathbf{A}$ is $\lambda^6 + 5\lambda^4 + 6\lambda^2 + 1 = 0$.

(2) Show that all eigenvalues are complex with real parts equal to zero. For this use the change of variable $z = -\lambda^2$ to get $z^3 - 5z^2 + 6z - 1 = 0$. This equation has roots in $(0, 1)$, $(1, 2)$ and $(3, 4)$.

(3) Use Part (ii) to describe the motion of the disks without computing w_1, w_2 and w_3.

5.29 Planck [22] postulated that the motion of a particle of negligible mass projected from the Earth, taking into account the effects on the particle caused by the rotation of the Earth, is governed by the following system

$$\frac{d^2x}{dt^2} = 2\omega \sin(\beta)\frac{dy}{dt}$$

$$\frac{d^2y}{dt^2} = -2\omega \left(\sin(\beta)\frac{dx}{dt} + \cos(\beta)\frac{dz}{dt} \right)$$

$$\frac{d^2z}{dt^2} = 2\omega \cos(\beta)\frac{dy}{dt} - g$$

$$x(0) = y(0) = z(0) = 0$$

$$x'(0) = u, \quad y'(0) = v, \quad z'(0) = w$$

where the positive x-axis represents south, the positive y-axis east, and the positive z-axis points in the direction opposite to the direction of the acceleration, g, due to gravity, β represents the latitude from the origin, and ω is the angular velocity of the Earth. Show that

$$x(t) = \frac{u}{2\omega}[2\omega t \cos^2(\beta) + \sin^2(\beta)\sin(2\omega t)] + \frac{v}{\omega}\sin(\beta)\sin^2(\omega t)$$
$$- \frac{w}{2\omega}\sin(\beta)\cos(\beta)[2\omega t - \sin(2\omega t)] + \frac{g}{2\omega^2}\sin(\beta)\cos(\beta)[\omega^2 t^2 - \sin^2(\omega t)]$$

$$y(t) = -\frac{u}{\omega}\sin(\beta)\sin^2(\omega t) + \frac{v}{2\omega}\sin(2\omega t) - \frac{w}{\omega}\cos(\beta)\sin^2(\omega t)$$
$$+ \frac{g}{4\omega^2}\cos(\beta)[2\omega t - \sin(2\omega t)]$$

$$z(t) = -\frac{u}{2\omega}\sin(\beta)\cos(\beta)[2\omega t - \sin(2\omega t)] + \frac{w}{2\omega}[2\omega t \sin^2(\beta) + \sin(2\omega t)\cos^2(\beta)]$$
$$+ \frac{v}{\omega}\cos(\beta)\sin^2(\omega t) - \frac{g}{2\omega^2}[\omega^2 t^2 \sin^2(\beta) + \cos^2(\beta)\sin^2(\omega t)].$$

Observe that $\lim_{\omega \to 0} x(t) = ut$, $\lim_{\omega \to 0} y(t) = vt$, $\lim_{\omega \to 0} z(t) = wt - \frac{1}{2}gt^2$, i.e., the solution reduces to the usual equations of motion when the Earth's rotation is ignored.

5.30 Let $p_n(t)$ be the probability that there are n species at time t (n is an integer and t is a real number). To compute $p_n(t + dt)$, Yule [31] assumed that:

1. If there are $(n - 1)$ species at time t, each species has a probability rdt of generating one new species between t and $(t + dt)$; in the limit $dt \to 0$, there will be n species at time $(t + dt)$ with a probability $(n - 1)rdt$.

2. If there are n species at time t, there will be $n + 1$ species at time $t + dt$ with a probability $nrdt$.

These assumptions lead to the following system of differential equations

$$\frac{dp_1}{dt} = -rp_1$$

$$\frac{dp_n}{dt} = (n-1)rp_{n-1} - nrp_n, \quad n \geq 2$$

with the initial conditions $p_n(0) = 0$, $n \geq 1$. Show that

(1) $p_n(t) = e^{-rt}\left(1 - e^{-rt}\right)^{n-1}$, i.e., at some fixed time t, the distribution of probabilities $\{p_n(t)\}$ is geometric with a ratio between two consecutive terms equal to $(1 - e^{-rt})$.

(2) $\sum_{n=1}^{\infty} np_n(t) = e^{rt}$, i.e., the expected number of species increase exponentially with time.

5.31 In the year 1966, Zimmerman [32] showed that if $C = [Ca^{++}]$ and $P = [HPO_4^{--}]$, then the system of differential equations

$$\frac{dC}{dt} = 10(K - kCP)$$

$$\frac{dP}{dt} = 6(K - kCP)$$

$$C(0) = P(0) = 0$$

governs how organic acids dissolve tooth enamel. Here, K and k are reaction rate constants. Show that $C = (5/3)P$ and

$$P(t) = \sqrt{\frac{3K}{5k} \frac{1 - e^{-\sqrt{240Kkt}}}{1 + e^{-\sqrt{240Kkt}}}}.$$

5.32 The populations of two species of animals are described by the nonlinear system of first-order differential equations

$$\frac{dx}{dt} = k_1 x(a - x)$$

$$\frac{dy}{dt} = k_2 xy.$$

Show that $x = aAe^{ak_1 t}/(1 + Ae^{ak_1 t})$, $y = B(1 + Ae^{ak_1 t})^{k_2/k_1}$.

5.33 A population model used in food science research (see [7]) is the following pair of initial value problems:

$$\frac{dP}{dt} = \mu \frac{q(t)}{1 + q(t)}\left[1 - \left(\frac{P(t)}{P_e}\right)^m\right]P(t)$$

$$\frac{dq}{dt} = \nu q(t)$$

$$P(0) = P_0, \quad q(0) = q_0.$$

(5.107)

Here, $P(t)$ represents the population of bacteria at time t; $q(t)$ represents the concentration of critical substance present; $\nu > 0$ represents the growth rate of the critical substance; μ accounts for the effects of environmental conditions, such as temperature, upon the growth rate of the bacteria; P_0 and q_0 represent the bacterial population and the amount of critical substance present at time $t = 0$, respectively; and m is an integer. Show that in the case μ and ν are constants the solution of (5.107) can be written as

$$q(t) = q_0 e^{\nu t}, \quad P(t) = P_e \left[1 + \left(\left[\frac{P_0}{P_e} \right]^{-m} - 1 \right) \left(\frac{1 + q_0}{1 + q_0 e^{\nu t}} \right)^{m\mu/\nu} \right]^{-1/m}.$$

5.34 A projectile is shot vertically intro the air with an initial velocity v_0 ft/s. Assuming that air resistance is proportional to the square of the instantaneous velocity, the motion is described by the pair of differential equations

$$m \frac{dv}{dt} = -mg - kv^2, \quad k > 0 \tag{5.108}$$

for positive y-axis up with origin at ground level so that $v = v_0$ when $y = 0$, and

$$m \frac{dv}{dt} = mg - kv^2, \quad k > 0 \tag{5.109}$$

for positive y-axis down with origin at the maximum height h, so that $v = 0$ at $y = h$. Clearly, Eqs. (5.108) and (5.109) respectively describe the motion of the projectile when rising and falling (see Fig. 5.15).

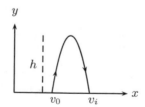

Fig. 5.15

Use $dv/dt = v\,dv/dy$ to obtain the solutions of Eqs. (5.108) and (5.109)

$$v^2 = \frac{mg + kv_0^2}{k} e^{-2ky/m} - \frac{mg}{k} \quad \text{and} \quad v^2 = \frac{mg}{k} \left(1 - e^{-2ky/m} \right).$$

Hence, deduce that
(1) $h = (m/2k) \ln(mg + kv_0^2)/mg$
(2) terminal velocity $= \sqrt{mg/k}$

(3) impact velocity $v_i = v_0/\sqrt{1 + (kv_0^2/mg)}$, which is less than the initial velocity v_0.

5.35 For the system (5.80)–(5.82) with $\alpha = \beta$ show that
(1) $S(t) < I(t)$ on the interval $t_{max} \leq t \leq T$, and hence $S(\infty) < I_*$. What happens to $S(\infty)$ if I_* decrease to zero?
(2) $t_{max} < 1/\alpha$.
(3) $I(1/\alpha) = 1/e$ and $S(1/\alpha) = S(0)/e$, and conclude that $I_{max} > 1/e$.
(4) The curve $I = I(t)$ has an infection point at $t = t_{max} + 1/\alpha$.
(5) T can be obtained from the expression

$$T = \frac{1}{\alpha}\left(\frac{I_*}{S(\infty)} - \frac{I(0)}{S(0)}\right),$$

and conclude that

$$T = t_{max} + \frac{1}{\alpha}\left(\frac{I_*}{S(\infty)} - 1\right).$$

What happens if I_* decreases to zero?

5.36 For the system (5.80)–(5.82) assume that $\alpha > \beta$ and $I(0) > I_*$.

(1) For $0 \leq t < T$, show that

$$I(t) = I(0)e^{-\beta t} + \frac{\alpha S(0)}{\alpha - \beta}\left(e^{-\beta t} - e^{-\alpha t}\right).$$

(2) Use (1) to show that

$$t_{max} = \frac{1}{\alpha - \beta}\ln\left[\left(\frac{\alpha S(0)}{\beta}\right)\left(\frac{\alpha}{\alpha - \beta I(0)}\right)\right]$$

and

$$I_{max} = \frac{\alpha}{\beta}S(t_{max}).$$

(3) Show that $I = I(t)$ has an infection point at

$$t = t_{max} + \frac{1}{\alpha - \beta}\ln\left(\frac{\alpha}{\beta}\right).$$

(4) Show that

$$I_{max} > \left(\frac{\beta}{\alpha}\right)^{\beta/(\alpha-\beta)}.$$

(5) Show that

$$S(\infty) = \left[\left(\frac{1}{S(0)} \right)^{\beta/\alpha} \left(\frac{\alpha - \beta I(0)}{\alpha} \right) - I_* \left(\frac{\alpha - \beta}{\alpha} \right) \right]^{\alpha/(\alpha - \beta)}.$$

References

1. M.L. Abell, J.P. Braselton, *Introductory Differential Equations* (Elsevier, Amsterdam, 2014)
2. E. Ackerman, L.C. Gatewood, J.W. Rosevear, G.D. Molnal, in *Concepts and Models of Biomathematics*, ed. by F. Heinmets (Marcel Dekker, New York, 1969)
3. R.P. Agarwal, D. O'Regan, *An Introduction to Ordinary Differential Equations* (Springer, New York, 2008)
4. R.G.D. Allen, *Mathematical Economics* (Macmillan, London, 1957)
5. W.F. Ames, *Nonlinear Ordinary Differential Equations in Transport Processes* (Academic, New York, 1968)
6. N.T.J. Bailey, *The Mathematical Theory of Epidemics* (Harper, New York, 1957)
7. J. Baranyi, T.A. Roberts, P. McClure, Food Microbiol. **10**, 43–59 (1993)
8. R. Bellman, *Modern Elementary Differential Equations* (Addison-Wesley, Reading, 1968)
9. H. Cohen (ed.) *The Mathematical Sciences*, COSRIMS (MIT Press, Cambridge, 1969)
10. L. Danziger, G.L. Elmergreen, Bull. Math. Biophys. **18**, 1–13 (1956)
11. L. Danziger, G.L. Elmergreen, Bull. Math. Biophys. **19**, 9–18 (1957)
12. J.H. Engel, Oper. Res. **2**, 163–171 (1954)
13. S. Grossberg, J. Theoret. Biol. **22**, 325–364 (1969)
14. G.J. Hueter, M.A. McClelland, L.A. Resner, M.G. Zevallos, Interface **5**, 15–25 (1978)
15. J.A. Jacquez, *Compartmental Analysis in Biology and Medicine* (Elsevier, Amsterdam, 1972)
16. M.D. Johnston, C.M. Edwards, W.F. Bodmer, P.K. Maini, S.J. Chapman, Proc. Nat. Acad. Sci. **104**, 4008–4013 (2007)
17. F.W. Lanchester, *Aircraft in Warfare, The Dawn of the Fourth Army* (Constable and Co., Ltd., London, 1916)
18. F.W. Lanchester, *Mathematics in Warfare, The World of Mathematics*, vol. 4 (Simon and Schuster Inc., New York, 1956)
19. W.O. Kermack, A.G. McKendrick, Proc. R. Soc. Lond. A **115**, 700–721 (1927)
20. D. Klaus, J. Roy. Stat. Soc. Ser. A. **130**, 505–528 (1967)
21. A.S. Perelson, A.U. Neumann, M. Markowitz, J.M. Leonard, D.D. Ho, Science, New Series **271**, 1582–1586 (1996)
22. B.M. Planck, *Einführung in die Allgemeine Mechanik*, 4th edn (S. Hirzel, Leipsig, 1928)
23. A. Rapoport, Bull. Math. Biophys. **14**, 171–183 (1952)
24. N. Rashevsky, *Mathematical Biophysics* (The University of Chicago Press, Chicago, 1938)
25. N. Rashevsky, Bull. Math. Biophys. **29**, 181–186 (1967)
26. A.T. Reid, Bull. Math. Biophys. **14**, 313–316 (1952)
27. T.L. Saaty, *Mathematical Models of Arms Control and Disarmament* (Wiley, New York, 1968)
28. S.H. Strogatz, Math. Mag. **61**, 35 (1988)
29. K.D. Tocher, *Mathematical Models in Metallurgical Process Development* (The Iron and Steel Institute, London, 1970)
30. P. Waltman, Deterministic threshold models in the theory of epidemics, vol. 1. *Lecture Notes in Biomathematics* (Springer, New York, 1974)
31. G.U. Yule, Phil. Trans. Roy. Soc. Lond. B **213**, 21–87 (1925)
32. S.O. Zimmerman, Bull. Math. Biophys. **28**, 417–432 (1966)

Chapter 6
Runge–Kutta Method

The class of differential equations for which explicit solutions can be obtained is rather small. In fact, in Chap. 3, we have already remarked that to find an explicit solution of the second-order linear differential equation (3.2) there does not exist any method. Further, the problem of finding solutions of nonlinear differential equations is most of the time impossible. It turns out that for such problems, the only alternative we have is to resort to numerical methods.

Although, in the literature several numerical techniques have been analyzed thoroughly, and now various codes which require almost no problem preparation are readily available, for the numerical computation of solutions of initial value problems of the type (5.29), we shall use only the fourth-order classical Runge–Kutta method (after Carl David Tolmé Runge 1856–1927, and Martin Wilhelm Kutta 1867–1944). Of course, for almost all the models, we shall consider it should be possible to apply other known less accurate or more sophisticated methods.

The first step of fourth-order classical Runge–Kutta method is to partition the interval $[t_0, b]$ on which the solution $\mathbf{u}(t) = (u_1(t), \cdots, u_n(t))$ of (5.29) is desired into a finite number of subintervals by the points $t_0 < t_1 < \cdots < t_N = b$. These points are called the *mesh points* or the *grid points*. The spacing between the points, i.e., $h_j = t_j - t_{j-1}$, $j = 1, 2, \cdots, N$ is called the *mesh spacing* or *step length*. For simplicity, we shall assume that the points are spaced uniformly, i.e., $h_j = h = (b - t_0)/N$ constant, $j = 1, 2, \cdots, N$. Thus, the mesh points are $t_j = t_0 + jh$, $j = 0, 1, \cdots, N$.

The second step of the method is to construct the sequence $\{\mathbf{u}_j\}_{j=0}^{N}$, where $\mathbf{u}_j = (u_{1,j}, \cdots, u_{n,j})$ approximates the exact (but unknown) $u(t_j)$, by employing the recurrence relation

$$\mathbf{u}_0 = \mathbf{u}(t_0) = \mathbf{u}^0$$
$$\mathbf{u}_{j+1} = \mathbf{u}_j + \frac{1}{6}(\mathbf{K}_1 + 2\mathbf{K}_2 + 2\mathbf{K}_3 + \mathbf{K}_4), \quad j = 0, 1, \cdots, N - 1,$$

where $\mathbf{K}_r = (k_{1,r}, \cdots, k_{n,r})^T$, $r = 1, 2, 3, 4$ and

© Springer Nature Switzerland AG 2019
R. P. Agarwal et al., *500 Examples and Problems of Applied Differential Equations*, Problem Books in Mathematics,
https://doi.org/10.1007/978-3-030-26384-3_6

$$k_{i,1} = h\mathbf{g}_i(t_j, u_{1,j}, u_{2,j}, \cdots, u_{n,j})$$
$$k_{i,2} = h\mathbf{g}_i(t_j + 0.5h, u_{1,j} + 0.5k_{1,1}, u_{2,j} + 0.5k_{2,1}, \cdots, u_{n,j} + 0.5k_{n,1})$$
$$k_{i,3} = h\mathbf{g}_i(t_j + 0.5h, u_{1,j} + 0.5k_{1,2}, u_{2,j} + 0.5k_{2,2}, \cdots, u_{n,j} + 0.5k_{n,2})$$
$$k_{i,4} = h\mathbf{g}_i(t_j + h, u_{1,j} + k_{1,3}, u_{2,j} + k_{2,3}, \cdots, u_{n,j} + k_{n,3}).$$

Example 6.1 In certain circumstances, how the malaria affects a community can be represented by a system of differential equations. The model we shall develop is due to Sir Ronald Ross (1857–1932), which seems to be the first attempt to apply differential equations to epidemiology. Ronald Ross received the 1902 Nobel Prize in Physiology and Medicine for his work on malaria. In what follows we let $P(t) =$ human population at time t, $H(t) =$ human population affected with malaria at time t, $F =$ infective rate, i.e., fraction of affected human population that is infective, $R =$ recovery rate, i.e., fraction of affected human population that completely recovers per unit of time, $M =$ human death rate. We shall also let p, h, f, r and m denote the corresponding quantities to mosquito population.

Let a mosquito bite a human on an average b times per unit of time, then fh infective mosquitoes will inflict bfh bites per unit of time, and a fraction of these bites $(P - H)/P$ occur on healthy people. We assume that once bitten, the person becomes affected. Then, the number of new infections per unit of time is $bfh(P - H)/P$. Similarly, if a human is bitten, on an average, B times per unit of time, then the number of new infections among the mosquitoes is $BFH(p - h)/p$. Clearly, the rate at which humans are bitten, BP must be the same rate as which the mosquitoes bite, bp. Hence, $B = bp/P$, so the number of new infections per unit of time among the mosquitoes is $bFH(p - h)/P$. Next, we assume that immigration and emigration in the area under consideration are negligible for both people and mosquitoes, then it follows that

$$\begin{array}{ccccc} \text{Rate of} & = & \text{Rate of} & - \text{ Death } & - \text{ Recovery} \\ \text{affection among humans} & & \text{new affections} & \text{rate} & \text{rate,} \end{array}$$

i.e.,

$$\frac{dH}{dt} = \frac{bfh}{P}(P - H) - MH - RH.$$

Similarly, the rate of affection among the mosquitoes satisfies the equation

$$\frac{dh}{dt} = \frac{bFH}{P}(p - h) - mh - rh.$$

Finally, we assume that
1. the birth rate is equal to the death rate for both the humans and the mosquitoes, so the populations P and p are constant,
2. M is negligible with respect to R, whereas r is negligible with respect to m, and
3. the humans have one unit of population and the mosquitoes have α units of population, i.e., $P = 1$, $p = \alpha$.

With these assumptions the above equations are simplified to the coupled nonlinear system

$$\frac{dH}{dt} = bfh(1 - H) - RH$$
$$\frac{dh}{dt} = bFH(\alpha - h) - mh. \tag{6.1}$$

In the year 1923, Lotka studied this system extensively, and provided the following experimental values of the constants:

$$\begin{array}{lll} \alpha = 19.418 & R = 0.231046 & f = 0.33333 \\ m = 3.2958 & F = 0.25 & b = 0.82396 \end{array} \tag{6.2}$$

and time is measured in months.

Suppose that initially no humans and ten percent of the mosquitoes are affected with malaria, i.e., for the system (6.1) we have the initial conditions

$$H(0) = 0, \quad h(0) = 1.9418. \tag{6.3}$$

We use Runge–Kutta method to find the numerical solution of the initial value problem (6.1)–(6.3) on the interval $[0, 200]$ with the step-size $h = 0.5$, i.e., two weeks. Some of the computational results are given in Table 6.1.

Table 6.1

j	H_j	h_j
1	0.11155425	0.68739955
3	0.15287432	0.22667077
5	0.16239231	0.19637038
7	0.16937272	0.20129841
9	0.17605928	0.20900219
15	0.19524775	0.23190164
20	0.20998105	0.24952382
30	0.23514173	0.27965990
40	0.25421765	0.30254371
50	0.26781779	0.31887773
60	0.27709391	0.33002771
80	0.28720989	0.34219576
100	0.29137086	0.34720341
150	0.29384328	0.35017964
200	0.29408185	0.35046686
250	0.29410468	0.35049434
300	0.29410686	0.35049697
350	0.29410707	0.35049722
400	0.29410709	0.35049724

From the computed data it appears that H_j and h_j in the limit as $j \to \infty$ approach to about 0.29410709 and 0.35049724, respectively. In fact, these limiting values are the constants H_0 and h_0 so that the functions $H(t) = H_0$ and $h(t) = h_0$ are the solutions (critical points, see next chapter) of the differential system (6.1). From this table it is also clear that the affected human population increases monotonically, whereas the affected mosquito population initially decreases, in fact reaches a minimum, and then increases monotonically.

Example 6.2 On October 18, 1968 Robert Beamon (born 1946), during Summer Olympic Games in Mexico City set a world record for the long jump with a first jump of 8.90 m (29 ft. $2\frac{1}{2}$ in.). When his teammate and coach Ralph Boston (born 1939) told him that he had broken the world record by nearly 2 ft., his legs gave way and an astonished and overwhelmed Beamon suffered a brief cataplexy attack brought on by the emotional shock, and collapsed to his knees, his body unable to support itself, placing his hands over his face. Many commentators referred the length of the jump to the thinness of the air at the Mexico City. In an article, Brearley [2] disclaimed these critics. He proposed the model

$$M\frac{du}{dt} = -\rho k u (u^2 + v^2)^{1/2}$$
$$M\frac{dv}{dt} = -Mg - \rho k v (u^2 + v^2)^{1/2}, \tag{6.4}$$

where $M = 80$ (kg) is the mass of the long jumper, g is the acceleration due to gravity ($g = 9.8 \text{ m/s}^2$), at time t after takeoff u and v represent the velocity components parallel to the x and y axes, $k = 0.182 \text{ m}^2$, $\rho = 0.984 \text{ kg/m}^3$, and the initial conditions are

$$u(0) = 9.45 \text{ m/s}, \quad v(0) = 4.15 \text{ m/s}. \tag{6.5}$$

For the initial value problem (6.4), (6.5), we use Runge–Kutta method with the time step-size $h = 0.04$ on the time interval [0, 1], which corresponds to 1 s, and obtain the following values for the horizontal u and vertical v velocity components (see Table 6.2).

If we assume the motion through the air is described by a parabola that is symmetric about its highest point, then when the vertical velocity v reaches zero ($t \approx 0.4215$ s for a time step-size of 5×10^{-6}), the jumper is at the highest point, therefore the time to fly will be double of the time of zero vertical velocity. The jump distance is given by the numerical integral of the horizontal velocity within the time of fly which in this case is 7.9 m.

If we repeat the calculations for the normal air density $k = 0.163 \text{ m}^2$ and $\rho = 1.225 \text{ kg/m}^3$, the highest point occurs earlier, $t \approx 0.4213$ s which results in a jump distance of 7.88 m which is 0.02 m (0.78 in.) difference from the jump in a thinner air.

Table 6.2

j	u	v
1	9.4500	4.1500
3	9.4328	3.3592
5	9.4161	2.5699
7	9.3999	1.7821
9	9.3839	0.9958
11	9.3681	0.2108
13	9.3524	−0.5729
15	9.3367	−1.3553
17	9.3208	−2.1363
19	9.3047	−2.9160
21	9.2883	−3.6941
23	9.2714	−4.4707
25	9.2541	−5.2456

Example 6.3 Continuing with Example 5.12, again we let $S(t)$ denote the percent of the population susceptible to a disease, $I(t)$ the percent of the population infected with the disease, and $R(t)$ the percent of the population unable to contract the disease. For example, $R(t)$ could represent the percent of persons who have had a particular disease, recovered, and have subsequently become immune to the disease. Now following Hethcote [3], we assume that

1. Susceptible and infected individuals die at a rate proportional to the number of susceptible and infected individuals with proportionality constant μ called the *daily death removal rate*; the number $1/\mu$ is the *average lifetime* or *life expectancy*.
2. The constant λ represents the *daily contact rate*. On average, an infected person will spread the disease to λ people per day.
3. Individuals recover from the disease at a rate proportional to the number infected with the disease with proportionality constant γ. The constant γ is called the *daily recovery removal rate; the average period of infectivity* is $1/\gamma$.
4. The *contact number* $\sigma = \lambda/(\gamma + \mu)$ represents the average number of contacts an infected person has with both susceptible and infected persons.

If a person becomes susceptible to a disease after recovering from it (like gonorrhea, meningitis, and streptococcal sore throat), then the percent of persons susceptible to becoming infected with the disease, $S(t)$, and the percent of people in the population infected with the disease, $I(t)$, can be modeled by the system

$$\begin{aligned} S' &= -\lambda SI + \mu - \mu S + \gamma I \\ I' &= \lambda SI - \gamma I - \mu I \end{aligned} \tag{6.6}$$

$$S(0) = S_0, \quad I(0) = I_0, \quad S(t) + I(t) = 1. \tag{6.7}$$

This model is called an *SIS* (susceptible-infected-susceptible) model since once an individual has recovered from the disease, the individual again becomes susceptible to the disease.

Since $S(t) = 1 - I(t)$, we can write $I'(t) = \lambda I S - \gamma I + \mu I$ as

$$I'(t) = \lambda I (1 - I) - \gamma I - \mu I$$

and hence we need to solve the initial value problem

$$I' = [\lambda - (\gamma + \mu)]I - \lambda I^2, \quad I(0) = I_0, \tag{6.8}$$

whose solution can be written as

$$I(t) = \begin{cases} \dfrac{e^{(\lambda+\mu)(\sigma-1)t}}{\sigma[e^{(\lambda+\mu)(\sigma-1)t} - 1]/(\sigma - 1) + 1/I_0} & \text{if } \sigma \neq 1 \\[2ex] \dfrac{1}{\lambda t + 1/I_0} & \text{if } \sigma = 1. \end{cases} \tag{6.9}$$

From (6.9) it follows that

$$\lim_{t \to \infty} I(t) = \begin{cases} (\sigma - 1)/\sigma, & \sigma > 1 \\ I_0, & \sigma = 1 \\ 0, & \sigma < 1. \end{cases} \tag{6.10}$$

We use Runge–Kutta method to find the numerical solution of the initial value problem (6.8) with $I_0 = 1/2$ and (i) $\lambda = 3.6, \gamma = 2, \mu = 1$, (ii) $\lambda = 3.6, \gamma = 2, \mu = 1.6$, and (iii) $\lambda = 3.6, \gamma = 2, \mu = 2$, on the interval [0, 20] with the step-size $h = 0.2$, and in Table 6.3 compare the results with (6.9) and (6.10). We also graph various solutions in Fig. 6.1.

Table 6.3

t	$\mu = 1$		$\mu = 1.6$		$\mu = 2$	
	I (numeric)	I (exact)	I (numeric)	I (exact)	I (numeric)	I (exact)
0	0.5000	0.5000	0.5000	0.5000	0.5000	0.5000
0.2	0.2085	0.2266	0.1087	0.1078	0.0646	0.0398
0.4	0.1774	0.1862	0.0610	0.0604	0.0220	0.0103
0.6	0.1698	0.1739	0.0424	0.0420	0.0089	0.0031
0.8	0.1676	0.1695	0.0325	0.0322	0.0038	0.0010
1.0	0.1669	0.1678	0.0263	0.0261	0.0017	0.0003
1.2	0.1667	0.1671	0.0221	0.0219	0.0008	0.0001
1.4	0.1667	0.1668	0.0191	0.0189	0.0003	0.0000
1.6	0.1667	0.1667	0.0168	0.0166	0.0002	0.0000
1.8	0.1667	0.1677	0.0150	0.0148	0.0000	0.0000
2.0	0.1667	0.1667	0.0135	0.0000	0.0000	0.0000

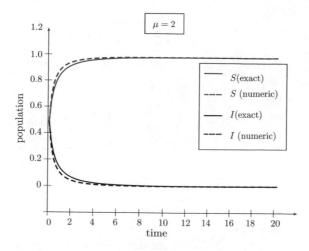

Fig. 6.1

The incidence of some diseases, like measles, rubella, and gonorrhea, oscillate seasonally. To model these diseases, we replace the constant contact rate, λ, by a periodic function $\lambda(t) = 5 - 2\sin(6t)$. Again, we use Runge–Kutta method to find the numerical solution of the initial value problem (6.8) with $I_0 = 1/2$ and (i) $\gamma = 1, \mu = 4$, and (ii) $\gamma = 1, \mu = 2$, on the interval $[0, 20]$ with the step-size $h = 0.2$, and in Table 6.4 present the numerical results. We also graph solutions in Fig. 6.2.

Table 6.4

t	$I\,(\mu = 4)$	$I\,(\mu = 2)$
0	0.5000	0.5000
0.2	0.1030	0.4697
0.4	0.0503	0.4476
0.6	0.0290	0.4084
0.8	0.0186	0.3656
1.0	0.0135	0.3336
1.2	0.0113	0.3211
1.4	0.0106	0.3306
1.6	0.0110	0.3591
1.8	0.0118	0.3989
2.0	0.0123	0.4383

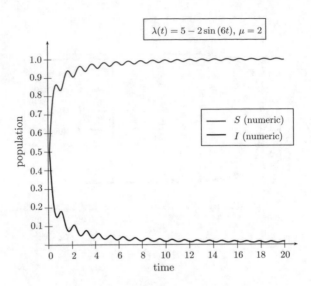

Fig. 6.2

Example 6.4 Following May [6] consider two prey populations H_1 and H_2 and two predator populations P_1 and P_2. If a_1, a_2, b_1, b_2, α_{ij}, and β_{ij} are positive parameters representing the interactions, then

$$
\begin{aligned}
H_1' &= H_1[a_1 - \alpha_{11}P_1 - \alpha_{12}P_2] \\
H_2' &= H_2[a_2 - \alpha_{21}P_1 - \alpha_{22}P_2] \\
P_1' &= P_1[-b_1 + \beta_{11}H_1 + \beta_{12}H_2] \\
P_2' &= P_2[-b_2 + \beta_{21}H_1 + \beta_{22}H_2].
\end{aligned}
\tag{6.11}
$$

One set of realistic vales are $a_1 = a_2 = 3$, $\alpha_{11} = \alpha_{22} = 2$, $\alpha_{12} = \alpha_{21} = 1$, $b_1 = 40$, $b_2 = 20$, $\beta_{11} = 3$, and $\beta_{12} = \beta_{21} = \beta_{22} = 1$, and $H_1(0) = 0.5$, $H_2(0) = 1$, $P_1(0) = P_2(0) = 1$. We use Runge–Kutta method with the time step-size $h = 0.01$ on the time interval $[0, 3]$ to find the numerical solution of the system (6.11). The numerical values are presented in Table 6.5.

Table 6.5

j	H_1	H_2	P_1	P_2
1	0.5000	1.0000	1.0000	1.0000
26	0.9508	1.8506	0.0001	0.0111
51	2.0114	3.9125	0.0000	0.0002
76	4.2581	8.2825	0.0000	0.0000
101	9.0144	17.5340	0.0000	0.0000
126	19.0828	37.1165	0.0000	0.0012
151	5.6893	1.5718	0.0264	22.3330
176	3.0207	0.2099	0.0000	0.3355
201	6.2661	0.4267	0.0000	0.0074
226	13.2575	0.9023	0.0000	0.0006
251	28.0627	1.9097	0.0000	0.0008
276	11.2360	1.7373	36.8127	0.1475
301	1.5385	0.9216	0.0075	0.0019

Based on the numerical results, we observe cyclic behavior with a time cycle ≈ 1.3 (see Fig. 6.3) for prey and predator populations with increasing number of prey species when both predator populations are small ($j \leq 126$) and rapid decrease in pray populations when at least one predator population exceeds both pray populations ($j = 151, 276$).

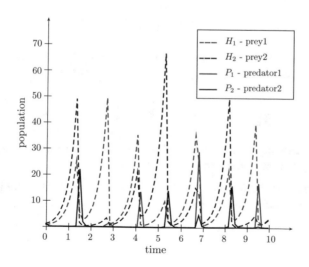

Fig. 6.3

Problems

6.1 The following initial value problem arises in hydraulics studies

$$y' = 2 + y^{1.5}, \quad y(0) = 0.7.$$

Use Runge–Kutta method to approximate the solution on $[0, 1.5]$ with the step-sizes $h = 0.1, \; 0.05$ and 0.01.

6.2 Consider the modified logistic population growth equation

$$P' = P(a - bP) + ce^{-kP}; \tag{6.12}$$

here, c and k are positive constants. In (6.12) the additional term ce^{-kP} represents the immigration. Clearly, the immigration is less when the population is large than when it is small. This decrease may be caused, for example, by the imposition of quotas, or by overcrowding of the region and a resulting deterioration of the favorable conditions that had attracted immigrants. Use Runge–Kutta method to find the solution of the initial value problem (6.12), $P(0) = P_0$ over the interval $[0, 10]$ when $a = b = k = 1$, $c = P_0 = 0.1$ and the step-size $h = 0.2$. Compare your result with the exact solution of (6.12), $P(0) = P_0$ with $a = b = 1$, $c = 0$, $P_0 = 0.1$.

6.3 If the magnetic characteristic of a coil wound on an iron core is of cubic form, then a sudden application of a periodic voltage $e = e_0 \sin \omega t$ across the coil gives rise to the initial value problem

$$e = iR + \frac{d\psi}{dt}, \quad \psi(0) = 0 \tag{6.13}$$

where $i = a\psi + b\psi^3$ and ψ denotes the magnetic flux. In (6.13) let $y = aR\psi/e_0$, $x = aRt$ and assume that $be_0^2 = 2a^3R^2$, $2aR = \omega$ to obtain

$$\frac{dy}{dx} = -y - 2y^3 + \sin(2x), \quad y(0) = 0. \tag{6.14}$$

Use Runge–Kutta method to find the solution of the initial value problem (6.14) on the interval $[-1, 1]$ with the step-size $h = 0.1$.

6.4 In nonlinear oscillation theory in the consideration of limit cycles (see [11]) the following differential equation arises

$$\frac{dv}{d\xi} = \alpha^2 \frac{(v - v^3/3) - \xi}{v}, \tag{6.15}$$

where v and ξ are space variables and α is a parameter. Use Runge–Kutta method to find the solution of (6.15) with $\alpha = 0.5$, $v(0) = 1$ on the interval $[0, 1]$ with the step-size $h = 0.1$.

6.5 In the study of the nonisothermal flow of a Newtonian fluid between parallel plates, the equation

$$\frac{d^2y}{dx^2} + x^2e^y = 0, \quad x > 0$$

was encountered. Using some clever substitutions this equation can be transformed into the first-order equation

$$\frac{dv}{du} = \frac{1}{2}\left[u(u+2)v^3 + (2u+5)v^2\right]. \tag{6.16}$$

Use Runge–Kutta method to find the solution $v(u)$ of (6.16) satisfying $v(2) = -4$ over the interval [2, 3] with the step-size $h = 0.01$.

6.6 Considering the nonlinear dissipative effects in the neck of a Helmholtz cavity resonator, an approximate differential equation for free oscillations of such a resonator takes the form (see [8])

$$M\frac{d^2x}{dt^2} + R\frac{dx}{dt}\left[1 + \beta\left(\frac{dx}{dt}\right)^2\right] + Kx = 0. \tag{6.17}$$

(1) Show that Eq. (6.17) can be written as

$$\frac{dv}{dx} = -\left[\left(\frac{K}{R}\right)x + v(1 + \beta v^2)\right]\bigg/\left(\frac{M}{R}\right)v, \tag{6.18}$$

where $v = dx/dt$.
(2) Use Runge–Kutta method to find the numerical solution of (6.18) when $K/R = 100$, $M/R = 9$, $\beta = 2 \times 10^{-5}$, and $v(0) = 700\,\text{cm/s}$ on [0, 2] with the step-size $h = 0.1$.

6.7 Use Runge–Kutta method to find the numerical solution of the simple pendulum Eq. (3.54) when $g = 32$, $L = 2$ and the initial conditions are $\theta(0) = 0.25$ (radian), $\theta'(0) = 0$ on [0, 2] with the step-size $h = 0.1$. Compare this numerical solution with the exact solution of the approximate Eq. (3.56).

6.8 In the design of a sewage treatment plant, the following initial value problem occurs (see [7]):

$$77.7\,H'' + 19.42\,H'^2 + H = 60, \quad H(0) = H'(0) = 0, \tag{6.19}$$

where $H(t)$ is the level of the fluid in an ejection chamber and t is the time in seconds. Use Runge–Kutta method to find the solution of (6.19) over the interval [0, 5] with the step-sizes $h = 0.5$ and 0.1.

6.9 For Van der Pol's equation (4.73) with $\mu = 1$ and initial conditions $y(0) = 0$, $y'(0) = 1$, use Runge–Kutta method to approximate the solution on [0, 5] with the step-sizes $h = 0.1$ and 0.05. Also compare this numerical solution with the series solution given in Problem 4.14.

6.10 Consider a weight attached in the middle of two springs and disturbed slightly from its equilibrium position (see Fig. 6.4). The small oscillations of the mass are governed by the following initial value problem:

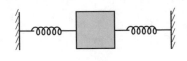

Fig. 6.4

$$y'' = -y^3, \quad y(0) = 0.2, \quad y'(0) = 0 \qquad (6.20)$$

It is known that the solution of (6.20) is a periodic oscillation of period approximately 40. Use Runge–Kutta method to approximate the solution of (6.20) on $[0, 80]$ with the step-size $h = 0.1$. Does the computed solution show periodic oscillation?

6.11 In a study of the variation of flux F in a large transformer the following non-linear differential equation occurs

$$\frac{d^2 F}{dt^2} + \alpha F + \beta F^3 = \frac{\omega}{N} E,$$

where the design parameters $\alpha = 74$, $\beta = 0.22$, the number of turns in the primary winding $N = 450$ and $E = 150 \sin \omega t$, $\omega = 120\pi$. Use Runge–Kutta method to find the numerical solution of this equation with the initial conditions $F(0) = 0.5$, $F'(0) = 0.1$ on $[0, 3]$ and the step-sizes $h = 0.1$ and 0.05.

6.12 When a charged surface is in contact with a liquid medium, it attracts ions of the opposite charge and repels ions of the same charge. In the study of such a situation in one dimension, the following Poisson–Boltzmann (Siméon Denis Poisson 1781–1840) equation arises

$$\frac{d^2 y}{dt^2} = P \sinh(Qy), \qquad (6.21)$$

where P and Q are positive constants (see [5]). Use Runge–Kutta method to approximate the solution of (6.21) with $P = Q = 1$ and initial conditions $y(0) = 1$, $y'(0) = 0$ on $[0, 2]$ with the step-sizes $h = 0.1$ and 0.05.

6.13 Enzymes have the ability to catalyze numerous chemical reactions in living organisms. One such model for the action of enzymes is

$$\frac{d^2 y}{dt^2} + \frac{ay}{b + y} = 0, \qquad (6.22)$$

where y is the concentration of the substance acted upon by the enzyme, and a, b are positive constants (see [12]). Use Runge–Kutta method to approximate the solution of (6.22) with $a = 1$, $b = 8$ and initial conditions $y(0) = 1$, $y'(0) = 0$ on $[0, 1]$ with the step-sizes $h = 0.1$ and 0.05.

6.14 Use Runge–Kutta method to find the numerical solution of the Emden equation (4.76) with the initial conditions $y(0.1) = 1$, $y'(0.1) = 0$ for $n = 2, 3, 4, 5$ on $[0.1, 10]$ and the step-size $h = 0.1$. Compare the numerical solution for $n = 5$ with the exact solution given in Problem 4.17.

6.15 In constructing the dikes to hold the North Sea water and reclaim the lowlands, the Dutch created Earth dams of the type shown in Fig. 6.5. The water from the low side is continually pumped back up to the high side by windmills. The sizing of the Earth dam involves solving the seepage problem to minimize the passage of water. In a particular application, it leads to solving the following initial value problem

$$y'' + 2f(x)y + y^{n/3} = 0, \quad y(0) = 0, \quad y'(0) = 0.75. \tag{6.23}$$

Use Runge–Kutta method to find the numerical solution of (6.23) with $f(x) = 1.5 \sin(2x)$, $n = 4$ and $f(x) = 4.6 \sin(3x)$, $n = 5$ on $[0, 2]$ with the step-size $h = 0.1$.

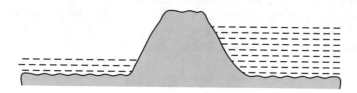

Fig. 6.5

6.16 When a fluid of low viscosity flows past a solid surface, the transition from the velocity of the surface to that of the stream is accomplished in a narrow layer near the surface (boundary-layer). In the year 1904 Ludwig Prandtl (1875–1953) used approximation techniques to develop boundary-layer theory. However, in 1908, Paul Richard Heinrich Blasius (1883–1970) formulated a problem to steady two-dimensional motion along a flat plate placed edgeways to the stream. When the origin is at the forward edge, the general equations reduce to the following third-order initial value problem

$$y''' = -yy'', \quad y(0) = y'(0) = 0, \quad y''(0) = 1.$$

Use Runge–Kutta method to approximate the solution on $[0, 2]$ with the step-sizes $h = 0.1$ and 0.05.

6.17 Let x be the segment of the total population N committing criminal acts and y as the segment engaged in crime prevention and law enforcement, then $(N - x - y)$ is the remaining portion, which may also have an influence. Let p be the influence by this largest group on the rate of change of x, and q be the influence on the rate of change of y. These changes may be produced by varying the degree of enforcement support, altering the parole policy, and enacting or rescinding legislation. Thus, the contemporary social problem of crime control can be investigated by the following system of nonlinear differential equations:

$$\frac{dx}{dt} = p + \alpha x - \beta xy$$

$$\frac{dy}{dt} = q - \gamma y + \delta xy.$$

In this system $\alpha > 0$ and $\gamma > 0$ imply that in the absence of a control force, the number of violators would increase, whereas, if there were no crime, prevention and enforcement personnel would decay in number. The interpretation of the signs of the product terms are the incarceration of violators and intrinsic growth of an organization. Use Runge–Kutta method to find the numerical solution of this system with $p = 1.6$, $\alpha = 0.1$, $\beta = 0.08$, $q = 0.7$, $\gamma = 0.4$, $\delta = 0.05$ with the initial conditions $x(0) = 40$, $y(0) = 6$ on the interval $[0, 4]$ with the step-size $h = 0.1$.

6.18 Consider an isolated intertidal marine community where two species live in the same environment and compete for the same food resources, but neither is a predator nor a prey for the other. Let $x(t)$ and $y(t)$ represent the number of each population present at time t. For this situation, Volterra proposed the following model:

$$\frac{dx}{dt} = [\alpha - \beta(x + y)]x$$

$$\frac{dy}{dt} = [\gamma - \delta(x + y)]y,$$

where the constants α, β, γ, δ are positive. The values of these parameters indicate the strength of the species, their eating habits, appetites, birth and death rates. Use Runge–Kutta method to find the numerical solution of this system with the initial conditions $x(0) = 200$, $y(0) = 100$ on the interval $[0, 4]$ with the step-size $h = 0.01$ for the following values of the constants, also in each case interpret your result.

(1) The food supply is limited compared to the species' appetites, e.g., $\alpha = 2$, $\beta = 0.2$, $\gamma = 1$, $\delta = 0.1$.

(2) Both species are tough enough to survive on what is available, e.g., $\alpha = 4$, $\beta = 0.1$, $\gamma = 2$, $\delta = 0.05$.

(3) $\alpha = 12$, $\beta = 0.1$, $\gamma = 2$, $\delta = 0.05$.

(4) $\alpha = 12$, $\beta = 0.1$, $\gamma = 2$, $\delta = 0.01$.

6.19 To accommodate more realistic situations Lotka–Volterra system (5.73) has been modified in several different ways. For example, in nature the responses and interactions between predators and prey rarely happen immediately; they usually involve time delays, which tend to produce destabilization. To involve this, it has been suggested that K should be replaced by $r\left[1 - (y/C)\right]$, where r is the proportionality constant and C is the carrying capacity set by environmental resources; and the terms Pxy and Lxy must be, respectively, changed to $Px(1 - e^{-cy})$ and $kx\left(1 - e^{-cy}\right)$, where k and c are constants. The system (5.73) then becomes

$$\frac{dx}{dt} = x\left[-M + P\left(1 - e^{-cy}\right)\right]$$
$$\frac{dy}{dt} = ry\left(1 - \frac{y}{C}\right) - kx\left(1 - e^{-cy}\right). \tag{6.24}$$

Use Runge–Kutta method to find the numerical solution of the system (6.24) with $M = r = k = 1$, $P = 1.5$, $c = 10^{-3}$, $C = 4000$, and the initial conditions $x(0) = 1$, $y(0) = 1.8$ on the interval $[0, 75]$ with the step-sizes $h = 1$ and 0.5.

6.20 In a study of the population cycles in rodents the following nonlinear system occurred:

$$\frac{dx}{dt} = x[a_1(t) - (b_1 - c_1)y - c_1(x + y)]$$
$$\frac{dy}{dt} = y[-a_2(t) + b_2 x], \tag{6.25}$$

where x represents the density (in animals per acre) of emigrants, y represents the density of tolerants, and t is the time in months. Clearly, these equations represent a possible interaction between two populations. The experimental values of the parameters are $b_1 = 1.75 \times 10^{-3}$, $b_2 = 1.5 \times 10^{-3}$, $c_1 = 1.0 \times 10^{-3}$, and initially $x(0) = y(0) = 0.5$. Now assuming that the net reproduction rate is minimal in late winter and maximal in late summer, it is appropriate to assume that $a_1(t) = 1 + 0.35 \sin(\pi t/6)$ and $a_2(t) = 0.14 - 0.075 \sin(\pi t/6)$, where $t = 0$ represents mid-June in the northern hemisphere. Use Runge–Kutta method to find the numerical solution of the system (6.25) on the interval $[0, 60]$ with the step-size $h = 1$.

6.21 Consider a first-order irreversible chemical reaction under nonisothermal conditions in a continuously stirred tank reactor. The control of this reactor is achieved by manipulating the flow of cooling fluid through a cooling coil inserted in the reactor. The dynamic-mass and heat-balance equations are

$$x' = -(1 - e^{\beta})x - 0.5(e^{\beta} - 1)$$
$$y' = e^{\beta}x - 8.9y^2 - 4.225 + 0.5(e^{\beta} - 1)$$
$$y(0) = -0.1111889, \quad z(0) = 0.0323358$$

where $\beta = 25y/(y + 2)$. This represents a self-sustained oscillation of an energy-mass system. It leads to a closed elliptic curve (*limit cycle*) in the xy-plane, i.e., if we start with the given initial conditions, the values of x and y will change, but eventually return to the same initial point. Use Runge–Kutta method to find the numerical solution of this system with the step-size $h = 0.001$ until one cycle is achieved.

6.22 The heart is an elastic muscle that contracts and relaxes in a regular rhythm regulated by the pacemaker. When the chemical control present in the tissue of the pacemaker reaches a specific value, called the threshold, an electrochemical wave is triggered, thus causing the heart to contract. As the chemical control decreases, the heart muscle relaxes. Based on some simplified assumptions (see [4]) it has been shown that the cycle of the heartbeat is governed by the following nonlinear system of differential equations:

$$\epsilon \frac{dx}{dt} = -(x^3 - Tx + b)$$
$$\frac{db}{dt} = (x - b_0),$$
$$\text{(6.26)}$$

where $x(t)$ represents the change in muscle fiber length and depends on time t, $b(t)$ represents a chemical control variable that governs the electrochemical wave, $T = 3x_0^2 - a > 0$ is the tension in the system, x_0 is the initial muscle fiber length, $b(0) = b_0 = 2x_0^3 - ax_0 < 0$, and $\epsilon > 0$ is a small positive constant. The constant a is chosen so that T and $b(0)$ are in the correct range. The resulting solution curves oscillate about $x = b = 0$ such that the minimum x-value corresponds to diastolic relaxation and while b is negative, x increases to a maximum representing systolic contraction.

Use Runge–Kutta method to find the numerical solution of the system (6.26) with $a = x_0 = 0.45$ and $\epsilon = 0.025$ on the interval $[0, 5]$ with the step-size $h = 0.1$. Find the approximate diastolic and systolic equilibrium points. Next, compute the solution when $\epsilon = 0.0125$. Do you find any changes at these equilibrium points?

6.23 The movement of substances through the gastral tract is usually influenced by the tension in the circumferential muscles. If the canal is a surface of revolution defined by $r = r(z)$ and hydrostatic forces are taken into account, then the equilibrium equations are

$$\frac{dr}{dz} = \tan(\phi)$$
$$T = p\cos(\phi)\left[r\sec^2(\phi) + \frac{r^2}{2}\frac{d}{dz}\tan(\phi)\right],$$

which on using the substitution $x = pz/T$, $y = pr/T$ leads to the system

$$\frac{d\phi}{dx} = \frac{2}{y}\left(\frac{\cos(\phi)}{y} - 1\right)$$

$$\frac{dy}{dx} = \tan(\phi).$$

Use Runge–Kutta method to find the numerical solution of this system with the initial conditions $y(0) = 2$, $\phi(0) = 0$ on $[0, 2]$ and the step-sizes $h = 0.1$ and 0.05.

6.24 Under certain assumptions, the *FitzHugh–Nagumo equation*, which arises in the study of the impulses in a nerve fiber, can be written as the system of differential equations

$$\frac{dV}{d\xi} = W$$

$$\frac{dW}{d\xi} = F(V) + R - uW \qquad\qquad (6.27)$$

$$\frac{dR}{d\xi} = \frac{\epsilon}{u}(bR - V - a),$$

where $F(V) = (1/3)V^3 - V$ (see [9, 10]). Use Runge–Kutta method to find the numerical solution of the system (6.27) with $\epsilon = 0.08$, $a = 0.7$, $b = 0$, $u = 1$ and the initial conditions $V(0) = 1$, $W(0) = 0$, $R(0) = 1$ on the interval $[0, 10]$ with the step-sizes $h = 1$ and 0.5.

6.25 Consider the transient heat flow in an electron tube filled with an inert gas (see Fig. 6.6). Assume that before the current is applied the whole system is at temperature t_1. At $t = 0$ the filament temperature is suddenly raised to $t_2 > t_1$ by the electric current. Then, heat is converted to the surrounding gas and radiated to the tube walls. The wall receives heat by convection from the gas and by radiation from the filament. Finally, the wall transfers heat by convection to the surrounding atmosphere, which is at temperature t_1. Let $t_2 = 2t_1$, the heat capacities C_1 and C_2 of the gas and wall be related by $C_2 = 2C_1$, and let the radiation coefficients be known. Under these assumptions the propagation problem in dimensionless variables is governed by the following system:

$$\begin{aligned}
y' &= -2y + z + 2 \\
z' &= 0.5y - z - 0.1z^4 + 2.1 \\
y(0) &= z(0) = 1.
\end{aligned}$$

Use Runge–Kutta method to find the numerical solution of this system on $[0, 2]$ with the step-sizes $h = 0.1$ and 0.05.

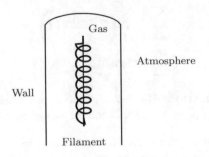

Fig. 6.6

6.26 Under various assumptions, the following system of nonlinear differential equations:

$$\frac{dX}{dt} = rX - \gamma XN - \beta XY$$
$$\frac{dI}{dt} = \beta XY - (\sigma + b + \gamma N)I$$
$$\frac{dY}{dt} = \sigma I - (\alpha + \beta + \gamma N)Y \qquad (6.28)$$
$$\frac{dN}{dt} = aX - (b + \gamma N)N - \alpha Y,$$

has been used to model a fox population in which rabies (a fatal viral disease which infects all warm-blooded animals and has killed humans and animals for hundreds of years) is present (see [1]). Here, $X(t)$ represents the population of foxes susceptible to rabies at time t; $I(t)$ the population that has contracted the rabies virus but is not yet ill; $Y(t)$ the population that has developed rabies; $N(t)$ the total population of the foxes; a represents the average per capita birth rate of foxes; $1/b$ denotes fox life expectancy (without resource limitations) which is typically in the range of 1.5–2.7 years; $r = a - b$ represents the intrinsic per capita population growth rate; $K = r/\gamma$ represents the fox carrying capacity of the defined area, which is typically in the range of 0.1–4 foxes per km^2; $1/\sigma$ represents the average latent period, i.e., the average time (in years) that a fox can carry the rabies virus but not actually be ill with rabies, typically $1/\sigma$ is between 28 and 30 days; α represents the death rate of foxes with rabies, $1/\alpha$ is the life expectancy (in years) of a fox with rabies and is typically between 3 and 10 days; and β represents a transmission coefficient, typically, $1/\beta$ is between 4 and 6 days. Use Runge–Kutta method to find the numerical solution of the system (6.28) with $a = 1$, $b = 0.5$, $r = 0.5$, $K = 2$, $\sigma = 12.1667$, $\alpha = 73$, $\beta = 80$ and the initial conditions $X(0) = 0.93$, $I(0) = 0.035$, $Y(0) = 0.035$, $N(0) = 1.0$ on the interval $[0, 40]$ with the step-sizes $h = 1$ and 0.5.

6.27 The motion of a moon moving in a planar orbit about a planet is governed by the equations

$$x'' = -G\frac{Mx}{r^3}, \quad y'' = -G\frac{My}{r^3},$$

where $r = (x^2 + y^2)^{1/2}$, G is the gravitational constant, and M is the mass of the moon. Assume that $GM = 1$. Use Runge–Kutta method to find numerical solution of this system on the interval $[0, 40]$ with the step-size $h = 0.1$
(1) with $x(0) = 1$, $x'(0) = 0$, $y(0) = 0$, $y'(0) = 1$, and observe that the motion is a circular orbit of radius $a = 1$ and period 2π (see Remark 3.1).
(2) with $x(0) = 1$, $x'(0) = 0$, $y(0) = 0$, $y'(0) = \sqrt{6}/2$, and observe that the motion is an elliptical orbit with major semiaxis $a = 2$ and period $4\pi\sqrt{2}$.
(3) with $x(0) = 1$, $x'(0) = 1$, $y(0) = 0$, $y'(0) = 1$, and investigate the orbit.

6.28 Russian scientists Nikolay Mitrofanovich Krylov (1879–1955) and Nikolay Nikolayevich Bogolyubov (1909–1992) used the method of variation of parameters to find a *first approximation* of the differential equation

$$y'' + \omega^2 y + \mu f(y, y') = 0$$

in the form

$$y(x) = \alpha(x) \sin(\omega x + \phi(x)),$$

where

$$\frac{d\alpha}{dx} = -\frac{\mu}{2\pi\omega} \int_0^{2\pi} f(a\sin(\theta), \omega a \cos(\theta)) \cos(\theta) d\theta,$$

$$\frac{d\phi}{dx} = \frac{\mu}{2\pi\omega a} \int_0^{2\pi} f(\alpha \sin(\theta), \omega a \cos(\theta)) \sin(\theta) d\theta.$$

(1) For the Duffing equation (4.75) with $\delta = 0$, $\alpha = \omega^2$, $\beta = \mu$, $\gamma = 0$, i.e.,

$$y'' + \omega^2 y + \mu y^3 = 0, \tag{6.29}$$

show that the Krylov–Bogolyubov first approximation is

$$y(x) = a_0 \sin\left[\left(\frac{3\mu a_0^2}{8\omega} + \omega\right)x + \phi_0\right], \tag{6.30}$$

where a_0 and ϕ_0 are arbitrary constants. Use Runge–Kutta method to find numerical solution of (6.29) with $\omega = \mu = 1$ on the interval $[0, 20]$ with the step-size $h = 0.1$ and initial conditions $y(0) = 1$, $y'(0) = 0$ and compare this numerical solution with the approximate solution (6.30).

(2) For the Van der Pol equation (4.73) show that the Krylov–Bogolyubov first approximation is

$$y(x) = \frac{a_0 e^{\mu x/2}}{\sqrt{1 + a_0^2(e^{\mu x} - 1)/4}} \sin(x + \phi_0), \tag{6.31}$$

where a_0 and ϕ_0 are arbitrary constants. If $a_0 = 2$, the first approximation (6.31) reduces to the *periodic* function $y(x) = 2\sin(x + \phi_0)$, and if $0 < a_0 < 2$ or $a_0 > 2$, then note that

$$\lim_{x \to \infty} \frac{a_0 e^{\mu x/2}}{\sqrt{1 + a_0^2(e^{\mu x} - 1)/4}} = 2.$$

Thus for any $a_0 > 0$ except $a_0 = 2$ the nonperiodic oscillation given by (6.31) tends to the periodic oscillation as $x \to \infty$. Use Runge–Kutta method to find numerical solution of (4.73) with $\mu = 1$ on the interval $[0, 20]$ with the step-size $h = 0.1$ and initial conditions $y(0) = 1$, $y'(0) = 0$ and compare this numerical solution with the approximate solution (6.31).

References

1. R.M. Anderson, H.C. Jackson, R.N. May, A.M. Smith, Nature **289**, 765–771 (1981)
2. M.N. Brearley, Math. Mag. **45**, 241–246 (1972)
3. H.W. Hethcote, in *Applied Mathematical Ecology*, ed. by S.A. Levin, T.G. Hallam, L.J. Gross (Springer, New York, 1989), pp. 119–143
4. D.S. Jones, M.J. Plank, B.D. Sleeman, *Differential Equations and Mathematical Biology*. Mathematical Biology and Medicine Series (CRC Press, Boca Raton, 2009)
5. M. Jones, *Biological Interfaces* (Elsevier, Amsterdam, 1975)
6. N.M. Krylov, N.N. Bogoliubov, *Introduction to Non-Linear Mechanics* (Princeton University Press, Princeton, 1950)
7. R.M. May, Math. Biosci. **12**, 59–79 (1971)
8. W. Milne, *Numerical Solution of Differential Equations* (Dover Publications Inc., New York, 1970)
9. P.M. Morse, K.U. Ingard, *Theoretical Acoustics* (McGraw-Hill, New York, 1968)
10. J.D. Murray, *Mathematical Biology* (Springer, New York, 1990)
11. A.C. Scott, Rev. Mod. Phys. **47**, 487–533 (1975)
12. J.J. Stoker, *Nonlinear Vibrations in Mechanical and Electrical Systems* (Interscience, New York, 1950)

Chapter 7
Stability Theory

The main aim of our previous chapters was to find analytic or numeric solutions of differential equations. However, most of the systems which appear in real-world applications cannot be solved analytically and their numerical solutions often fail to provide exactly the required qualitative information. Fortunately, for many such systems certain properties of the solutions can be established directly. In this chapter, mainly we shall study stability of the solutions by linearizing the nonlinear differential systems [3]. For this, in what follows we shall assume that for the initial value problem (5.29) there always exists a solution $u(t) = u(t, t_0, u^0)$ on the interval $[t_0, \infty)$. We begin with the following definitions.

A solution $u(t, t_0, u^0)$ of the initial value problem (5.29) existing on the interval $[t_0, \infty)$ is said to be *stable* if small changes in u^0 bring only small changes in the solutions of (5.29) for all $t \geq t_0$. Otherwise, we say that the solution $u(t, t_0, u^0)$ is *unstable*. Precisely, we have

Definition 7.1 A solution $u(t) = u(t, t_0, u^0)$ of the initial value problem (5.29) is said to be *stable*, if for each $\epsilon > 0$ there is a $\delta = \delta(\epsilon, t_0) > 0$ such that $\|\Delta u^0\| < \delta$ implies that $\|u(t, t_0, u^0 + \Delta u^0) - u(t, t_0, u^0)\| < \epsilon$.

Definition 7.2 A solution $u(t) = u(t, t_0, u^0)$ of the initial value problem (5.29) is said to be *unstable* if it is not stable.

Definition 7.3 A solution $u(t) = u(t, t_0, u^0)$ of the initial value problem (5.29) is said to be *asymptotically stable* if it is stable and there exists a $\delta_0 > 0$ such that $\|\Delta u^0\| < \delta_0$ implies that

$$\|u(t, t_0, u^0 + \Delta u^0) - u(t, t_0, u^0)\| \to 0 \quad \text{as} \quad t \to \infty.$$

Theorem 7.1 *All solutions of the differential system (5.3) are stable if and only if they are bounded.*

© Springer Nature Switzerland AG 2019
R. P. Agarwal et al., *500 Examples and Problems of Applied Differential Equations*, Problem Books in Mathematics,
https://doi.org/10.1007/978-3-030-26384-3_7

Corollary 7.1 *If the real parts of the multiple eigenvalues of the matrix A are negative, and the real parts of the simple eigenvalues of the matrix A are nonpositive, then all solutions of the differential system (5.12) are stable.*

Corollary 7.2 *If the real parts of the eigenvalues of the matrix A are negative, then all solutions of the differential system (5.12) are asymptotically stable.*

Theorem 7.2 (Adolf Hurwitz's (1859–1919) Theorem) *A necessary and sufficient condition for the negativity of the real parts of all zeros of the polynomial*

$$x^n + a_1 x^{n-1} + \cdots + a_{n-1} x + a_n$$

with real coefficients is the positivity of all the leading principal minors of the $n \times n$ Hurwitz matrix

$$\begin{pmatrix} a_1 & 1 & 0 & 0 & 0 & \cdots & 0 \\ a_3 & a_2 & a_1 & 1 & 0 & \cdots & 0 \\ a_5 & a_4 & a_3 & a_2 & a_1 & \cdots & 0 \\ & \cdots & & & & \\ 0 & 0 & 0 & 0 & 0 & \cdots & a_n \end{pmatrix},$$

i.e.,

$$a_1 > 0, \quad \begin{vmatrix} a_1 & 1 \\ a_3 & a_2 \end{vmatrix} > 0, \quad \begin{vmatrix} a_1 & 1 & 0 \\ a_3 & a_2 & a_1 \\ a_5 & a_4 & a_3 \end{vmatrix} > 0, \cdots .$$

If the function $\mathbf{g}$ is independent of time, then the system $\mathbf{u}' = \mathbf{g}(t, \mathbf{u})$ reduces to an *autonomous* system

$$\mathbf{u}' = \mathbf{g}(\mathbf{u}). \tag{7.1}$$

Solutions of $\mathbf{g}(\mathbf{u}) = \mathbf{0}$ are called *points of equilibrium, fixed points, stationary points, rest points, singular points,* or *critical points* of (7.1). Clearly, critical points are the constant solutions of the system (7.1), i.e., if u starts at a critical point, it remains at this point. The set of all critical points is called the *critical point set.*

Theorem 7.3 *If $u(t)$ is a solution of the differential system (7.1) in the interval (α, β), then for any constant c the function $v(t) = u(t + c)$ is also a solution of (7.1) in the interval $(\alpha - c, \beta - c)$.*

Obviously, the above property does not hold for nonautonomous differential systems, e.g., a solution of $u_1' = u_1, u_2' = t u_1$ is $u_1(t) = e^t, u_2(t) = t e^t - e^t$, and $u_2'(t + c) = (t + c) e^{t+c} \neq t u_1(t + c)$ unless $c = 0$.

If the matrix $\mathbf{A}$ is nonsingular, then for the system (5.12) origin is the only critical point. In general, system (7.1) can have a finite or infinite number of fixed points. For example, the simple pendulum equation (3.54), $\theta'' + \omega^2 \sin\theta = 0$, $\omega^2 = g/L$ has an infinite number of critical points at $\theta = n\pi$, $n = 0, \pm 1, \pm 2, \cdots$.

Consider the first-order autonomous differential equation

$$y' = f(y), \tag{7.2}$$

where $f(y)$ is continuously differentiable function of y. This condition ensures that for any given t_0 and y_0 the initial value problem (7.2), $y(t_0) = y_0$ has a unique solution. From (7.2), it is clear that if $f(a) > 0$, then starting at point $y = a$, y will increase. On the other hand, if $f(a) < 0$, then starting with $y = a$, y will decrease. This increase or decrease will continue until a critical point is reached. An easy way to determine the stability of a critical point, say, c is to check the sign of $f(y)$ for $y < c$ and $y > c$. If $y_1 < c$ and $f(y_1) > 0$, and if $y_2 > c$ and $f(y_2) < 0$, then the solution is always moving toward c. In this case we conclude that c is a stable critical point. However, if $f(y_1) < 0$ or $f(y_2) > 0$, the critical point is unstable.

Now, we shall consider two-dimensional autonomous systems of the form

$$\begin{aligned} u_1' &= g_1(u_1, u_2) \\ u_2' &= g_2(u_1, u_2). \end{aligned} \tag{7.3}$$

Throughout, we shall assume that the functions g_1 and g_2 together with their first partial derivatives are continuous in some domain D of the $u_1 u_2$-plane. Thus, for all $(u_1^0, u_2^0) \in D$ the differential system (7.3) together with $u_1(t_0) = u_1^0, u_2(t_0) = u_2^0$ has a unique solution in some interval J containing t_0. The main interest in studying (7.3) is twofold:

1. A large number of dynamic processes in applied sciences are governed by such systems.
2. The qualitative behavior of its solutions can be illustrated through the geometry in the $u_1 u_2$-plane.

In the domain D of the $u_1 u_2$-plane, any solution of the differential system (7.3) may be regarded as a parametric curve given by $(u_1(t), u_2(t))$ with t as the parameter. This curve $(u_1(t), u_2(t))$ is called a *trajectory* or an *orbit* or a *path* of (7.3), and the $u_1 u_2$-plane is called the *phase plane*. Thus, from Theorem 7.3, for any constant c both $(u_1(t), u_2(t))$, $t \in (\alpha, \beta)$ and $(u_1(t + c), u_2(t + c))$, $t \in (\alpha - c, \beta - c)$ represent the same trajectory. For the trajectories of the differential system (7.3), the following property is very important.

Theorem 7.4 *Through each point* $(u_1^0, u_2^0) \in D$, *there passes one and only one trajectory of the differential system* (7.3).

For the differential system $u_1' = u_2, u_2' = -u_1$, there are an infinite number of solutions $u_1(t) = \sin(t + c), u_2(t) = \cos(t + c)$, $0 \le c < 2\pi$, $-\infty < t < \infty$. However, they represent the same trajectory, i.e., the circle $u_1^2 + u_2^2 = 1$. Thus, it is important to note that a trajectory is a curve in D that is represented parametrically by more than one solution. Hence, in conclusion, $\mathbf{u}(t) = (u_1(t), u_2(t))$ and $\mathbf{v}(t) = (u_1(t+c), u_2(t+c))$, $c \neq 0$ represent distinct solutions of (7.3), but they represent the same curve parametrically. If (c_1, c_2) is a critical point of (7.3), then obviously $u_1(t) = c_1$, $u_2(t) = c_2$ is a solution of (7.3), and from Theorem 7.4 no trajectory can pass through the point (c_1, c_2). A critical point (c_1, c_2) is said to be *isolated* if there exists no other critical point in some neighborhood of (c_1, c_2). By a critical point, we shall hereafter mean an isolated critical point. If (c_1, c_2) is a critical point of (7.3), then the substitution $v_1 = u_1 - c_1$, $v_2 = u_2 - c_2$ transforms (7.3) into an equivalent system with $(0, 0)$ as a critical point, thus without loss of generality we can assume that $(0, 0)$ is a critical point of the system (7.3).

An effective technique in studying the differential system (7.3) near the critical point $(0, 0)$ is to approximate it by a linear system of the form

$$\mathbf{u}' = \begin{pmatrix} u_1' \\ u_2' \end{pmatrix} = \begin{pmatrix} a_{11}u_1 + a_{12}u_2 \\ a_{21}u_1 + a_{22}u_2 \end{pmatrix} = \begin{pmatrix} a_{11} & a_{12} \\ a_{12} & a_{22} \end{pmatrix} \begin{pmatrix} u_1 \\ u_2 \end{pmatrix} = \mathbf{Au}, \qquad (7.4)$$

where $a_{11}a_{22} - a_{21}a_{12} \neq 0$, so that $(0, 0)$ is the only critical point of (7.4). It is expected that a "good" approximation (7.4) will provide solutions which themselves are "good" approximations to the solutions of the system (7.3). For example, if the system (7.3) can be written as

$$\begin{aligned} u_1' &= a_{11}u_1 + a_{12}u_2 + h_1(u_1, u_2) \\ u_2' &= a_{21}u_1 + a_{22}u_2 + h_2(u_1, u_2), \end{aligned} \qquad (7.5)$$

where $h_1(0, 0) = h_2(0, 0) = 0$ and

$$\lim_{u_1, u_2 \to 0} \frac{h_1(u_1, u_2)}{\sqrt{u_1^2 + u_2^2}} = \lim_{u_1, u_2 \to 0} \frac{h_2(u_1, u_2)}{\sqrt{u_1^2 + u_2^2}} = 0, \qquad (7.6)$$

then the following result holds.

Theorem 7.5 *(i) If the zero solution of the system (7.4) is asymptotically stable, then the zero solution of the system (7.5) is asymptotically stable.*
(ii) If the zero solution of the system (7.4) is unstable, then the zero solution of the system (7.5) is unstable.
(iii) If the zero solution of the system (7.4) is stable, then the zero solution of the system (7.5) may be asymptotically stable, stable, or unstable.

Of course, if the functions $g_1(u_1, u_2)$ and $g_2(u_1, u_2)$ possess continuous second-order partial derivatives in the neighborhood of the critical point $(0, 0)$, then by Taylor's formula differential system (7.3) can always be written in the form (7.5) with

$$a_{11} = \frac{\partial g_1}{\partial u_1}(0,0), \quad a_{12} = \frac{\partial g_1}{\partial u_2}(0,0), \quad a_{21} = \frac{\partial g_2}{\partial u_1}(0,0), \quad a_{22} = \frac{\partial g_2}{\partial u_2}(0,0).$$

The picture of all trajectories of a system is called the *phase portrait* of the system. Since the solutions of (7.4) can be determined explicitly, a complete description of its phase portrait can be given. However, the nature of the solutions of (7.4) depends on the eigenvalues of the matrix A, i.e., the roots of the equation

$$\lambda^2 - (a_{11} + a_{22})\lambda + a_{11}a_{22} - a_{21}a_{12} = 0. \tag{7.7}$$

The phase portrait of (7.4) depends almost entirely on the roots λ_1 and λ_2 of (7.7). For this, there are several different cases which must be studied separately.

Case 1. λ_1 **and** λ_2 **are real, distinct, and of the same sign.** If v^1, v^2 are the corresponding eigenvectors of A, then from (5.20) the general solution of (7.4) can be written as

$$\begin{pmatrix} u_1(t) \\ u_2(t) \end{pmatrix} = c_1 \begin{pmatrix} v_1^1 \\ v_2^1 \end{pmatrix} e^{\lambda_1 t} + c_2 \begin{pmatrix} v_1^2 \\ v_2^2 \end{pmatrix} e^{\lambda_2 t}, \tag{7.8}$$

where c_1 and c_2 are arbitrary constants.

For simplicity, we can always assume that $\lambda_1 > \lambda_2$. Thus, if $\lambda_2 < \lambda_1 < 0$ then all solutions of (7.4) tend to $(0,0)$ as $t \to \infty$. Therefore, the critical point $(0,0)$ of (7.4) is asymptotically stable. In case $c_1 = 0$ and $c_2 \neq 0$, we have $u_2 = (v_2^2/v_1^2)u_1$, i.e., the trajectory is a straight line with slope v_2^2/v_1^2. Similarly, if $c_1 \neq 0$ and $c_2 = 0$ we obtain the straight line $u_2 = (v_2^1/v_1^1)u_1$. To obtain other trajectories, we assume that c_1 and c_2 both are different from zero. Then, since

$$\frac{u_2(t)}{u_1(t)} = \frac{c_1 v_2^1 e^{\lambda_1 t} + c_2 v_2^2 e^{\lambda_2 t}}{c_1 v_1^1 e^{\lambda_1 t} + c_2 v_1^2 e^{\lambda_2 t}} = \frac{c_1 v_2^1 + c_2 v_2^2 e^{(\lambda_2 - \lambda_1)t}}{c_1 v_1^1 + c_2 v_1^2 e^{(\lambda_2 - \lambda_1)t}}, \tag{7.9}$$

which tends to v_2^1/v_1^1 as $t \to \infty$, all trajectories tend to $(0,0)$ with the slope v_2^1/v_1^1. Similarly, as $t \to -\infty$ all trajectories become asymptotic to the line with the slope v_2^2/v_1^2. This situation is illustrated in Fig. 7.1 for two different values of the slope v_2^1/v_1^1 and v_2^2/v_1^2. Here the critical point $(0,0)$ is called a *stable node*.

If $\lambda_1 > \lambda_2 > 0$, then all nontrivial solutions tend to infinity as t tends to ∞. Therefore, the critical point $(0,0)$ is unstable. The trajectories are same as for $\lambda_2 < \lambda_1 < 0$ except that the direction of the motion is reversed as depicted in Fig. 7.2. As $t \to -\infty$, the trajectories tend to $(0,0)$ with the slope v_2^2/v_1^2, and as $t \to \infty$ trajectories become asymptotic to the line with the slope v_2^1/v_1^1. Here the critical point $(0,0)$ is called an *unstable node*.

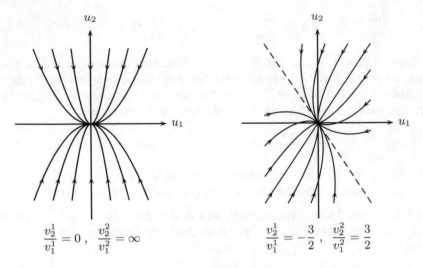

$$\frac{v_2^1}{v_1^1} = 0 , \quad \frac{v_2^2}{v_1^2} = \infty \qquad\qquad \frac{v_2^1}{v_1^1} = -\frac{3}{2} , \quad \frac{v_2^2}{v_1^2} = \frac{3}{2}$$

Fig. 7.1

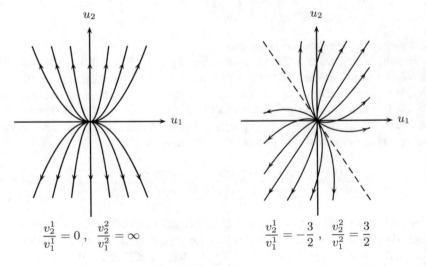

$$\frac{v_2^1}{v_1^1} = 0 , \quad \frac{v_2^2}{v_2^2} = \infty \qquad\qquad \frac{v_2^1}{v_1^1} = -\frac{3}{2} , \quad \frac{v_2^2}{v_1^2} = \frac{3}{2}$$

Fig. 7.2

Case 2. λ_1 **and** λ_2 **are real with opposite signs**. Of course, the general solution of the differential system (7.4) remains the same (7.8). Let $\lambda_1 > 0 > \lambda_2$. If $c_1 = 0$ and $c_2 \neq 0$, then as in Case 1 we have $u_2 = (v_2^2/v_1^2)u_1$, and as $t \to \infty$ both $u_1(t)$ and $u_2(t)$ tend to zero. If $c_1 \neq 0$ and $c_2 = 0$, then $u_2 = (v_2^1/v_1^1)u_1$ and both $u_1(t)$ and $u_2(t)$ tend to infinity as $t \to \infty$, and approach zero as $t \to -\infty$. If c_1 and c_2 both are different from zero, then from (7.9) it follows that u_2/u_1 tends to v_2^1/v_1^1 as $t \to \infty$. Hence, all trajectories are asymptotic to the line with slope v_2^1/v_1^1 as $t \to \infty$.

Similarly, as $t \to -\infty$ all trajectories are asymptotic to the line with slope v_2^2/v_1^2. Also, it is obvious that both $u_1(t)$ and $u_2(t)$ tend to infinity as $t \to \pm\infty$. This type of critical point is called a *saddle point*. Obviously, saddle point displayed in Fig 7.3 is an unstable critical point of the system.

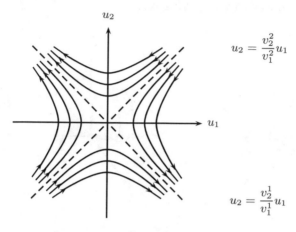

$$u_2 = \frac{v_2^2}{v_1^2} u_1$$

$$u_2 = \frac{v_2^1}{v_1^1} u_1$$

Fig. 7.3

Case 3. λ_1 and λ_2 are equal, i.e., $\lambda_1 = \lambda_2 = \lambda$. In this case from (5.22), the general solution of the differential system (7.4) can be written as

$$\begin{pmatrix} u_1(t) \\ u_2(t) \end{pmatrix} = c_1 \begin{pmatrix} 1 + (a_{11} - \lambda)t \\ a_{21}t \end{pmatrix} e^{\lambda t} + c_2 \begin{pmatrix} a_{12}t \\ 1 + (a_{22} - \lambda)t \end{pmatrix} e^{\lambda t}, \qquad (7.10)$$

where c_1 and c_2 are arbitrary constants.

If $\lambda < 0$, both $u_1(t)$ and $u_2(t)$ tend to 0 as $t \to \infty$ and hence the critical point $(0, 0)$ of (7.4) is asymptotically stable. Further, from (7.10), it follows that

$$\frac{u_2}{u_1} = \frac{c_2 + [a_{21}c_1 + (a_{22} - \lambda)c_2]t}{c_1 + [a_{12}c_2 + (a_{11} - \lambda)c_1]t}. \qquad (7.11)$$

Thus, in particular, if $a_{12} = a_{21} = 0$, $a_{11} = a_{22} \neq 0$, then Eq. (7.7) gives $\lambda = a_{11} = a_{22}$, and (7.11) reduces to $u_2/u_1 = c_2/c_1$. Therefore, all trajectories are straight lines with slope c_2/c_1. The phase portrait in this case is illustrated in Fig. 7.4a. Here, the origin is called *stable proper (star-shaped) node*. In the general case as $t \to \pm\infty$, (7.11) tends to

$$\frac{a_{21}c_1 + (a_{22} - \lambda)c_2}{a_{12}c_2 + (a_{11} - \lambda)c_1}.$$

But, since $(a_{11} - \lambda)(a_{22} - \lambda) = a_{12}a_{21}$ this ratio is the same as $a_{21}/(a_{11} - \lambda)$. Thus, as $t \to \pm\infty$, all trajectories are asymptotic to the line $u_2 = (a_{21}/(a_{11} - \lambda))u_1$. The origin $(0, 0)$ here is called *stable improper node*, see Fig. 7.4b.

If $\lambda > 0$, all solutions tend to ∞ as $t \to \infty$ and hence the critical point $(0, 0)$ of (7.4) is unstable. The trajectories are the same as for $\lambda < 0$ except that the direction of the motion is reversed (see Fig. 7.5a, b).

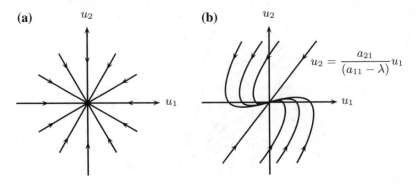

(a) u_2 **(b)** u_2

$$u_2 = \frac{a_{21}}{(a_{11} - \lambda)}u_1$$

Fig. 7.4

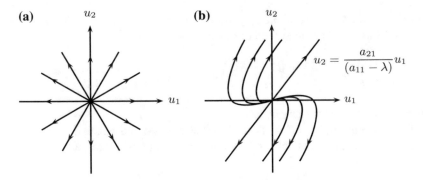

(a) u_2 **(b)** u_2

$$u_2 = \frac{a_{21}}{(a_{11} - \lambda)}u_1$$

Fig. 7.5

Case 4. λ_1 **and** λ_2 **are complex conjugates.** Let $\lambda_1 = \mu + i\nu$ and $\lambda_2 = \mu - i\nu$, where we can assume that $\nu > 0$. If $v = v^1 + iv^2$ is the eigenvector of A corresponding to the eigenvalue $\lambda_1 = \mu + i\nu$, then a solution of (7.4) can be written as

$$u(t) = e^{(\mu+i\nu)t}(v^1 + iv^2) = e^{\mu t}[\cos(\nu t) + i\sin(\nu t)](v^1 + iv^2)$$
$$= e^{\mu t}[v^1 \cos(\nu t) - v^2 \sin(\nu t)] + ie^{\mu t}[v^1 \sin(\nu t) + v^2 \cos(\nu t)].$$

Therefore, it follows that

$$u^1(t) = e^{\mu t}[v^1\cos(\nu t) - v^2\sin(\nu t)] \quad \text{and} \quad u^2(t) = e^{\mu t}[v^1\sin(\nu t) + v^2\cos(\nu t)]$$

are two real-valued linearly independent solutions of (7.4), and every solution $u(t)$ of (7.4) is of the form $u(t) = c_1 u^1(t) + c_2 u^2(t)$. This expression can easily be rewritten as

$$\begin{aligned}u_1(t) &= r_1 e^{\mu t}\cos(\nu t - \delta_1)\\ u_2(t) &= r_2 e^{\mu t}\cos(\nu t - \delta_2),\end{aligned} \tag{7.12}$$

where $r_1 \geq 0$, $r_2 \geq 0$, δ_1 and δ_2 are some constants.

If $\mu = 0$, then both $u_1(t) = r_1\cos(\nu t - \delta_1)$ and $u_2(t) = r_2\cos(\nu t - \delta_2)$ are periodic of period $2\pi/\nu$. The function $u_1(t)$ varies between $-r_1$ and r_1, while $u_2(t)$ varies between $-r_2$ and r_2. Thus, each trajectory beginning at the point (u_1^*, u_2^*) when $t = t^*$ will return to the same point when $t = t^* + 2\pi/\nu$. Thus, the trajectories are closed curves, and the phase portrait of (7.4) has the form described in Fig. 7.6a. In this case the critical point $(0, 0)$ is stable but not asymptotically stable, and is called a *center*.

If $\mu < 0$, then the effect of the factor $e^{\mu t}$ in (7.12) is to change the simple closed curves of Fig. 7.6a into the spirals of Fig. 7.6b. This is because the point

$$(u_1(2\pi/\nu), u_2(2\pi/\nu)) = \exp(2\pi\mu/\nu)(u_1(0), u_2(0))$$

is closer to the origin $(0, 0)$ than $(u_1(0), u_2(0))$. In this case the critical point $(0, 0)$ is asymptotically stable, and is called a *stable focus*.

If $\mu > 0$, then all trajectories of (7.4) spiral away from the origin $(0, 0)$ as $t \to \infty$ and are illustrated in Fig. 7.6c. In this case the critical point $(0, 0)$ is unstable, and is named an *unstable focus*.

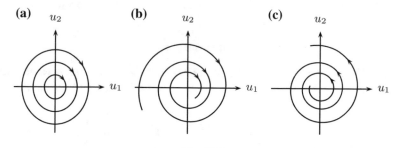

Fig. 7.6

We summarize the above analysis in the following theorem.

Theorem 7.6 *For the differential system (7.4), let λ_1 and λ_2 be the eigenvalues of the matrix A. Then, the behavior of its trajectories near the critical point $(0, 0)$ is as follows:*

(i)   stable node, if λ_1 and λ_2 are real, distinct, and negative,
(ii)   unstable node, if λ_1 and λ_2 are real, distinct, and positive,
(iii) saddle point (unstable), if λ_1 and λ_2 are real, distinct, and of opposite sign,
(iv)   stable node, if λ_1 and λ_2 are real, equal, and negative,
(v)   unstable node, if λ_1 and λ_2 are real, equal, and positive,
(vi)   stable center, if λ_1 and λ_2 are pure imaginary,
(vii) stable focus, if λ_1 and λ_2 are complex conjugates, with negative real part,
(viii) unstable focus, if λ_1 and λ_2 are complex conjugates with positive real part.

The behavior of the linear system (7.4) near the origin also determines the nature of the trajectories of the nonlinear system (7.5) near the critical point $(0, 0)$. For this, we state the following result.

Theorem 7.7 *For the differential system (7.4), let λ_1 and λ_2 be the eigenvalues of the matrix A. Then,*
(a) the nonlinear system (7.5) has the same type of critical point at the origin as the linear system (7.4) whenever:
(i) $\lambda_1 \neq \lambda_2$ and $(0, 0)$ is a node of the system (7.4),
(ii) $(0, 0)$ is a saddle point of the system (7.4),
(iii) $\lambda_1 = \lambda_2$ and $(0, 0)$ is not a star-shaped node of the system (7.4),
(iv) $(0, 0)$ is a focus of the system (7.4)
(b) the origin is not necessarily the same type of critical point for the two systems:
(i)   if $\lambda_1 = \lambda_2$ and $(0, 0)$ is a star-shaped node of the system (7.4), then $(0, 0)$ is either a node or a focus of the system (7.5),
(ii) if $(0, 0)$ is a center of the system (7.4), then $(0, 0)$ is either a center or a focus of the system (7.5).

Remark 7.1 If the general nonlinear system (7.3) does not contain linear terms, then an infinite number of critical points are possible. Further, the nature of these points depend on the nonlinearity in (7.3), and hence it is rather impossible to classify these critical points.

Now we state a general result for the system (7.1).

Theorem 7.8 *Let $0 = (0, 0, \cdots, 0)$ be a critical point of the autonomous system (7.1), and let $g(u)$ be twice continuously differentiable. Further, let the Jacobian matrix*

$$J(u) = \left(\frac{\partial g_i}{\partial u_j}\right) = \begin{pmatrix} \dfrac{\partial g_1}{\partial u_1} & \dfrac{\partial g_1}{\partial u_2} & \cdots & \dfrac{\partial g_1}{\partial u_n} \\ \vdots & \vdots & \cdots & \vdots \\ \dfrac{\partial g_n}{\partial u_1} & \dfrac{\partial g_n}{\partial u_2} & \cdots & \dfrac{\partial g_n}{\partial u_n} \end{pmatrix}$$

at 0 be nonsingular. Then, if all the eigenvalues of $J(0)$ have negative real parts, 0 is asymptotically stable. If any eigenvalue has positive real part, then 0 is unstable. Further, the local phase portrait of the linear and nonlinear system are the same except perhaps when 0 is a center.

Definition 7.4 A closed trajectory (closed periodic solution) of the differential system (7.3) which is approached spirally from either the inside or the outside by a nonclosed trajectory of (7.3) either as $t \to +\infty$ or as $t \to -\infty$ is called a *limit cycle* of (7.3) (see Fig 7.7).

The limit cycle solution is said to be a *self-excited oscillation* because even the slightest disturbance from the equilibrium point at the origin results in a motion that grows and inevitably approaches the periodic limit cycle motion as $t \to \infty$.

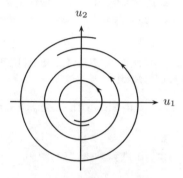

Fig. 7.7

The following result due to Poincaré and Ivar Otto Bendixson (1861–1935) provides sufficient conditions for the existence of limit cycles of the differential system (7.3).

Theorem 7.9 (Poincaré–Bendixson Theorem) *Suppose that a solution $u(t) = (u_1(t), u_2(t))$ of the differential system (7.3) remains in a bounded region of the u_1u_2-plane which contains no critical points of (7.3). Then, its trajectory must spiral into a simple closed curve, which itself is the trajectory of a periodic solution of (7.3).*

The Poincaré–Bendixson Theorem is false for systems of dimension 3 or more. The next result provides sufficient conditions for the nonexistence of closed trajectories, and hence, in particular, limit cycles of the differential system (7.3).

Theorem 7.10 (Bendixson's Theorem) *If*

$$\frac{\partial g_1(u_1, u_2)}{\partial u_1} + \frac{\partial g_2(u_1, u_2)}{\partial u_2}$$

has the same sign throughout the domain D, then the differential system (7.3) has no closed trajectory in D.

The following result provides sufficient conditions for the small periodic solutions of a special two-dimensional system.

Theorem 7.11 (Poincaré) *Suppose the functions* $g_i(u_1, u_2)$, $i = 1, 2$ *of the system*

$$u_1' = \beta u_2 + g_1(u_1, u_2)$$
$$u_2' = -\beta u_1 + g_2(u_1, u_2) \tag{7.13}$$

satisfy the following symmetry conditions $g_1(u_1, -u_2) = -g_1(u_1, u_2)$ *and* g_2 $(u_1, -u_2) = g_2(u_1, u_2)$, *then all solutions* $u(t) = (u_1(t), u_2(t))$ *of (7.13) with* $\beta \neq 0$ *and* $\|u(t)\|$ *sufficiently small are periodic. Furthermore, the orbits of these solutions encircle the origin.*

Now, we shall state a result of Alfred Marie Liénard (1869–1958), Norman Levinson (1912–1975), and Oliver K. Smith.

Theorem 7.12 (Liénard–Levinson–Smith Theorem) *Consider the Liénard differential equation*

$$y'' + f(y)y' + g(y) = 0, \tag{7.14}$$

where we assume that
(i) f *is even and continuous for all* y,
(ii) *there exists a number* $y_0 > 0$ *such that*

$$F(y) = \int_0^y f(t)dt < 0$$

for $0 < y < y_0$, *and* $F(y) > 0$ *and monotonically increasing for* $y > y_0$, *also* $F(y) \to \infty$ *as* $y \to \infty$,
(iii) g *is odd, has a continuous derivative for all* y, *and is such that* $g(y) > 0$ *for* $y > 0$,
(iv) $G(y) = \int_0^y g(t)dt \to \infty$ *as* $y \to \infty$.
Then, the differential equation (7.14) possesses an essentially unique nontrivial periodic solution.

By "essentially unique" in the above result we mean that if $y = y(t)$ is a nontrivial periodic solution of (7.14), then all other nontrivial periodic solutions of (7.14) are of the form $y = y(t - t_1)$ where t_1 is a real number. This, of course, implies that the equivalent system $u_1' = u_2$, $u_2' = -f(u_1)u_2 - g(u_1)$ has a unique closed trajectory in the u_1u_2-plane.

Example 7.1 Comparing Verhulst's model (2.4) with (7.2), we have $f(P) = P(a - bP)$. Thus, for (2.4), $P = 0$ and $P = a/b$ are the stationary points. Since for a small $\epsilon > 0$, $f(\epsilon) = \epsilon(a - b\epsilon) > 0$, the stationary point $P = 0$ is unstable. Further, since $f\left(\frac{a-\epsilon}{b}\right) = \left(\frac{a-\epsilon}{b}\right)\epsilon > 0$ and $f\left(\frac{a+\epsilon}{b}\right) = \left(\frac{a+\epsilon}{b}\right)(-\epsilon) < 0$, the stationary point $P = a/b$ is stable. To show the importance of stability, we note that the exact values of a and b in (2.4) are unlikely to be known. Thus, if we replace these with close approximate values, say, $\tilde{a}$ and $\tilde{b}$, then the approximate limiting value $\tilde{P} = \tilde{a}/\tilde{b}$ will be close to the actual limiting value $P = a/b$. In conclusion, the actual limiting population of (2.4) is stable with respect to small perturbations.

Example 7.2 Comparing Gompertz's model (2.29) with (7.2), we have $f(P) = P[a - b\ln(P)]$. Here, we assume that a and b are positive constants. Thus, for (2.29), $P = 0$ and $P = e^{a/b}$ are the stationary points. Since for a small $\epsilon > 0$, $f(\epsilon) = \epsilon[a - b\ln(\epsilon)] > 0$, the stationary point $P = 0$ is unstable. Further, since $f\left(e^{(1-\epsilon)a/b}\right) = e^{(1-\epsilon)a/b}(\epsilon a) > 0$ and $f\left(e^{(1+\epsilon)a/b}\right) = e^{(1+\epsilon)a/b}(-\epsilon a) < 0$, the stationary point $P = e^{a/b}$ is stable.

Example 7.3 In system form, Eq. (3.23) can be written as

$$u_1' = u_2$$
$$u_2' = -\frac{k}{M}u_1 - \frac{a}{M}u_2 \tag{7.15}$$

for which $(0, 0)$ is the only critical point. For (7.15), Eq. (7.7) takes the form

$$\lambda^2 + \frac{a}{M}\lambda + \frac{k}{M} = 0,$$

which has the solutions

$$\lambda = \frac{1}{2M}(-a \pm \sqrt{a^2 - 4Mk})$$

the nature of which depends on the discriminant $(a^2 - 4Mk)$. We have the following cases:

Case 1. $a = 0$, no damping. Since M and k are positive $\lambda = \pm i\sqrt{k/M}$, this yields Theorem 7.6(vi), i.e., stable center. Physically, this means the system (7.15) undergoes stable sinusoidal oscillation about its equilibrium point.

Case 2. $a^2 - 4Mk = 0$. In this case $\lambda = -a/2M$ and we have Theorem 7.6(iv), i.e., stable node. Physically, this means the system (7.15) is critically damped and the motion is not oscillatory, but returns to its equilibrium point.

Case 3. $a^2 - 4Mk < 0$, $a \neq 0$. In this case, the eigenvalues are complex conjugates with negative real parts, so we have Theorem 7.6(vii), stable focus. Physically, this means the motion of the system (7.15) is a damped oscillation which returns to the equilibrium position as $t \to \infty$.

Case 4. $a^2 - 4Mk > 0$. In this case the eigenvalues are real, unequal, and negative, so we have Theorem 7.6(i), stable node. Physically, this means the displacement and velocity of the system (7.15) go to zero without oscillation. This is the overdamped situation.

Example 7.4 Consider the equation (3.46) of the form

$$LI'' + RI' + \frac{1}{C}I + h(I, I') = 0, \tag{7.16}$$

where the nonlinear term h, in particular, represents the induced currents from other sources. In what follows, we assume that

$$h(0, 0) = \left.\frac{\partial h}{\partial u_1}\right|_{(0,0)} = \left.\frac{\partial h}{\partial u_2}\right|_{(0,0)} = 0.$$

In system form, Eq. (7.16) can be written as

$$\begin{aligned} u_1' &= u_2 \\ u_2' &= -\frac{1}{CL}u_1 - \frac{R}{L}u_2 - h(u_1, u_2) \end{aligned} \tag{7.17}$$

for which $(0, 0)$ is the only critical point. The associated linear system for (7.17) is

$$\begin{aligned} u_1' &= u_2 \\ u_2' &= -\frac{1}{CL}u_1 - \frac{R}{L}u_2. \end{aligned} \tag{7.18}$$

Clearly, (7.15) is the same as (7.18) with $M = L$, $k = 1/C$, and $a = R$. Thus, the cases 2–4 of Example 7.3 for the system (7.18) also hold (the Case 1 does not exist). Therefore, from Theorem 7.7, if $R^2 - 4L/C \geq 0$ the critical point $(0, 0)$ of the system (7.17) is a stable node, and if $R^2 - 4L/C < 0$ it is a stable focus. In all the cases, the critical point $(0, 0)$ of (7.17) is asymptotically stable, hence in analyzing the circuit Eq. (7.16), we can safely ignore the nonlinear terms for small values of I and I'.

Example 7.5 In system form, Van der Pol's equation (4.73) can be written as

$$\begin{aligned} u_1' &= u_2 = g_1(u_1, u_2) \\ u_2' &= \mu u_2 - u_1 - \mu u_1^2 u_2 = g_2(u_1, u_2), \end{aligned} \tag{7.19}$$

for which $(0, 0)$ is the only critical point. The nonlinear term in the second equation of (7.19) is $h_2(u_1, u_2) = -\mu u_1^2 u_2$, and since

$$\lim_{u_1, u_2 \to 0} \frac{-\mu u_1^2 u_2}{\sqrt{u_1^2 + u_2^2}} = 0,$$

i.e., condition (7.6) is satisfied, in view of Theorem 7.7 the system (7.19) can be linearized, i.e., we can consider the system

$$
\begin{aligned}
u_1' &= u_2 \\
u_2' &= \mu u_2 - u_1.
\end{aligned}
\tag{7.20}
$$

For (7.20), Eq. (7.7) takes the form

$$
\lambda^2 - \mu\lambda + 1 = 0,
$$

which has the roots

$$
\lambda = \frac{1}{2}\left(\mu \pm \sqrt{\mu^2 - 4}\right).
$$

Now from Theorems 7.6 and 7.7, it follows that the critical point $(0, 0)$ of the system (7.19) is unstable focus if $0 < \mu < 2$, unstable node if $\mu \geq 2$, stable focus if $-2 < \mu < 0$, and stable node if $\mu \leq -2$.

From the system (7.19), we have $\partial g_1/\partial u_1 + \partial g_2/\partial u_2 = \mu(1 - u_1^2)$. Thus, it follows from Theorem 7.10 that closed trajectories, if there are any, are not contained in the strip $|u_1| < 1$.

Example 7.6 In system form, simple pendulum equation (3.54) can be written as

$$
\begin{aligned}
u_1' &= u_2 \\
u_2' &= -\omega^2 \sin(u_1), \quad \omega^2 = g/L.
\end{aligned}
\tag{7.21}
$$

For this system, we have already mentioned that there are infinite number of critical points $(0, 0)$ and $(n\pi, 0)$, $n = \pm 1, \pm 2, \cdots$. To discuss the stability of the critical point $(0, 0)$, we expand $\sin(u_1)$ about $u_1 = 0$, i.e.,

$$
\sin(u_1) = u_1 - \frac{u_1^3}{3!} + \cdots ,
$$

from which and Theorem 7.7 it is clear that the system (7.21) can be lineaized to

$$
\begin{aligned}
u_1' &= u_2 \\
u_2' &= -\omega^2 u_1.
\end{aligned}
\tag{7.22}
$$

For this system, eigenvalues are $\lambda = \pm i\omega$. Thus, from Theorem 7.6(vi) the critical point $(0, 0)$ of the system (7.22) is a stable center. But, then Theorem 7.7 concludes that the critical point $(0, 0)$ of (7.21) is either a center or a focus. We recall that the critical point $(0, 0)$ corresponds physically to the pendulum hanging vertically downwards. A small displacement from this position leads to only small oscillations about it. Thus, we can conclude that $(0, 0)$ is a stable critical point of (7.21). In fact, near this critical point trajectories are simple closed curves. Hence, in conclusion $(0, 0)$ is a stable center of (7.21). We can also conclude the same result from the fact that the simple pendulum is a conservative system, and hence there is no gain or loss of energy. Now since at the critical points $(2n\pi, 0)$, $n = \pm 1, \pm 2, \cdots$ also the pendulum is hanging vertically downwards, all these points are stable center of (7.21). Mathematically, if we use the substitution $U_1 = u_1 \mp 2n\pi$ in (7.21) and use the fact that $\sin(U_1 \pm 2n\pi) = \sin(U_1)$, we find that at the critical points $(\pm 2n\pi, 0)$, $n = 1, 2, \cdots$, the system (7.21) remains the same.

Now we shall discuss the critical point $(\pi, 0)$. For this, we expand $\sin u_1$ about the point $u_1 = \pi$, which yields

$$\sin u_1 = (\pi - u_1) - \frac{1}{3!}(\pi - u_1)^3 + \cdots .$$

Thus, the corresponding linear system is

$$\begin{aligned} u_1' &= u_2 \\ u_2' &= -\omega^2(\pi - u_1) = \omega^2(u_1 - \pi). \end{aligned} \tag{7.23}$$

In (7.23), we make the substitution $U_1 = u_1 - \pi$, to get the system

$$\begin{aligned} U_1' &= u_2 \\ u_2' &= \omega^2 U_1. \end{aligned} \tag{7.24}$$

Notice that a direct substitution $U_1 = u_1 - \pi$ in (7.21) leads to the system

$$\begin{aligned} U_1' &= u_2 \\ u_2' &= \omega^2 \cos(U_1), \end{aligned} \tag{7.25}$$

whose linearization is the same as (7.24). For the system (7.24), the eigenvalues are $\lambda_1 = \omega$, $\lambda_2 = -\omega$, and hence from Theorem 7.6(iii), the critical point $(0, 0)$ of (7.24) is an unstable saddle point. Thus, from Theorem 7.7(ii), for the system (7.21) the critical point $(\pi, 0)$ is an unstable saddle point. It is clear that the critical points $((2n + 1)\pi, 0)$, $n = \pm 1, \pm 2, \cdots$ correspond to the pendulum pointing vertically upwards. Therefore, all these critical points are unstable saddle points. In fact, in (7.21) the substitution $U_1 = u_1 - (2n + 1)\pi$, $n = \pm 1, \pm 2, \cdots$ leads to the same system (7.25).

In conclusion, the critical points of (7.21) are alternately stable centers and unstable saddle points.

Now we consider the case when a simple pendulum is under the resistance of the angular velocity, i.e., we consider the general *damped* oscillations model

$$\theta'' + c\theta' + \omega^2 \sin(\theta) = 0, \quad c > 0. \tag{7.26}$$

In system form, (7.26) can be written as

$$u_1' = u_2$$
$$u_2' = -\omega^2 \sin(u_1) - cu_2 \qquad (7.27)$$

for which also $(n\pi, 0)$, $n = 0, \pm 1, \pm 2, \cdots$ are the only critical points. At the the critical point $(0, 0)$, the associated linear system is

$$u_1' = u_2$$
$$u_2' = -\omega^2 u_1 - cu_2 \qquad (7.28)$$

for which the eigenvalues are

$$\lambda = \frac{1}{2}\left(-c \pm \sqrt{c^2 - 4\omega^2}\right), \quad c > 0.$$

Thus, from Theorem 7.6(i) and (iv) the critical point of (7.28) is a stable node if $c \geq 2\omega$, and from Theorem 7.6(vii) stable focus if $c < 2\omega$. From Theorem 7.7 this result carries over for the system (7.28). Finally, we remark that the critical point $(\pi, 0)$ of (7.28) is still unstable saddle point.

Example 7.7 Consider the Lotka–Volterra system (5.73), where recall that $x(t)$, $y(t)$, respectively, represent the size of predator and prey at time t. Clearly, for this system $(0, 0)$ and $(K/L, M/P)$ are the only critical points. For the critical point $(0, 0)$ we compare (5.73) with (7.5), to obtain $a_{11} = -M, a_{12} = 0, h_1(x, y) = Pxy, a_{21} = 0, a_{22} = K, h_2(x, y) = -Lxy$, so that $a_{11}a_{22} - a_{21}a_{12} = -MK \neq 0$, and h_1 and h_2 satisfy the conditions (7.6). Thus, in view of Theorem 7.7, the system (5.73) can be linearized, i.e., we can consider the system

$$x' = -Mx$$
$$y' = Ky,$$

for which the eigenvalues are $\lambda_1 = -M < 0$ and $\lambda_2 = K > 0$. Therefore, from Theorems 7.6 ans 7.7, the critical point $(0, 0)$ of (5.73) is a saddle point. Clearly, the critical point $(0, 0)$ of the system (5.73) is not of physical interest.

For the critical point $(K/L, M/P)$, we make the change of variables $u_1 = x - K/L, u_2 = y - M/P$, to obtain the system

$$u_1' = \frac{PK}{L}u_2 + Pu_1u_2$$
$$u_2' = -\frac{LM}{P}u_1 - Lu_1u_2. \qquad (7.29)$$

Since for this system $(0, 0)$ is a critical point, we can follow the above arguments, and linearize (7.29) to the system

$$u'_1 = \frac{PK}{L} u_2$$

$$u'_2 = -\frac{LM}{P} u_1$$

(7.30)

for which the eigenvalues are $\lambda = \pm i \sqrt{KM}$. Thus, from Theorem 7.6(vi) the critical point $(0, 0)$ of the system (7.30) is a stable center. But, then Theorem 7.7 tells us that the critical point $(K/L, M/P)$ of (5.73) is either a center or a focus. However, fortunately for the system (5.73) in Example 5.10 we have seen that the trajectories are closed curves in the xy-plane. Thus, the critical point $(K/L, M/P)$ of (5.73) is a stable center. Therefore, each solution $(x(t), y(t))$ of (5.73) with the initial condition $(x(0), y(0))$ in the first quadrant is a periodic function of t with period, say, ω. Then, the average values of the populations $x(t)$ and $y(t)$ can be given by

$$\overline{x} = \frac{1}{\omega} \int_0^\omega x(t)dt, \quad \overline{y} = \frac{1}{\omega} \int_0^\omega y(t)dt.$$

Now integrating the first equation of (5.73) over the interval $[0, \omega]$, we find

$$0 = \ln x(\omega) - \ln x(0) = -M\omega + P \int_0^\omega y(t)dt,$$

which immediately gives $\overline{y} = M/P$, and similarly the second equation of (5.73) gives $\overline{x} = K/L$. This means that the average sizes of the populations $x(t)$ and $y(t)$ which interact according to the system (5.73) are the same as equilibrium values.

Now assume that the prey population $y(t)$ is harvested in a small amount, so that both prey and predators will decrease at the rates $\epsilon y(t)$ and $\epsilon x(t)$, where $\epsilon > 0$ is small. In this case, the system (5.73) takes the form

$$x' = -(M + \epsilon)x + Pxy$$
$$y' = (K - \epsilon)y - Lxy.$$

(7.31)

For this system, it is easy to find that $\overline{x} = (K - \epsilon)/L$ and $\overline{y} = (M + \epsilon)/P$, i.e, the average size of the prey is higher than before harvesting, whereas the average size of the predator is smaller than before harvesting.

Example 7.8 The egg of a certain parasite deposited on a host when hatched kills the host. We shall study the variation of host and parasite populations with time. For this, let $h(t)$ and $p(t)$ denote the number of hosts and parasites at any time t. The number of eggs deposited per unit time depends on the probability of hosts and parasites coming together and is proportional to the population product, i.e., Khp. If b is the birth rate of the hosts and d is their death rate when no parasites are present, and if d_1 is the death rate of parasites, then the changes in the host-population and parasite-population are described by the differential equations

$$\frac{dh}{dt} = bh(t) - dh(t) - Kh(t)p(t)$$

and

$$\frac{dp}{dt} = Kh(t)p(t) - d_1 p(t),$$

, respectively. Thus, we have the system

$$h' = \alpha h - Khp$$
$$p' = Khp - d_1 p, \tag{7.32}$$

where $\alpha = b - d$. Since h and p are nonzero, an equilibrium point for the system (7.32) is

$$h = \frac{d_1}{K}, \quad p = \frac{\alpha}{K} = \frac{b-d}{K}.$$

Thus, the equilibrium host-population size depends on the death rate of parasites, whereas the equilibrium parasite-population depends on birth and death rates of the host.

Example 7.9 Following [5], also see [1, 27], we consider the continuous-flow stirred tank reactor (see Fig. 7.8). In this reaction, a stream of chemical C flows into the tank of volume V, and produces as well as residue flow out of the tank at a constant rate q. Because the reaction is continuous, the composition and temperature of the contents of the tank are constant and are the same as the composition and temperature of the stream flowing out of the tank. Suppose that a concentration c_{in}, of the chemical C flows into the tank while a concentration c flows out of the tank. In addition, suppose that the chemical C changes into products at a rate proportional to the concentration c of C, where the constant of proportionality $k(T)$ depends on the temperature T, $k(T) = Ae^{-B/T}$. Because

$$\begin{array}{cccc} \text{Rate of change} & = \text{Rate in} & - \text{ Rate out} & - \text{ Rate that } C\text{disappears} \\ \text{of amount of} C & \text{of } C & \text{of } C & \text{by the reaction} \end{array}$$

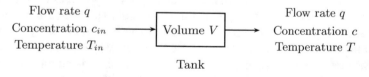

Flow rate q Flow rate q

Concentration c_{in} Volume V Concentration c

Temperature T_{in} Temperature T

Tank

Fig. 7.8

We have the differential equation

$$\frac{d}{dt}(Vc) = qc_{in} - qc - Vk(T)c. \tag{7.33}$$

Similarly, we balance the heat of the reaction with

Rate of change = Rate in − Rate out − Heat removed + Heat produced
of heat content   of heat   of heat   by cooling   by reaction.

Let C_p be the specific heat so that the heat content per unit volume of the reaction mixture at temperature T is $C_p T$. If H is the rate at which heat is generated by the reaction and $VS(T)$ is the rate at which heat is removed from the system by a cooling system, then we have the differential equation

$$VC_p \frac{dT}{dt} = qC_p T_{in} - qC_p T - VS(T) + HVk(T)c, \qquad (7.34)$$

where T_{in} is the temperature at which the chemical flows into the tank.

Assuming that the volume V is constant, and there is no cooling system, i.e., $VS(T) = 0$, then the equilibrium points of the nonlinear system (7.33), (7.34) satisfy the equations

$$c_{in} - c = \frac{V}{q} ck(T), \qquad (7.35)$$

$$T - T_{in} = \frac{HVc_{in}}{qC_p + VC_p k(T)} k(T). \qquad (7.36)$$

Solutions of (7.36) are called *steady-state temperatures*.

Notice that $y = T - T_{in}$ describes heat removal while $y = \frac{HVc_{in}}{qC_p + VC_p k(T)} \times k(T)$ represents heat production. Thus, if the slope of the heat production curve is greater than that of the heat removal curve, the steady state is unstable. On the other hand, if the slope of the heat production curve is less than or equal to that of the heat removal curve, then the steady state is stable.

Example 7.10 A study of McLaren and Peterson [22] in 1994 involves balsam fir tree, moose, and wolf. The fir trees are eaten by the moose, moose are eaten by wolves, and the change in the wolf population affects the trees. In what follows at time t, we denote by $x(t), y(t), z(t)$, respectively, the populations of trees, moose, and wolves. We assume that these populations in isolation can be modeled with a logic growth and the effect of intersection between species is proportional to the product of the populations. Further, for simplicity, we assume that all the parameters, namely, the growth rate, the carrying capacity, and the the effect of iteration, on which the behavior of solutions depend are 1. This leads to the following system of equations:

$$\begin{aligned} x' &= x(1 - x) - xy \\ y' &= y(1 - y) + xy - yz \\ z' &= z(1 - z) + yz. \end{aligned} \qquad (7.37)$$

For this system, the critical points are $(0, 0, 0), (1, 0, 0), (0, 1, 0), (0, 0, 1), (1, 0, 1)$, and $(2/3, 1/3, 4/3)$. Clearly, among all these critical points the only point at which all the three species coexist is $(2/3, 1/3, 4/3)$. To study stability of this critical point, in (7.37) we make the substitution $x = X + 2/3$, $y = Y + 1/3$, $z = Z + 4/3$ and obtain the Jacobian matrix at the point $(0, 0, 0)$, which appears as

$$\begin{pmatrix} -2/3 & -2/3 & 0 \\ 1/3 & -1/3 & -1/3 \\ 0 & 4/3 & -4/3 \end{pmatrix}. \tag{7.38}$$

For this matrix, the characteristic polynomial equation is

$$\lambda^3 + \frac{7}{3}\lambda^2 + \frac{20}{9}\lambda + \frac{8}{9} = 0.$$

Now from Theorem 7.2, it immediately follows that all eigenvalues of (7.38) have negative real parts, in fact, the eigenvalues are $\lambda_1 = -1$ and $\lambda_{2,3} = 2(-1 \pm i)/3$. Thus, from Theorem 7.8, it follows that the critical point $(2/3, 1/3, 4/3)$ of (7.37) is asymptotically stable, i.e., all solutions of (7.37) with initial conditions sufficiently close to this critical point will tend toward this point.

Example 7.11 Let E and s be positive constants and let $f(x)$ be a continuous odd function that approaches a finite limit as $x \to \infty$, is increasing, and is concave down for $x > 0$. In 1949, Androkov and Chaiken showed that the voltages over the deflection plate in a sweeping circuit for an oscilloscope are determined by the system

$$\frac{dV_1}{dt} = -sV_1 + f(E - V_2)$$

$$\frac{dV_2}{dt} = -sV_2 + f(E - V_1).$$

Clearly, from Theorem 7.10 this system has no limit cycles.

Example 7.12 In Van der Pol's equation (4.73), $f(y) = \mu(y^2 - 1)$ and $g(y) = y$. Thus, it is easy to check the following:
(i) $f(-y) = \mu(y^2 - 1) = f(y)$, the function f is even and continuous for all y,
(ii) $F(y) = \mu(y^3/3 - y) < 0$ for $0 < y < \sqrt{3}$, $F(y) > 0$ and monotonically increasing for $y > \sqrt{3}$, also $F(y) \to \infty$ as $y \to \infty$,
(iii) $g(-y) = -y = -g(y)$, the function g is odd; $dg/dy = 1$, and the derivative of g is continuous for all y; $g(y) > 0$ for $y > 0$,
(iv) $G(y) = y^2/2 \to \infty$ as $y \to \infty$.
Hence, all the conditions of Theorem 7.12 for the differential equation (4.73) are satisfied. In conclusion, we find that the d.e. (4.73) possesses an essentially unique nontrivial periodic solution. In other words, the equivalent system (7.19) has a unique closed trajectory in the $u_1 u_2$-plane. For $\mu = 0.1$, 1 and 10, these trajectories are illustrated in Fig. 7.9.

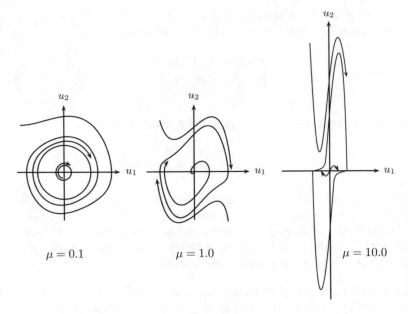

Fig. 7.9

Problems

7.1 The flow of chemically reacting mixtures of gases plays a fundamental role in studying such diverse problems as the solar atmosphere and the atmosphere of other stars, and the gas flow in the combustion chamber of a rocket engine. It can be shown that for certain types of gases the propagation of small disturbances through the gas at time t varies is described by the differential equation

$$y''' + ay'' + by' + cy = 0,$$

where the given constants a, b, and c are all positive. The dependent variable y is proportional to the gas pressure. The coefficients a, b, and c are related to the physical properties and the temperature of the gas. In particular, the constants b and c are usually called the frozen and equilibrium sound speeds of the gas, respectively. From the physical properties, it is known that $b > c$. If the differential equation is asymptotically stable, then all disturbances to the gas will eventually disappear because they are dissipated by the chemical reactions. If the differential equation is not asymptotically stable, then there are disturbances which do not decay as $t \to \infty$. Then, shock waves may form in the gas.

(1) Use the Hurwitz test (Theorem 7.2) to determine conditions on the coefficients on the constants a, b, and c for which the differential equation is asymptotically stable.

(2) Is the differential equation asymptotically stable if $c = 0$?

7.2 For the differential equation (2.28) find critical points and discuss their stability properties. (The case $a^2 - 4bH < 0$ gives complex critical points, which are unrealistic).

7.3 The differential equation

$$P' = bP^2 - aP - H,$$

where $a, b, H > 0$ describes the harvesting of an unsophisticated population such as alligators. Show that for this differential equation the critical point $P_1 = (a + \sqrt{a^2 + 4bH})/(2b) > 0$ is unstable, whereas the critical point $P_2 = (a - \sqrt{a^2 + 4bH})/(2b) < 0$ is stable. (P_2 is an unrealistic critical point). Further, show that if $P(0) = P_0 > P_1$, then $P(t) \to \infty$ in a finite period of time, but if $P_0 < P_1$, then $P(t) \to 0$ in a finite period of time. Thus, $P_0 = P_1$ is a threshold value that increases as the harvesting rate H increases—in other words, for more alligators to be killed annually, the initial number must be large to prevent extinction.

7.4 Gilpin and Ayala's [10] model of a single-species population is described by the equation

$$P' = rP \left(1 - (P/K)^\theta\right),$$

where r, K, and θ are positive constants. Find critical points of this equation and discuss their stability properties.

7.5 Schoener's [29] model of a single-species population with a constant energy input is

$$P' = rP \left(\frac{I}{P} - C - bP\right),$$

where r, I, C, and b are positive constants. Find critical points of this equation and discuss their stability properties.

7.6 *Allee effects* are broadly defined as a decline in individual fitness at low population size or density, that can result in critical population thresholds below which populations crash to extinction. The following Volterra's [34] model describes this type of situation

$$P' = P(-d + aP - bP^2),$$

where a, b, and d are positive constants. Find critical points of this equation and discuss their stability properties.

7.7 A model of a single-species of a population which experiences an Allee effect is given by the equation (see [11, 25])

$$P' = P\left(b + P\frac{a - P}{1 + cP}\right), \tag{7.39}$$

where a, b, and c are positive constants. Find critical points of this equation and discuss their stability properties.

7.8 If in model (7.39) an effort is applied per unit time to catch the animals in a population is fixed, then this leads to the differential equation (see [11])

$$P' = P\left(b + P\frac{a - P}{1 + cP}\right) - EP, \tag{7.40}$$

where a, b, c, and E are positive constants. If $d = E - b > 0$, show that two positive critical points of (7.40) are

$$P_1 = \left((a - cd) - \sqrt{(a - cd)^2 - 4d}\right)/2,$$

$$P_2 = \left((a - cd) + \sqrt{(a - cd)^2 - 4d}\right)/2.$$

The critical point P_1 is unstable, whereas P_2 is stable. If $d < 0$, then there is a unique positive stable critical point.

7.9 Duffing's equation (4.75) with $\alpha > 0, \beta \geq 0, \delta \geq 0$, and $\gamma = 0$ in system form appears as

$$u_1' = u_2$$
$$u_2' = -\alpha u_1 - \delta u_2 - \beta u_1^3.$$

Show that for this system the critical point $(0, 0)$ is a stable node if $\delta^2 \geq 4\alpha$, a stable focus if $\delta^2 < 4\alpha$, and a stable center if $\delta = 0$. (Note that if $\delta = 0$, then the trajectories are $2y^2 + 2\alpha x^2 + \beta x^4 = c^2$. Further, for this system with $\alpha = 1$ and $\delta = 0$ conditions of Theorem 7.11 are satisfied.)

7.10 Rayleigh's equation (4.74) in system form appears as

$$u_1' = u_2$$
$$u_2' = -\frac{k}{m}u_1 + \frac{a}{m}u_2 - \frac{b}{m}u_2^3.$$

Show that for this system the critical point $(0, 0)$ is an unstable saddle point if $a^2 > 4km$, unstable node if $a^2 = 4km$, and unstable focus if $a^2 < 4km$.

7.11 A mass m is attached to a spring and encounters friction as it moves along a surface. The motion is governed by

$$mx'' + \mu \operatorname{sign}(x') + kx = 0,$$

where μ is a positive constant depending on the mass of the object and the coefficient of friction between mass and the surface, k is the spring constant, and $\operatorname{sign}(x')$ is $(+1)$

when $x' > 0$ and (-1) when $x' < 0$. Give a phase plane analysis for this equation and interpret the results.

7.12 The motion of an object falling through the air is governed by the equation

$$x'' = g - \frac{g}{V^2} x'|x'|,$$

where x is the distance fallen and V is the terminal velocity. Give a phase plane analysis for this equation and interpret the results.

7.13 Suppose that a satellite is in flight on the line between a planet of mass m_1 and its moon of mass m_2 which are a constant distance r apart. The distance x between the satellite and the planet satisfies the nonlinear differential equation

$$x'' = -\frac{gm_1}{x^2} + \frac{gm_2}{(r-x)^2}.$$

Give a phase plane analysis for the linearized equation and interpret the results. $(r\sqrt{m_1}/(\sqrt{m_1} - \sqrt{m_2}), 0)$, saddle point.

7.14 Consider a mass hanging on a massless rod from a support point. The support point is allowed to oscillate in the vertical direction. Let x denote the angle in the positive (counterclockwise) direction which the pendulum makes with the straight down position. Then, x satisfies the equation (see [16])

$$b^2 x'' + rx' + (1 + a \sin(\omega t)) \sin(x) = 0, \tag{7.41}$$

where b denotes the ratio of the forcing frequency to the free frequency of the pendulum, ω is typically taken to be 2π, and a is the amplitude of the acceleration applied to the support point. Show that the critical points $(0, 0)$ and $(\pi, 0)$ of (7.41) are asymptotically stable as long as $r > 0$.

7.15 For the system (5.78) show that $(0, 0)$, $(-M/Q, 0)$, $(0, K/R)$, and $\left(\frac{KP-MR}{PL+QR}, \frac{ML+KQ}{PL+QR}\right)$ are all critical points. Study the stability of each critical point and determine whether it is a node, saddle, or focus. (Note that from biological point of view negative critical points are not of interest; however, in particular applications the constants in (5.78) may be negative).

7.16 Let $x(t)$ represent the income of a company and $y(t)$ the amount of consumer spending. Also suppose that z represents the rate of company expenditures. The differential system which models this situation is

$$x' = x - ay, \quad 1 < a < \infty$$
$$y' = b(x - y - z), \quad b \geq 1.$$

(1) If $z = c$ is constant, find and classify the critical points of the system. Also, consider the special case when $b = 1$.

(2) If the expenditure depends on income linearly, i.e., $z = c + dx$ ($d > 0$), find and classify the critical points of the system.

7.17 To describe lasers with an inertialess bleachable filter, Samson [28] used the following system:

$$u' = \nu u[y - 1 - \beta/(1 + \alpha u)]$$
$$y' = y_0 - (1 + u)y. \tag{7.42}$$

Here y, y_0, β are the ratios of the gain, the unsaturated gain, and the unsaturated loss factor of the filter to the radiation-independent loss factor; u is the ratio of probability of an induced transition in the lasing channel to the probability of relaxation of the active material (proportional to the density of the generated radiation); α is a dimensionless nonlinearity parameter; ν is proportional to the velocity of light, the filling factor of the active material in the resonator, and the loss factor, and is independent of u and inversely proportional to the relaxation probability. The derivatives u' and y' are calculated with respect to the dimensionless variable τ, equal to the product of t and the probability of relaxation of the active material. The parameters ν, β, α, y_0 in (7.42) can vary over quite wide ranges. Find the critical points of the system (7.42) and study their stability properties.

7.18 Consider the Lotka–Volterra type of system

$$u_1' = -Mu_1 + P \left(1 - e^{-bu_1}\right) u_2$$
$$u_2' = -Ku_2 + L \left(1 - e^{-bu_1}\right) u_2,$$

where all the constants are positive and $K \neq L$. Show that for this system $(0, 0)$ and $\left(\frac{1}{b} \ln \left[\frac{L}{|L-K|}\right], \frac{ML}{bKP} \ln \left[\frac{L}{|L-K|}\right]\right)$ are the critical points. Classify these critical points.

7.19 Consider Bonhoeffer–Van der Pol's oscillator (this system and name was suggested by FitzHugh in 1961)

$$u_1' = u_1 - u_1^3/3 - u_2 + I(t)$$
$$u_2' = c(u_1 + a - bu_2).$$

Find and classify the critical points (numerically) of this system when $I(t) = 0$, $a = b = c = 1$.

7.20 (1) If $T_{in} = 1$, $HVC_{in} = VC_p = qC_p = 1$, and $k(T) = e^{-1/T}$ graph $y = T - T_{in}$ and $y = \frac{HVC_{in}}{qC_p + VC_p k(T)} k(T)$ to determine the number of roots of Eq. (7.36).
(2) If $T_{in} = 0.15$, $HVC_{in} = 1$, $VC_p = 0.25$, $qC_p = 0.015$, and $k(T) = e^{-3/T}$ graph $y = T - T_{in}$ and $y = \frac{HVC_{in}}{qC_p + VC_p k(T)} k(T)$ to determine the number of roots of Eq. (7.36).
(3) Determine which of the temperature found in (1) and (2) are stable and which are unstable.

7.21 Let P be the price of a single item on the market. Let Q be the quantity of the item available on the market. Both P and Q are functions of time. It we consider price and quantity as two interacting species, then the following economic model can be proposed:

$$\frac{dP}{dt} = aP\left(\frac{b}{Q} - P\right)$$
$$\frac{dQ}{dt} = cQ(fP - Q),$$

where a, b, c, and f are positive constants. Find and classify critical points of this system.

7.22 Let in Australia $k(t)$ be the number of kangaroos at time t (the prey) and $p(t)$ be the number of predators at time t. This prey and predators situation can be described by the system

$$k' = \alpha k - \beta k^2 - \gamma kp$$
$$p' = -\sigma p + \lambda kp,$$

where $\alpha, \beta, \gamma, \sigma$, and λ are constants. Find and classify critical points of this system.

7.23 The following system describes the rotational motion of a book tossed into the air (see [6]):

$$x' = yz$$
$$y' = -2xz \qquad\qquad (7.43)$$
$$z' = xy$$

(1) Show that $x^2 + y^2 + z^2 = c^2$. Thus, if we choose $c = 1$, then the trajectories of (7.43) lie on the surface of a sphere of radius 1.
(2) Examine the stability of (7.43) at the critical points $(\pm 1, 0, 0)$, $(0, \pm 1, 0)$, and $(0, 0, \pm 1)$.

7.24 In system (7.37), assume that a disease affects the population of wolves that kills a small fraction of population each year. We accommodate this by adding the term $(-\gamma z)$ in the equation z', where γ is a small positive parameter. Then, the model takes the form

$$x' = x(1 - x) - xy$$
$$y' = y(1 - y) + xy - yz$$
$$z' = z(1 - z) - \gamma z + yz.$$

Show that for this system $(x(\gamma), y(\gamma), z(\gamma)) = ((2 - \gamma)/3, (1 + \gamma)/3, (4 - 2\gamma)/3))$ is the only critical point with all three coordinates nonzero, and it is asymptotically stable. Further, from $dz(\gamma)/d\gamma = -2/3, dy(\gamma)/d\gamma = 1/3, dx(\gamma)/d\gamma = -1/3$ conclude that an increase in γ decreases the number of wolves, good for the moose population, and decreases the population of trees.

7.25 The equations that describe a simple food chain in a chemostat (a system in which the chemical composition is kept at a controlled level, especially for the culture of microorganisms) are (see [30])

$$\frac{dS}{dt} = 1 - S - \frac{m_1 y S}{a_1 + S}$$

$$\frac{dx}{dt} = \frac{m_2 x y}{a_2 + y} - x$$

$$\frac{dy}{dt} = \frac{m_1 y S}{a_1 + S} - y - \frac{m_2 x y}{a_2 + y}$$

$$S(0) = S_0 \geq 0, \quad x(0) = x_0 > 0, \quad y(0) = y_0 > 0.$$

(7.44)

In the system (7.44), x (the predator) consumes y (the prey), y consumes the nutrient S, and the constants are positive.

(1) Let $\sum = 1 - S - x - y$ to show that $\sum' = -\sum$. Thus, $\sum = \sum_0 e^{-t} \to 0$ as $t \to \infty$, and hence $S = 1 - x - y$. Therefore, the system (7.44) can be written as

$$\frac{dx}{dt} = \frac{m_2 x y}{a_2 + y} - x$$

$$\frac{dy}{dt} = \frac{m_1 y (1 - x - y)}{1 + a_1 - x - y} - y - \frac{m_2 x y}{a_2 + y}$$

$$x(0) = x_0 > 0, \quad y(0) = y_0 > 0.$$

(7.45)

(2) Show that $E_0(0, 0)$ and $E_1 \left(0, 1 - \frac{a_1}{m_1 - 1} \right)$ are equilibrium points of (7.45). Besides E_0 and E_1 are there other equilibrium points?

(3) Find conditions so that E_0 is unstable.

(4) Show that E_1 is stable if $\lambda_1 + \lambda_2 > 1$ and unstable saddle point if $\lambda_1 + \lambda_2 < 1$, where $\lambda_i = a_i/(m_i - 1)$, $i = 1, 2$.

7.26 In Problem 5.19, if we assume that the population of stem or semi-differentiated cells increases the (per-capita) rate at which they differentiate increases in proportion, i.e., α_2 and β_2 are, respectively, replaced by $(\alpha_2 + k_0 N_0)$ and $(\beta_2 + k_1 N_1)$, where k_0 and k_1 are positive constants, then the system takes the form (see [18])

$$N_0' = \alpha N_0 - k_0 N_0^2$$

(7.46)

$$N_1' = \beta N_1 - k_1 N_1^2 + N_0(\alpha_2 + k_0 N_0)$$

(7.47)

$$N_2' = -\gamma N_2 + N_1(\beta_2 + k_1 N_1).$$

(7.48)

(1) Show that $E_0 = (0, 0, 0)$ and $E_1 = (N_0^*, N_1^*, N_2^*)$ are the stationary points of the system (7.46)–(7.48), where

$$N_0^* = \frac{\alpha}{k_0}, \ N_1^* = \frac{1}{2k_1}\left[\beta+\sqrt{\beta^2+4k_1 N_0^*(\alpha_2+k_0 N_0^*)}\right], \ N_2^* = \frac{N_1^*}{\gamma}(\beta_2+k_1 N_1^*).$$

(2) Discuss stability of the critical point E_0.

(3) Compare (7.46) with (2.4) to conclude that the stem cells exhibit logistic growth, with a carrying capacity of N_0^* (see Example 2.1). If $\alpha > 0$ all solutions of (7.46) are attracted to this stable steady state. If $\alpha < 0$ all solutions of (7.46) are attracted to the equilibrium point $N_0 = 0$. Also, conclude that there are no values of the parameters except $k_0 > 0$ that allow unbounded growth of $N_0(t)$. Thus, if $\alpha_3 > \alpha_1 + \alpha_2$ the stem cell population is able to sustain itself and reaches a nonzero steady state.

(4) Show that for the semi-differentiated cells $N_1(t)$ the equilibrium point N_1^* is stable.

7.27 In Problem 5.19, we assume that α_2 and β_2 are, respectively, increased by $\alpha_2 + k_0 N_0/(1 + m_0 N_0)$ and $\beta_2 + k_1 N_1/(1 + m_1 N_1)$, where m_0 and m_1 are positive constants, then the system takes the form (see [18])

$$N_0' = \alpha N_0 - \frac{k_0 N_0^2}{1 + m_0 N_0} \tag{7.49}$$

$$N_1' = \beta N_1 - \frac{k_1 N_1^2}{1 + m_1 N_1} + \alpha_2 N_0 + \frac{k_0 N_0^2}{1 + m_0 N_0} \tag{7.50}$$

$$N_2' = -\gamma N_2 + \beta_2 N_1 + \frac{k_1 N_1^2}{1 + m_1 N_1}. \tag{7.51}$$

(1) Show that $E_0 = (0, 0, 0)$ is a stationary point of the system (7.49)–(7.51). Discuss stability of this critical point.

(2) Show that for Eq. (7.49), the point $N_0 = 0$ is unstable for $\alpha > 0$ and globally attracting for $\alpha < 0$.

(3) Show that for the stem cells *homeostasis* (the ability to maintain a constant internal environment in response to environmental changes) exists when $0 < \alpha < k_0/m_0$, and is given by $N_0^* = \alpha/(k_0 - m_0\alpha)$. If $\alpha > k_0/m_0$, no steady state exists and the cell population grows unboundely.

7.28 When diseases persist in a population for long period of time, in SIR model (see Example 5.12) births and deaths must be taken into consideration. If a person becomes immune to a disease after recovering from it and births and deaths in the population are taken into account, then the percent of persons susceptible to becoming infected with the disease, $S(t)$, and the percent of people in the population infected with the disease, $I(t)$, can be modeled by the system (see [4, 14, 23])

$$\begin{aligned} S' &= -\lambda S I + \mu - \mu S \\ I' &= \lambda S I - \gamma I - \mu I \end{aligned} \tag{7.52}$$

with the conditions (5.91).

The model (7.52) is called an *SIR model with vital dynamics*. This model might be used to model diseases that are *endemic* to a population; those diseases that persist in a population for long periods of time (10 or 20 years). Smallpox is an example of a disease that was endemic until it was eliminated in 1977.

(1) Show that the equilibrium points of the system (7.52) are $(1, 0)$ and $E^* = \left(\dfrac{\gamma + \mu}{\lambda}, \dfrac{\mu[\lambda - (\gamma + \mu)]}{\lambda(\gamma + \mu)} \right)$.

(2) Since $S(t) + I(t) + R(t) = 1$, we can use the fact that $S(t) + I(t) \leq 1$ to determine conditions on γ, μ, and λ so that the system (7.52) has the equilibrium point E^*. In this case, classify the equilibrium point.

(3) Use the fact that $S(t) + I(t) \leq 1$ to determine conditions on γ, μ, and λ so that the system (7.52) does not have the equilibrium point E^*.

7.29 Consider the situation in which the disease is periodic. The symbolic representation is $S \rightarrow I \rightarrow R \rightarrow S$, with vital dynamics, and for which $S(t) + I(t) + R(t) = 1$. We assume that there is a temporary immunity with the rate of loss of immunity per immune individual per unit time denoted by α. This situation can be modeled by

$$S' = -\lambda SI + \mu - \mu S + \alpha R$$
$$I' = \lambda SI - \gamma I - \mu I \tag{7.53}$$

In the first equation, the terms in the right side are, respectively, due to: the decrease in S due to individuals' contracting the disease, the births into class S, the deaths from class S, and the influx to S of persons who lost their immunity. In the second equation, the terms in the right side are, respectively, due to: those coming from S by having contracted the disease, the outflow that is due to recovery from the disease, and those dying while in class I. Observe that in view of $S(t) + I(t) + R(t) = 1$ the system (7.53) can be written as

$$S' = -\lambda SI + (\mu + \alpha) - (\mu + \alpha)S - \alpha I$$
$$I' = \lambda I \left(S - \frac{\gamma + \mu}{\lambda} \right). \tag{7.54}$$

Find and classify the critical points of the system (7.54).

7.30 Following [2] (also see [9]) assume that the fox population is divided into susceptibles and infectives with viral disease rabies (see Problem 6.26), with their population densities denoted by S and I, respectively. The rate of increase of infective foxes (rabid) is assumed to be proportional to the number of contacts between rabid and healthy foxes. Since rabid foxes never recover and death is certain, it is assumed that the death rate from rabies is proportional to the number of infective foxes. These assumptions lead to the normalized system

$$S' = S[b - (1 + b)I - bS]$$
$$I' = I[(1 - b)S - r - bI], \tag{7.55}$$

where $b = 0.022$/year and $r = 0.456$/year represent the normalized birth and death rates, respectively. Show that for the system (7.55) the critical points are $E_1 = (0, 0)$, $E_2 = (1, 0)$, $E_3 = (S^*, I^*)$, $E_4 = (0, -r/b)$, where $S^* = b^2 + br + r$ and $I^* = b - br - b^2$. Further, show that E_1 is unstable; if $r + b \leq 1$ then E_2 is unstable, otherwise it is stable; if $r + b < 1$ and $(b^2 + br)^2 \geq 4(b^2 + br + r)(b - br - b^2)$ then E_3 is a stable node; if $r + b < 1$ and $(b^2 + br)^2 < 4(b^2 + br + r)(b - br - b^2)$ then E_3 is a stable spiral; if $r + b = 1$ then E_3 is the same as E_2, and when $r + b > 1$, I^* is negative and thus E_3 is meaningless. Critical point E_4 is also meaningless.

7.31 In modeling of porcine reproduction and respiratory syndrome with time dependent infection rate, Suksamran et al. [31] obtained the following system:

$$
\begin{aligned}
u_1' &= u_2 \\
u_2' &= -(b_U - d_U)u_1 - cu_2 - \frac{\gamma b_U v}{\alpha}u_3 + \frac{cb_U v}{\alpha}u_4 + \beta u_1 u_3 \\
u_3' &= u_4 \\
u_4' &= \frac{\gamma(b_V - d_V)}{c^2}u_3 + \frac{(\gamma - b_V + d_V)}{c}u_4 + \frac{\alpha\beta}{c^2}u_1 u_3,
\end{aligned}
\tag{7.56}
$$

where all constants are positive. Show that

(1) The system (7.56) has two equilibrium points $E_0 = (0, 0, 0, 0)$, and $E_1 = \left(\frac{-\gamma(b_V - d_V)}{\alpha\beta}, 0, \frac{(b_U - d_U)(b_V - d_V)}{\beta(b_{UV} + b_V - d_V)}, 0\right)$ where $b_V - d_V < 0$ and $\frac{b_U - d_U}{b_{UV} + b_V - d_V} < 0$.

(2) The equilibrium point E_0 is unstable. Further, E_0 is a saddle point if $b_U - d_U < 0$ or $b_V - d_V > 0$.

(3) The equilibrium point E_1 is unstable (use Theorem 7.2).

7.32 In [7], Dobson et al. have proposed the model

$$
\begin{aligned}
\frac{dF}{dt} &= sU - dPF \\
\frac{dA}{dt} &= dPF + bU - aA \\
\frac{dU}{dt} &= aA - (b + s)U \\
\frac{dP}{dt} &= rP\frac{A - hP}{A}.
\end{aligned}
\tag{7.57}
$$

These equations provide a framework which compare the impact on natural habitats of agricultural expansions, ecological restoration, improvements in agricultural efficiency, and human population growth. The model (7.57) considers the rate at which an original area of pristine forest habitat (of area F, initially assumed to be the entire forest in a country or a patch of forest connected to other patches in a spatial array) is converted first to agriculture land (area A), which after a period of time $1/a$, becomes unused land (area U), which in turn recovers through natural succession or ecological restoration to become forest after a time interval $1/s$. The efficiency with which humans convert natural habitat to agricultural land is assumed to be d. Unused land may also be restored to agriculturally viable land after a time interval $1/b$. Rates of habitat conversion are assumed to be a simple function of the number of humans P using the land at any time. Initially, this parameter is assumed to be a constant. It is assumed that human population growth r occurs at a maximum rate of 4% and can be modeled as a simple logistic function with carrying capacity, given by the minimum amount of land h required to support an individual human. The initial assumption is that it remains constant. The basic parameters (a, b, d, h, r, s) can be estimated from the studies of tropical and temperate forests. Show that the critical point of (7.57) is

$$A^* = hP^*, \quad U^* = \frac{aA^*}{s+b}, \quad F^* = h\frac{a}{d}\frac{s}{s+b}, \quad P^* = \frac{1}{h}\frac{F_0 - h\frac{a}{d}}{\frac{a}{s+b}+1}\frac{s}{s+b}, \quad (7.58)$$

where F_0 is the initial total area. Assuming that the critical point (7.58) of the system (7.57) is stable, what conclusions you can deduce.

7.33 For the differential system (6.11) show that (for details see [24])
(1) $E_1 = (0, 0, 0, 0)$ is a critical point and it is unstable.
(2) $E_2 = (0, b_2/\beta_{22}, 0, a_2/\alpha_{22})$ is a critical point and it is stable if $A_1 = (a_1/\alpha_{12}) - (a_2/\alpha_{22}) < 0$, $B_1 = (b_1/\beta_{12}) - (b_2/\beta_{22}) > 0$, and unstable otherwise.
(3) $E_3 = (b_1/\beta_{11}, 0, a_1/\alpha_{11}, 0)$ is a critical point and it is stable if $A_2 = (a_2/\alpha_{21}) - (a_1/\alpha_{11}) < 0$, $B_2 = (b_2/\beta_{21}) - (b_1/\beta_{11}) > 0$, and unstable otherwise.
(4) $E_4 = (0, b_1/\beta_{12}, a_2/\alpha_{21}, 0)$ is a critical point and it is stable if $A_2 > 0$, $B_1 < 0$, and unstable otherwise.
(5) $E_5 = (b_2/\beta_{21}, 0, 0, a_1/\alpha_{12})$ is a critical point and it is stable if $A_1 > 0$, $B_2 < 0$, and unstable otherwise.
(6) $E_6 = (\beta_{12}\beta_{22}B_1/B_3, \beta_{11}\beta_{21}B_2/B_3, \alpha_{12}\alpha_{22}A_1/A_3, \alpha_{11}\alpha_{21}A_2/A_3)$, where $A_3 = \alpha_{11}\alpha_{22} - \alpha_{12}\alpha_{21}$, $B_3 = \beta_{11}\beta_{22} - \beta_{12}\beta_{21}$ is a critical point.
(7) If $A_1 > 0$, $A_2 > 0$, then $A_3 > 0$; if $A_1 < 0$, $A_2 < 0$, then $A_3 < 0$; if $B_1 > 0$, $B_2 > 0$, then $B_3 > 0$; and, if $B_1 < 0$, $B_2 < 0$, then $B_3 < 0$.
(8) If $A_1 > 0$, $A_2 > 0$ and $B_1 > 0$, $B_2 > 0$, or $A_1 < 0$, $A_2 < 0$ and $B_1 < 0$, $B_2 < 0$ then the critical point E_6 is stable. Further, if $A_3 > 0$, $B_3 < 0$, or $A_3 < 0$, $B_3 > 0$, then E_6 is unstable.
(9) For the numerical values of the constants given in Example 6.4, $E_6 = (10, 10, 1, 1)$ and it is stable.

7.34 The fifth-order system

$$y_i' = y_{i+1}y_{i+2} + y_{i-1}y_{i-2} - 2y_{i+1}y_{i-1}, \quad i = 1, 2, \cdots, 5$$

where $y_i = y_{i+5}$ for all i, has the energy integral $\frac{d}{dt}\sum_{i=1}^{5} y_i^2 = 0$, so the orbits are confined to the surface of a ball in five-dimensional phase space. Show that this system has critical points and they are all unstable (see [6]). Further, plot the graph of $y_1(t)$ verses $y_2(t)$ and draw the conclusions.

7.35 Show that if $\delta \neq 0$ and $\gamma = 0$, Duffing's equation (4.75) has no limit cycles.

7.36 In chemical kinetics (see [8, 17]) the following system appears:

$$
\begin{aligned}
x' &= s^2(x + y - xy - qx^2) \\
y' &= (-y - xy + z) \\
z' &= sw(x - z),
\end{aligned}
\tag{7.59}
$$

where the parameters have range $s \sim 100$, $q \sim 10^{-5}$, $sw \sim 10$. Show that
(1) When $1 < x \leq q^{-1}$, $z > 0$ the system (7.59) has a critical point $(\theta, \theta/(1 - \theta), \theta)$,
where $\theta = -\dfrac{1}{2} + \sqrt{\dfrac{1}{4} + \dfrac{2}{q}}$.
(2) In the range of parameters, the system (7.59) has a limit cycle.

7.37 Edward Norton Lorenz (1917–2008) was an American meteorologist. In 1963 he studied the system

$$
\begin{aligned}
x' &= \sigma(-x + y) \\
y' &= rx - y - xz \\
z' &= -bz + xy.
\end{aligned}
\tag{7.60}
$$

This system (now known as *Lorenz equations*) arises in fluid dynamics and in the study of meteorology. (Actually, the system (7.60) is much simpler than originally used for modeling the weather, it cannot tell us even tomorrow's temperature. However, by examining this system, Lorenz made a scientific revolution in the field called *Chaos Theory*.) In (7.60), σ, r, and b are parameters.
(1) Locate the critical points of (7.60) and determine their local behavior.
(2) The solutions of (7.60) are extremely sensitive to parameters and perturbations in the initial conditions. As an example, find the numerical solution $(x(t), y(t), z(t))$ of (7.60) with the parameters $\sigma = 10, r = 28, b = 8/3$ and the initial conditions $x(0) = -8, y(0) = 8, z(0) = 27$, and draw its projection in xz-plane.
(3) With the same values of the parameters as in (2) find the numerical solutions of (7.60) with the initial conditions $x(0) = 0, y(0) = 1, z(0) = 0$ and $x(0) = 0, y(0) = 1.001, z(0) = 0$. Draw the solution curves in xyz-space and xt-plane, and respectively, compare the curves. Do you find some strange patterns?

7.38 Similar to Lorenz equations another extensively studied system was introduced in 1970 by Otto Eberhard Rössler (born in 1940)

$$
\begin{aligned}
x' &= -y - z \\
y' &= x + ay \\
z' &= b + z(x - c).
\end{aligned}
\tag{7.61}
$$

(1) Locate the critical points of (7.61) and determine their local behavior.

(2) The solutions of (7.61) are extremely sensitive to parameters a, b, c. Find the numerical solutions $(x(t), y(t), z(t))$ of (7.61) with the parameters $a = 0.2, a = 0.3, a = 0.35, a = 0.375$ and $b = 2, c = 4$ in all cases, and draw their projections in xy-plane.

7.39 Many physical systems involve some physical parameters which vary continuously. In some of these systems, there exist certain critical values of the parameters such that the system's qualitative behavior changes abruptly and dramatically as a parameter passes through such a critical value. Such a critical point is called *bifurcation* (generally refers to something splitting apart) point. (The term bifurcation was introduced by Poincaré in 1885. He also classified various types of bifurcation points.) For example, as the amount of one of the chemicals in a certain mixture is increased, spiral wave patterns of varying color suddenly emerge in an originally quiescent fluid. In several cases the mathematical modeling finally leads to the equation of the form

$$x' = (R - R_c)x - ax^3, \tag{7.62}$$

where a and R_c are positive constants, and R can assume various values. As an example, R measures the amount of chemical and x measures the chemical reaction. (In fluid mechanics Eq. (7.62) occurs in the study of the transition from laminar to turbulent flow, and is known as Landau's equation, after the Russian Nobel laureate Lev Davidovic Landau (1908–1968).) For Eq. (7.62) show that

(1) If $R < R_c$, then $x = 0$ is the only critical point and it is asymptotically stable.

(2) If $R > R_c$, then there are three critical points $x = 0$ and $x = \pm\sqrt{(R - R_c)/a}$, where $x = 0$ is unstable and the other two are asymptotically stable.

Thus, $R = R_c$ is a bifurcation point. The manner in which the solution branches at R_c is called a *pitchfork bifurcation*. This name is justified from the graph in the Rx-plane showing all the three critical points.

7.40 Consider the buckling of the mechanical system (see Fig. 7.10) involving two massless rigid rods of length ℓ attached to a mass m and a lateral spring of stiffness k. We assume that $x = 0$, when the spring is neither stretched nor compressed and the roads are aligned vertically. As we increase the downward load P nothing happens until we reach a critical value P_{cr}, at which value x increases (to one side or the other, which is difficult to predict) and the system collapses. For the displacement $x(t)$, an application of Newton's second law of motion leads to the differential equation

$$mx'' - \frac{2Px}{\ell}\left[1 - \left(\frac{x}{l}\right)^2\right]^{-1/2} + kx = 0. \tag{7.63}$$

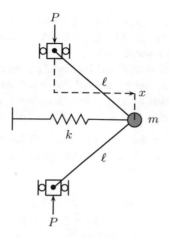

Fig. 7.10

Show that for the Eq. (7.63) the origin is a center if $P < k\ell/2$ and saddle point if $P > k\ell/2$, and $P = P_{cr} = k\ell/2$ is a bifurcation point.

7.41 The system

$$x' = 3\left(x + y - \frac{1}{3}x^3 - c\right)$$
$$y' = -\frac{1}{3}(x + 0.8y - 0.7) \tag{7.64}$$

is a particular case of the FitzHugh–Nagumo equations. In (7.64), the parameter c is the external stimulus.

(1) For $c = 0$ show that the system (7.64) has one critical point. Find this point and show that it is an asymptotically stable spiral.

(2) For $c = 0.5$ show that the system (7.64) has one critical point. Find this point and show that it is an unstable spiral.

(3) Find the critical value (bifurcation point) $c = c_0$ where asymptotic stability changes to instability of the system (7.64). Draw a phase portrait for the system for $c = c_0$.

7.42 The nonlinear system

$$x' = -cx + y$$
$$y' = \frac{x^2}{1 + x^2} - y \tag{7.65}$$

arises in molecular biology (see [12]), where x and y are proportional to protein and messenger RNA concentrations, and c is a positive empirical constant associated with the "death rate" of protein in the absence of the messenger RNA. Show that

(1) $S_0 = (0, 0)$, $S_+ = (x_+, y_+)$ and $S_- = (x_-, y_-)$, where $x_\pm = (1 \pm \sqrt{1 - 4c^2})/$
$(2c)$, $y_\pm = (1 \pm \sqrt{1 - 4c^2})/2$ and $c \le 1/2$ are the critical points of (7.65).
(2) The critical point S_0 is a stable node.
(3) For $c < 1/2$, S_- is a saddle point, whereas S_+ is an unstable node.
(4) For $c = 1/2$, S_- and S_+ merge and form a singularity of some other type.
(5) For $c > 1/2$ the singularity disappears completely, and the origin is the only
stable node, i.e., bifurcation occurs at $c = 1/2$.

7.43 An extensively studied system which presents the famous *Hopf bifurcation* (a
critical point where a system's stability switches and a periodic solution arises) is
the following almost linear system:

$$x' = \epsilon x + y - x(x^2 + y^2)$$
$$y' = -x + \epsilon y - y(x^2 + y^2). \tag{7.66}$$

(1) Use polar coordinates $x = r \cos \theta$, $y = r \sin \theta$ to transform the system (7.66) to
$r' = r(\epsilon - r^2)$, $\theta' = -1$.
(2) Separate the variables and integrate directly to show that if $\epsilon \le 0$ then $r(t) \to 0$
as $t \to \infty$, thus in this case the origin is a stable spiral.
(3) Show that if $\epsilon > 0$, then $r(t) \to \sqrt{\epsilon}$ as $t \to \infty$, thus in this case the origin is an
unstable spiral point.
(4) The circle $r(t) = \sqrt{\epsilon}$ itself is a closed periodic solution or limit cycle. Thus, a
limit cycle of increasing size is spawned as the parameter ϵ increases through the
critical value 0.

7.44 Lotka–Volterra system (5.73) displays closed curves in the phase plane. However, the sizes of these curves keep on increasing as we move the initial conditions
away from the critical point. This dependence is not expected in a model which is supposed to represent a real situation. In [26, 33] authors studied a predator–prey model
in which the interaction term was replaced by Holling type II functional response
(Holling [15] in 1965 described three types of functional responses and Haldane
[13] in 1930 used another type which is now called Holling type-IV response.).
Their system is governed by the equations

$$\frac{dx}{dt} = -Mx + \frac{Pxy}{y + D}$$
$$\frac{dy}{dt} = Ky\left(1 - \frac{y}{L}\right) - \frac{Lxy}{y + D}.$$

More realistic/complex systems have been examined in [19, 20, 32]. Letting $y = LY$, $x = KX$, $t = T/K$ this system reduces to

$$\frac{dX}{dT} = -bX + \frac{eXY}{Y + a}$$
$$\frac{dY}{dT} = Y(1 - Y) - \frac{XY}{Y + a}, \tag{7.67}$$

where $a = D/L, b = M/K, e = P/K$. Show that

(1) System (7.67) has three critical points $E_0 = (0, 0)$, $E_1(0, 1)$, and $E_2(x^*, y^*)$, where

$$x^* = \frac{ae}{e - b}\left(1 - \frac{ab}{e - b}\right), \quad y^* = \frac{ab}{e - b}.$$

(2) E_0 is a saddle point which repels in the x-direction and attracts in y-direction.

(3) E_1 is a saddle point which attracts in the x-direction and repels in y-direction if $e - b > ab$, and if $e - b < ab$ then E_1 is asymptotically stable. Further, if $e < b$, E_1 is *globally* asymptotically stable, i.e., every trajectory $(X(T), Y(T))$ of (7.67) tends to E_1 as $t \to \infty$.

(4) If $(1 - a)/2 < ab/(e - b) < 1$, then E_2 is asymptotically stable. Further, find criteria for E_2 to be globally asymptotically stable.

(5) The system (7.67) undergoes a Hopf-bifurcation at $y^* = ab/(e - b) = (1 - a)/2$ if $0 < ab/(e - b) < 1$ and $a < 1$.

7.45 In [21] (also see [35]) the following three species food chain model having Holling type II functional response was examined

$$x' = -d_1 x + \frac{a_1 xy}{1 + b_1 y} - \frac{a_2 xz}{1 + b_2 x}$$

$$y' = y(1 - y) - \frac{a_1 xy}{1 + b_1 y} \quad\quad (7.68)$$

$$z' = -d_2 z + \frac{a_2 xz}{1 + b_2 x},$$

where x, y, z are the densities of predator, prey, and top predator at time t, respectively, and $a_1, a_2, b_1, b_2, d_1, d_2$ are positive constants. Show that

(1) The system (7.68) has at most five nonnegative critical points: $E_0(0, 0, 0)$, $E_1(0, 1, 0)$, $E_2((a_1 - b_1 d_1 - d_1)/(a_1 - b_1 d_1)^2, d_1/(a_1 - b_1 d_1), 0)$, $E_3(x_1^*, y_1^*, z_1^*)$, and $E_4(x_2^*, y_2^*, z_2^*)$, where

$$x_i^* = \frac{d_2}{a_2 - b_2 d_2}, \quad y_i^* = \frac{b_1 - 1}{2b_1} + (-1)^{i-1}\frac{\sqrt{(b_1 + 1)^2 - 4a_1 b_1 d_2/(a_2 - b_2 d_2)}}{2b_1},$$

$$z_i^* = \frac{(a_1 - b_1 d_1)y_i^* - d_1}{(a_2 - b_2 d_2)(1 + b_1 y_i^*)}, \quad i = 1, 2.$$

(2) Analyze the complex dynamics in the neighborhood of critical points where the species in the food chain is absent, by illustrating the bifurcation diagrams.

References

1. M.L. Abell, J.P. Braselton, *Introductory Differential Equations* (Elsevier, Amsterdam, 2014)
2. M.R. Abo Elrish, E.H. Twizell, Int. J. Comput. Math. **81**, 1027–1038 (2004)
3. R.P. Agarwal, D. O'Regan, *An Introduction to Ordinary Differential Equations* (Springer, New York, 2008)
4. R.M. Anderson, R.M. May, Science **215**, 1053–1060 (1982)
5. Applications in Undergraduate Mathematics in Engineering, *The Mathematical Association of America* (Macmillan Co., New York, 1967), pp. 122–125
6. C.M. Bender, S.A. Orszag, *Advanced Mathematical Methods for Scientists and Engineers* (McGraw-Hill, Singapore, 1978)
7. A.P. Dobson, A.D. Bradshaw, A.J.M. Baker, Science **277**, 515–522 (1997)
8. R.J. Field, R.M. Noyes, J. Chem. Phys. **60**, 1877–1884 (1974)
9. G.A. Gardner, L.R.T. Gardner, J. Cunningham, Z. Naturforsch **45c**, 1230–1240 (1990)
10. M.E. Gilpin, F.J. Ayala, Proc. Nat. Acad. Sci. (USA) **70**, 3590–3593 (1973)
11. B.S. Goh, *Management and Analysis of Biological Populations* (Elsevier Scientific Publishing Company, Amsterdam, 1980)
12. J.S. Griffith, *Mathematical Neurobiology* (Academic Press, New York, 1971)
13. J.B.S. Haldane, *Enzymes* (Longmans, London, 1930)
14. H.W. Hethcote, in *Applied Mathematical Ecology*, ed. by S.A. Levin, T.G. Hallam, L.J. Gross, (Springer, 1989), pp. 119–143
15. C.S. Holling, Mem. entomol. Soc. Can. **46**, 1–60 (1965)
16. F.C. Hoppensteadt, Computer studies of nonlinear oscillations. Lect. Appl. Math. **17**, 131–139 (1979)
17. L.N. Howard, Nonlinear oscillations. Lect. Appl. Math. **17**, 1–67 (1979)
18. M.D. Johnston, C.M. Edwards, W.F. Bodmer, P.K. Maini, S.J. Chapman, Proc. Nat. Acad. Sci. **104**, 4008–4013 (2007)
19. T.K. Kar, A. Batabyal, R.P. Agarwal, J. Korean Soc. Ind. Appl. Math. **14**, 1–16 (2010)
20. T.K. Kar, S.K. Chattopadhyay, R.P. Agarwal, Commun. Appl. Anal. **14**, 21–38 (2010)
21. A. Klebanoff, A. Hastings, J. Math. Biol. **32**, 427–451 (1994)
22. B.E. McLaren, R.O. Peterson, Science **266**, 1555–1558 (1994)
23. J.D. Murray, *Mathematical Biology* (Springer, New York, 1990)
24. S.P. Neupane, D.B. Gurung, Neural Parallel Sci. Comput. **25**, 435–455 (2017)
25. E.R. Pianka, *Evolutionary Ecology* (Harper and Row Publisher, New York, 1974)
26. V. Rai, V. Kumar, L.K. Pandey, I.E.E.E. Trans, Syst. Man Cybern. **21**, 261–263 (1991)
27. W.F. Ramirez, *Computational Methods for Process Simulation* (Butterworths Series in Chemical Engineering, Boston, 1989), pp. 175–182
28. A.M. Samson, J. Appl. Spectrosc. **42**, 115–125 (1985)
29. T.W. Schoener, Theor. Popul. Biol. **4**, 56–84 (1973)
30. H.L. Smith, P. Waltman, *The Theory of the Chemostat: Dynamics of Microbial Competition* (Cambridge University Press, Cambridge, 1995)
31. J. Suksamran, Y. Lenbury, P. Satiracoo, C. Rattanakul, Adv. Differ. Equ. **2017**, 215 (2017). https://doi.org/10.1186/s13662-017-1282-3
32. P. Turchin, I. Hanski, Am. Nat. **149**, 842–874 (1997)
33. R.K. Upadhyay, S.R.K. Iyengar, Nonlinear Anal. Real World Appl. **6**, 509–530 (2005)
34. V. Volterra, Hum. Biol. **10**, 1–11 (1938)
35. S. Zhang, L. Chen, Chaos. Solitons Fractals **24**, 1269–1278 (2005)

Chapter 8
Linear Boundary Value Problems

We begin this chapter by showing that solutions of second-order linear boundary value problems exist (and analogously for the higher order) if and only if corresponding systems of algebraic equations of second (higher) order have solutions. Then, we shall use the method of *separation of variables* also known as the *Fourier method* (after Jean Baptiste Joseph Fourier (1768–1830)) to solve some basic partial differential equations. An important feature of this method is that it leads to finding the solutions of boundary value problems for ordinary differential equations. We shall also consider problems in infinite domains which can be effectively solved by finding the Fourier transform, or the Fourier sine or cosine transform of the unknown function.

Consider the second-order linear differential equation

$$p_0(t)y'' + p_1(t)y' + p_2(t)y = r(t), \tag{8.1}$$

where the functions $p_0, p_1, p_2,$ and r are continuous in the interval $J = [a, b]$. Together with the differential equation (8.1) we shall consider the boundary conditions of the form

$$\begin{aligned} \ell_1[y] &= a_0 y(a) + a_1 y'(a) + b_0 y(b) + b_1 y'(b) = A \\ \ell_2[y] &= c_0 y(a) + c_1 y'(a) + d_0 y(b) + d_1 y'(b) = B, \end{aligned} \tag{8.2}$$

where $a_i, b_i, c_i, d_i, i = 0, 1$ and A, B are given constants. In what follows, we shall assume that these are essentially two conditions, i.e., there does not exist a constant c such that $(a_0 \, a_1 \, b_0 \, b_1) = c(c_0 \, c_1 \, d_0 \, d_1)$. Boundary value problem (8.1), (8.2) is called a nonhomogeneous two point boundary value problem, whereas the homogeneous differential equation

$$p_0(t)y'' + p_1(t)y' + p_2(t)y = 0, \tag{8.3}$$

together with the homogeneous boundary conditions

© Springer Nature Switzerland AG 2019
R. P. Agarwal et al., *500 Examples and Problems of Applied Differential Equations*, Problem Books in Mathematics,
https://doi.org/10.1007/978-3-030-26384-3_8

$$\ell_1[y] = 0, \quad \ell_2[y] = 0 \tag{8.4}$$

will be called a homogeneous boundary value problem.

Boundary conditions (8.2) are quite general and in particular include the
(i) first boundary conditions also known as *Dirichlet conditions* (after Peter Gustav Lejeune Dirichlet (1805–1859))

$$y(a) = A, \quad y(b) = B; \tag{8.5}$$

(ii) second boundary conditions

$$y(a) = A, \quad y'(b) = B, \tag{8.6}$$

or

$$y'(a) = A, \quad y(b) = B; \tag{8.7}$$

(iii) separated boundary conditions akso known as *Sturm–Liouville conditions* (after Jacques Charles Francois Sturm (1803–1855) and Joseph Liouville (1809–1882))

$$\begin{aligned} a_0 y(a) + a_1 y'(a) &= A \\ d_0 y(b) + d_1 y'(b) &= B, \end{aligned} \tag{8.8}$$

where $a_0^2 + a_1^2 \neq 0$ and $d_0^2 + d_1^2 \neq 0$; and
(iv) periodic boundary conditions

$$y(a) = y(b), \quad y'(a) = y'(b). \tag{8.9}$$

Boundary value problem (8.1), (8.2) is called *regular* if both a and b are finite, and the function $p_0(t) \neq 0$ for all t in J. If $a = -\infty$ and/or $b = \infty$ and/or $p_0(t) = 0$ for at least one point t in J, then problems (8.1), (8.2) are said to be *singular*. By a *solution* of the boundary value problem (8.1), (8.2), we mean a solution of the differential equation (8.1) satisfying the boundary conditions (8.2). The existence and uniqueness theory even for the regular boundary value problems is more difficult than that of initial-value problems. In fact, in the case of boundary value problems a slight change in boundary conditions can lead to significant change in the behavior of the solutions. For example, the initial-value problem $y'' + y = 0$, $y(0) = c_1$, $y'(0) = c_1$ has a unique solution $y(t) = c_1 \cos(t) + c_2 \sin(t)$ for any set of values c_1, c_2. However, the boundary value problem $y'' + y = 0$, $y(0) = 0$, $y(\pi) = \epsilon(\neq 0)$ has no solution; the problem $y'' + y = 0$, $y(0) = 0$, $y(b) = \epsilon$, $0 < b < \pi$ has a unique solution $y(t) = \epsilon \sin(t)/\sin(b)$; while the problem $y'' + y = 0$, $y(0) = 0$, $y(\pi) = 0$ has an infinite number of solutions $y(t) = c \sin(t)$, where c is an arbitrary constant.

Obviously, for the homogeneous problem (8.3), (8.4) trivial solution always exists. However, from the above example, it follows that besides trivial solution homogeneous boundary value problems may have nontrivial solutions also. To find a necessary and sufficient condition so that the problem (8.3), (8.4) has only the trivial solution, we note that any solution $y(t)$ of (8.3) in terms of two linearly independent solutions $y_1(t)$, $y_2(t)$ of (8.3) can be written as

$$y(t) = c_1 y_1(t) + c_2 y_2(t).$$

This is a solution of the problem (8.3), (8.4) if and only if

$$\ell_1[c_1 y_1 + c_2 y_2] = c_1 \ell_1[y_1] + c_2 \ell_2[y_2] = 0 \\ \ell_2[c_1 y_1 + c_2 y_2] = c_1 \ell_2[y_1] + c_2 \ell[y_2] = 0. \tag{8.10}$$

Clearly, the system (8.10) has only the trivial solution if and only if

$$\Delta = \begin{vmatrix} \ell_1[y_1] & \ell_1[y_2] \\ \ell_2[y_1] & \ell_2[y_2] \end{vmatrix} \neq 0. \tag{8.11}$$

We summarize this simple observation in the following theorem.

Theorem 8.1 *Let* $y_1(t)$, $y_2(t)$ *be any two linearly independent solutions of the differential equation (8.3). Then, the homogeneous boundary value problem (8.3), (8.4) has*
(i) only the trivial solution if and only if $\Delta \neq 0$,
(ii) an infinite number of nontrivial solutions if and only if $\Delta = 0$.

Now again let $y_1(t)$, $y_2(t)$ be any two linearly independent solutions of the homogeneous differential equation (8.3) and $v(t)$ be a particular solution of the nonhomogeneous differential equation (8.1), then the general solution of (8.1) can be written as

$$y(t) = c_1 y_1(t) + c_2 y_2(t) + v(t). \tag{8.12}$$

Clearly, this is a solution of the problem (8.1), (8.2) if and only if

$$\ell_1[c_1 y_1 + c_2 y_2 + v] = c_1 \ell_1[y_1] + c_2 \ell_1[y_2] + \ell_1[v] = A \\ \ell_2[c_1 y_1 + c_2 y_2 + v] = c_1 \ell_2[y_1] + c_2 \ell_2[y_2] + \ell_2[v] = B. \tag{8.13}$$

The nonhomogeneous system (8.13) has a unique solution if and only if $\Delta \neq 0$, *i.e., if and only if the homogeneous system (8.10) has only the trivial solution. From Theorem 8.1,* $\Delta \neq 0$ *is equivalent to the homogeneous boundary value problem (8.3), (8.4) having only the trivial solution. We state this result in the following theorem.*

Theorem 8.2 *The nonhomogeneous boundary value problem (8.1), (8.2) has a unique solution if and only if the homogeneous boundary value problem (8.3), (8.4) has only the trivial solution.*

To apply the method of separation of variables for partial differential equations, we need the following concepts.

Definition 8.1 A function $f(x)$ is called *piecewise continuous* in an interval $\alpha < x < \beta$ if there are finitely many points $\alpha = x_0 < x_1 < \cdots < x_n = \beta$ such that

(i) $f(x)$ is continuous in each subinterval $(x_0, x_1), (x_1, x_2) \cdots , (x_{n-1}, x_n)$, and
(ii) on each subinterval (x_{k-1}, x_k) both right-hand limit $f(x_{k-1}+)$ and left-hand limit $f(x_k-)$ exist, i.e., are finite.

Note that the function $f(x)$ need not be defined at the points x_k.

Definition 8.2 A function $f(x)$ in an interval $\alpha < x < \beta$ is said to be *piecewise smooth* if both $f(x)$ and $f'(x)$ are piecewise continuous in $\alpha < x < \beta$.

Note that $f'(x)$ is piecewise smooth means that $f'(x)$ is continuous except at $s_0, \cdots , s_m$ (these points include $x_0, \cdots , x_n$ where $f(x)$ is not continuous) and in each subinterval (s_{k-1}, s_k) both $f'(s_{k-1}+)$ and $f'(s_k-)$ exist.

Any function $f(x)$ defined in $-\pi < x < \pi$ can be extended to a periodic function $F(x)$ in $\mathbb{R}$ with period 2π. The function $F(x)$ is called the *periodic extension* of $f(x)$. If a function $f(x)$, $-\pi \leq x \leq \pi$ can be extended to a periodic function with period 2π, we must have $f(-\pi) = f(\pi)$. If a function $g(x)$ is *odd*, i.e., $g(-x) = -g(x)$ then $\int_{-\alpha}^{\alpha} g(x)dx = 0$, and if $g(x)$ is *even*, i.e., $g(-x) = g(x)$ then $\int_{-\alpha}^{\alpha} g(x)dx = 2\int_{0}^{\alpha} g(x)dx$. Further, if $g(x)$ is defined only in $(0, \alpha)$ we can extend it as follows:

$$g_o(x) = \begin{cases} g(x), & 0 < x < \alpha \\ -g(-x), & -\alpha < x < 0. \end{cases}$$

Clearly, $g_o(x)$ is an odd function, and hence it is called an *odd extension* of $g(x)$. We can also extend $g(x)$ as

$$g_e(x) = \begin{cases} g(x), & 0 < x < \alpha \\ g(-x), & -\alpha < x < 0. \end{cases}$$

Since, $g_e(x)$ is even it is called an *even extension* of $g(x)$.

Clearly, the expression

$$\frac{a_0}{2} + \sum_{n=1}^{\infty} [a_n \cos(nx) + b_n \sin(nx)] \tag{8.14}$$

is periodic of period 2π. For a given piecewise continuous function $f(x)$ in $-\pi < x < \pi$, the series (8.14) with the constants

$$a_n = \frac{1}{\pi} \int_{-\pi}^{\pi} f(t) \cos(nt)dt, \quad n \geq 0$$

$$b_n = \frac{1}{\pi} \int_{-\pi}^{\pi} f(t) \sin(nt)dt, \quad n \geq 1. \tag{8.15}$$

is called the *Fourier series*, and until the equality is proved the correspondence is written as

$$f(x) \sim \frac{a_0}{2} + \sum_{n=1}^{\infty} [a_n \cos(nx) + b_n \sin(nx)], \tag{8.16}$$

Extending a given piecewise continuous function $f(x)$ in $0 < x < \pi$ to an even function in $-\pi < x < \pi$, the series (8.16) reduces to the *Fourier cosine series*

$$f(x) \sim \frac{a_0}{2} + \sum_{n=1}^{\infty} a_n \cos(nx), \tag{8.17}$$

where

$$a_n = \frac{2}{\pi} \int_0^{\pi} f(t) \cos(nt)dt, \quad n \geq 0. \tag{8.18}$$

Similarly, extending a given piecewise continuous function $f(x)$ in $0 < x < \pi$ to an odd function in $-\pi < x < \pi$, the series (8.16) reduces to the *Fourier sine series*

$$f(x) \sim \sum_{n=1}^{\infty} b_n \sin(nx), \tag{8.19}$$

where

$$b_n = \frac{2}{\pi} \int_0^{\pi} f(t) \sin(nt)dt, \quad n \geq 1. \tag{8.20}$$

Theorem 8.3 *Let $f(x)$ be piecewise smooth in the interval $[-\pi, \pi]$. Then, the Fourier series of $f(x)$ converges to $[f(x+) + f(x-)]/2$ at each point in the open interval $(-\pi, \pi)$ and at $x = \pm\pi$ the series converges to $[f(-\pi+) + f(\pi-)]/2$.*

Theorem 8.4 *Let $f(x)$ be piecewise smooth in the interval $[0, \pi]$. Then, the Fourier cosine series of $f(x)$ converges to $[f(x+) + f(x-)]/2$ at each point in the open interval $(0, \pi)$ and at $x = 0$ and π the series converges respectively to $f(0+)$ and $f(\pi-)$.*

Theorem 8.5 *Let $f(x)$ be piecewise smooth in the interval $[0, \pi]$. Then, the Fourier sine series of $f(x)$ converges to $[f(x+) + f(x-)]/2$ at each point in the open interval $(0, \pi)$ and at $x = 0$ and π the series converges to 0.*

Theorem 8.6 *If the series $\sum_{n=1}^{\infty}(|a_n| + |b_n|)$ converges, then the Fourier series of $f(x)$ converges uniformly in the interval $-\pi \leq x \leq \pi$ to $f(x)$ and, in fact, in the whole interval $-\infty < x < \infty$.*

Theorem 8.7 *If $f(x)$ is given in $-\pi < x < \pi$, and if $f(x)$ is continuous and bounded and has a piecewise continuous derivative, and if $f(-\pi+) = f(\pi-)$, then the Fourier series of $f(x)$ converges uniformly to $f(x)$ in the interval $-\pi \leq x \leq \pi$. The series converges to $f(\pi-) = f(-\pi+)$ at $x = \pm\pi$.*

Theorem 8.8 *If $f(x)$ is given in $0 < x < \pi$, and if $f(x)$ is continuous and bounded and has a piecewise continuous derivative, then the Fourier cosine series of $f(x)$ converges uniformly to $f(x)$ in the interval $0 \leq x \leq \pi$. The series converges to $f(0+)$ at $x = 0$ and to $f(\pi-)$ at $x = \pi$.*

Theorem 8.9 *If $f(x)$ is given in $0 < x < \pi$, if $f(x)$ is continuous and bounded and has a piecewise continuous derivative, and if $f(0+) = f(\pi-) = 0$, then the Fourier sine series of $f(x)$ converges uniformly to $f(x)$ in the interval $0 \leq x \leq \pi$. The series converges to 0 at $x = 0$ and $x = \pi$.*

For the differentiation and integration of Fourier trigonometric series, we have the following results.

Theorem 8.10 *Suppose $f(x)$ is continuous in $-\pi < x < \pi$, $f(-\pi) = f(\pi)$ and $f'(x)$ is piecewise continuous. Then,*

$$f'(x) = \sum_{n=1}^{\infty} [-na_n \sin(nx) + nb_n \cos(nx)]$$

at each point $x \in (-\pi, \pi)$ where $f''(x)$ exists.

Theorem 8.11 *Suppose $f(x)$ is piecewise continuous in $-\pi < x < \pi$ and the correspondence (8.16) holds. Then, for $x \in [-\pi, \pi]$*

$$\int_{-\pi}^{x} f(t)dt = \int_{-\pi}^{x} \frac{a_0}{2} dt + \sum_{n=1}^{\infty} \int_{-\pi}^{x} [a_n \cos(nt) + b_n \sin(nt)] dt$$

$$= \frac{a_0}{2}(x + \pi) + \sum_{n=1}^{\infty} \frac{1}{n} \left[a_n \sin(nx) - b_n \left[\cos(nx) - (-1)^n \right] \right].$$

If the piecewise continuous function $f(x)$ is defined in $-a < x < a$, instead of $-\pi < x < \pi$, then its Fourier trigonometric series (8.16) takes the form

$$f(x) \sim \frac{a_0}{2} + \sum_{n=1}^{\infty} \left[a_n \cos\left(\frac{n\pi x}{a}\right) + b_n \sin\left(\frac{n\pi x}{a}\right) \right], \tag{8.21}$$

where

$$\begin{aligned} a_n &= \frac{1}{a} \int_{-a}^{a} f(t) \cos\left(\frac{n\pi t}{a}\right) dt, \quad n \geq 0 \\ b_n &= \frac{1}{a} \int_{-a}^{a} f(t) \sin\left(\frac{n\pi t}{a}\right) dt, \quad n \geq 1. \end{aligned} \tag{8.22}$$

Theorem 8.12 *Let $f(x)$, $-\infty < x < \infty$ be piecewise continuous on each finite interval, and $\int_{-\infty}^{\infty} |f(x)| < \infty$. Then, the complex Fourier transform of $f(x)$ defined by*

$$F(\omega) = \frac{1}{\sqrt{2\pi}} \int_{-\infty}^{\infty} f(x) e^{-i\omega x} dx \tag{8.23}$$

exists. Further, at each x,

$$\frac{1}{\sqrt{2\pi}} \int_{-\infty}^{\infty} F(\omega) e^{i\omega x} d\omega = \frac{1}{2} [f(x+0) + f(x-0)].$$

The representation

$$f(x) = \frac{1}{\sqrt{2\pi}} \int_{-\infty}^{\infty} F(\omega)e^{i\omega x} d\omega \tag{8.24}$$

is called the *inverse Fourier transform* of $F(\omega)$. Equations (8.23) and (8.24) are called *Fourier transform pair*. The following properties of Fourier transform are immediate.

(P1) (Linearity Property). For arbitrary functions $f(x)$ and $g(x)$ whose Fourier transforms exist and arbitrary constants a and b, $F(af + bg) = aF(f) + bF(g)$.

(P2) (Transform of the Derivative). Let $f(x)$ be continuous on the x-axis and $f(x) \to 0$ as $|x| \to \infty$. Furthermore, let $f'(x)$ be absolutely integrable on the x-axis. Then, $F(f'(x)) = i\omega F(f)$. It is clear that $F(f'') = i\omega F(f') = (i\omega)^2 F(f) = -\omega^2 F(f)$.

(P3) (Convolution Theorem). Suppose that $f(x)$ and $g(x)$ are piecewise continuous, bounded, and absolutely integrable functions on the x-axis. Then,

$$F(f * g) = \sqrt{2\pi}F(f)F(g), \tag{8.25}$$

where $f * g$ is the convolution of functions f and g defined as

$$(f * g)(x) = \int_{-\infty}^{\infty} f(t)g(x - t)dt = \int_{-\infty}^{\infty} f(x - t)g(t)dt. \tag{8.26}$$

By taking the inverse Fourier transform on both sides of (8.25) and writing $F(f) = \hat{f}(\omega)$ and $F(g) = \hat{g}(\omega)$, and noting that $\sqrt{2\pi}$ and $1/\sqrt{2\pi}$ cancel each other, we get

$$(f * g)(x) = \int_{-\infty}^{\infty} \hat{f}(\omega)\hat{g}(\omega)e^{i\omega x} d\omega. \tag{8.27}$$

We will also need the Fourier transform of $f(x) = e^{-ax^2}$, $a > 0$. For this, we have

$$\begin{aligned}
F(\omega) &= \frac{1}{\sqrt{2\pi}} \int_{-\infty}^{\infty} e^{[-ax^2 - i\omega x]} dx \\
&= \frac{1}{\sqrt{2\pi}} \int_{-\infty}^{\infty} \exp\left[-\left(\sqrt{a}x + \frac{i\omega}{2\sqrt{a}}\right)^2 + \left(\frac{i\omega}{2\sqrt{a}}\right)^2\right] dx \\
&= \frac{1}{\sqrt{2\pi}} \exp\left(-\frac{\omega^2}{4a}\right) \int_{-\infty}^{\infty} \exp\left[-\left(\sqrt{a}x + \frac{i\omega}{2\sqrt{a}}\right)^2\right] dx \quad (8.28) \\
&= \frac{1}{\sqrt{2\pi}} \exp\left(-\frac{\omega^2}{4a}\right) \int_{-\infty}^{\infty} \exp\left(-v^2\right) \frac{dv}{\sqrt{a}}, \quad \sqrt{a}x + \frac{i\omega}{2\sqrt{a}} = v \\
&= \frac{1}{\sqrt{2\pi}} \exp\left(-\frac{\omega^2}{4a}\right) \frac{1}{\sqrt{a}}\sqrt{\pi} = \frac{1}{\sqrt{2a}}e^{-\omega^2/4a}.
\end{aligned}$$

From (8.23), (8.24) the *Fourier cosine transform pair* (taking $f(x)$ as even)

$$F_c(\omega) = \sqrt{\frac{2}{\pi}} \int_0^\infty f(x) \cos(\omega x) dx \tag{8.29}$$

$$f(x) = \sqrt{\frac{2}{\pi}} \int_0^\infty F_c(\omega) \cos(\omega x) d\omega \tag{8.30}$$

and the *Fourier sine transform pair* (taking $f(x)$ as odd)

$$F_s(\omega) = \sqrt{\frac{2}{\pi}} \int_0^\infty f(x) \sin(\omega x) dx \tag{8.31}$$

$$f(x) = \sqrt{\frac{2}{\pi}} \int_0^\infty F_s(\omega) \sin(\omega x) d\omega \tag{8.32}$$

follow immediately.

Example 8.1 One of the most commonly used elements in structures such as aircraft, buildings, ships, and bridges is the elastic beam. A beam is a long slender rod that is supported at either or both ends and is subjected to external forces or displacements along its length and possibly at its ends. Beams are used for girders and for wall and floor supports in building construction. A modern airplane contains thousands of beams in its fuselage, wing, and tail structures. The rails of the railroad track are examples of beams that are supported along their entire length by the roadbed (the ties are considered as parts of the roadbed). Differential equations considered in Problems 3.35, 3.38, 3.39 appear in beam analysis. In fact, for simplicity, we consider a mechanical model which involves the deflection of a long beam that is supported at one or both ends, as shown in Fig. 8.1.

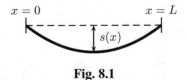

Fig. 8.1

Assuming that in its undeflected form the beam is horizontal, then the deflection of the beam can be expressed as a function of x. Suppose that the shape of the beam when it is deflected is given by the graph of the function $y(x) = -s(x)$, where x is the distance from the left end of the beam and s the measure of the vertical deflection from the equilibrium position. The boundary value problem that models this situation is derived as follows.

Let $m(x)$ equal the turning moment of the force relative to the point x, and $w(x)$ represent the weight distribution of the beam. These two functions are related by the equation

$$\frac{d^2 m}{dx^2} = w(x).$$

Also, the turning moment is proportional to the curvature of the beam. Hence,

$$m(x) = \frac{EI}{\left(1 + \left(\frac{dy}{dx}\right)^2\right)^{3/2}} \frac{d^2 y}{dx^2},$$

where E and I are constants related to the composition of the beam and the shape and size of a cross section of the beam, respectively. Notice that this equation is nonlinear. This difficulty is overcome with an approximation. For small values of y, the denominator of the right-hand side of the equation can be approximated by the constant 1. Therefore, the equation is simplified to

$$m(x) = EI \frac{d^2 y}{dx^2}.$$

This equation is linear and can be differentiated twice to obtain

$$\frac{d^2 m}{dx^2} = EI \frac{d^4 y}{dx^4},$$

which is then used with the equation above relating $m(x)$ and $w(x)$ to obtain the single fourth-order linear nonhomogeneous differential equation

$$EI \frac{d^4 y}{dx^4} = w(x). \tag{8.33}$$

Boundary conditions for this problem may vary. In most cases, two conditions are given for each end of the beam. Some of these conditions that are specified in pairs are

Fixed (embedded) end: $y = 0$, $y' = 0$
Free end: $y'' = 0$, $y''' = 0$
Simple support (pinned) end: $y = 0$, $y'' = 0$
Sliding clamped end: $y' = 0$, $y''' = 0$.
In particular, if the beam is embedded at both ends and a load w_0 is uniformly distributed along its length, i.e., in (8.33), $w(x) = w_0$, $0 < x < L$, then it leads to the boundary value problem

$$EI\frac{d^4y}{dx^4} = w_0 \tag{8.34}$$
$$y(0) = 0, \quad y'(0) = 0, \quad y(L) = 0, \quad y'(L) = 0.$$

The general solution of the differential equation in (8.34) can be written as

$$y(x) = c_1 + c_2 x + c_3 x^2 + c_4 x^3 + \frac{w_0}{24EI} x^4.$$

This solution satisfies the boundary conditions if and only if

$$c_1 = 0$$
$$c_2 = 0$$
$$c_3 L^2 + c_4 L^3 + \frac{w_0}{24EI} L^4 = 0$$
$$2c_3 L + 3c_4 L^2 + \frac{w_0}{6EI} L^3 = 0,$$

which gives $c_1 = 0$, $c_2 = 0$, $c_3 = w_0 L^2/(24EI)$, $c_4 = -w_0 L/(12EI)$. Thus, the solution of the problem (8.34) is

$$y(x) = \frac{w_0 L^2}{24EI} x^2 - \frac{w_0 L}{12EI} x^3 + \frac{w_0}{24EI} x^4 = \frac{w_0}{24EI} x^2 (L - x)^2.$$

Hence, the maximum deflection of the beam is $y(L/2) = w_0 L^4/(384EI)$.

Now assume that the beam is embedded at both ends and a load w_0 is uniformly distributed on its central half length. In this case, the differential equation takes the form

$$EI\frac{d^4y}{dx^4} = \begin{cases} 0, & 0 \le x < L/4 \\ w_0, & L/4 \le x \le 3L/4 \\ 0, & 3L/4 < x \le L. \end{cases}$$

The solution of this differential equation satisfying the embedded conditions appears as

$$y(x) = \frac{1}{EI} \begin{cases} C_1 x^2 + D_1 x^3, & 0 \le x \le L/4 \\ A_2 + B_2 x + C_2 x^2 + D_2 x^3 + w_0 x^4/24, & L/4 \le x \le 3L/4 \\ C_3 (L - x)^2 + D_3 (L - x)^3, & 3L/4 \le x \le L. \end{cases}$$

In this solution first the unknowns A_2, B_2, C_2, D_2 are determined by using the conditions $y(L/4) = y(3L/4)$, $y'(L/2) = 0$, $y'(L/4) = -y'(3L/4)$, $y''(L/4) = y''(3L/4)$ and then the unknowns C_1, D_1, C_3, D_3 are calculated by using the fact that the solution is continuously differentiable at $L/4$ and $3L/4$. This gives

$$y(x) = \frac{w_0}{24EI} \begin{cases} \dfrac{11}{16}L^2x^2 - Lx^3, & 0 \le x \le L/4 \\[2mm] \dfrac{L^4}{256} - \dfrac{1}{16}L^3x + \dfrac{17}{16}L^2x^2 - 2Lx^3 + x^4, & L/4 \le x \le 3L/4 \\[2mm] \dfrac{11}{16}L^2(L-x)^2 - L(L-x)^3, & 3L/4 \le x \le L. \end{cases}$$

Thus, the maximum deflection of the beam is $y(L/2) = 13w_0L^4/(6144EI)$.

Finally, we note that if the beam is resting on an elastic foundation, then the differential equation (8.33) must be modified to

$$EIy^{(4)} + ky = w(x), \tag{8.35}$$

where k is called the spring constant of the elastic foundation. If the beam is subjected to an axial force T, then the differential equation is

$$EIy^{(4)} - Ty'' = w(x). \tag{8.36}$$

Here T is positive if the force acts to stretch the beam.

Example 8.2 In applications the eigenvalues and eigenfunctions have important physical interpretations. For example, in quantum mechanics eigenvalues denote the possible energy states of a system and the eigenfunctions are the wave functions. In buckling problems, the eigenvalues represent the critical compressive loads that a column can withstand without buckling and the eigenfunctions provide the possible buckling configurations of the column. Further, in vibration problems the eigenvalues are proportional to the squares of the natural frequencies of vibration, while the eigenfunctions give the natural configuration modes of the system.

Columns were used extensively in Greek and Roman structures. They were designed according to empirical formulas developed by the architects of these ancient civilizations. It was Euler in the eighteenth century whose study of columns that engineers considered rational and efficient procedure for their design. Euler considered the buckling of a long and slender elastic rod that is deformed by compressive forces P acting axially on the ends of the rod; see Fig. 8.2. Slender rods are conventionally used as columns in civil, aircraft, and marine structures. For example, in a steel-framed building the vertical columns support the weight and loads of the structure above them. This weight induces the force P on the ends of the column. If P is small, then the column deflects only slightly from its natural straight state. For a sufficiently large value of P, the column suddenly bows out of the straight state, with deflections of large amplitude. This is called buckling. Once the column buckles it is unable to support the load. Clearly, the buckling of a single column in a building can have a disastrous effect on the entire structure.

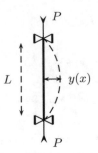

Fig. 8.2

If the ends of the column are simply supported then the eigenvalue problem derived by Euler to describe the buckling of the rod is

$$y'' + \lambda y = 0 \tag{8.37}$$

$$y(0) = y(1) = 0. \tag{8.38}$$

Here $y(x)$ is the lateral displacement of the column, see Fig. 8.2, and the constant

$$\lambda = \frac{P}{EI}. \tag{8.39}$$

We have taken the length of the column $L = 1$ for simplicity. Here, $E > 0$ is the elastic modulus (Young's modulus) of the rod and $I > 0$ is the moment of inertia of the rod's cross section. In Eq. (8.39), the applied force P and, consequently, λ are positive for compression and negative for tension. Compression means that the column is squeezed by the force. Tension stretches the column. We wish to determine the values of P and, consequently, of λ (eigenvalues) for which the problem (8.37), (8.38) has nonzero solutions (eigenfunctions). These nonzero solutions will be the bowed shapes of the column.

If $\lambda = 0$, then the general solution of (8.37) (reduced to $y'' = 0$) is $y(x) = c_1 + c_2 x$ and this solution satisfies the boundary conditions (8.38) if and only if $c_1 = c_2 = 0$, i.e., $y(x) \equiv 0$ is the only solution of (8.37), (8.38). Hence, $\lambda = 0$ is not an eigenvalue of the problem (8.37), (8.38).

If $\lambda \neq 0$, it is convenient to replace λ by μ^2, where μ is a new parameter, for mathematical generality, assumed not necessarily real. In this case the general solution of (8.37) is $y(x) = c_1 e^{i\mu x} + c_2 e^{-i\mu x}$, and this solution satisfies the boundary conditions (8.38) if and only if

$$c_1 + c_2 = 0$$
$$c_1 e^{i\mu} + c_2 e^{-i\mu} = 0. \tag{8.40}$$

The system (8.40) has a nontrivial solution if and only if

$$e^{-i\mu} - e^{i\mu} = 0. \tag{8.41}$$

If $\mu = a + ib$, where a and b are real, condition (8.41) reduces to

$$e^{b}[\cos(a) - i \sin(a)] - e^{-b}[\cos(a) + i \sin(a)] = 0,$$

or

$$(e^{b} - e^{-b}) \cos(a) - i(e^{b} + e^{-b}) \sin(a) = 0,$$

or

$$2 \sinh(b) \cos(a) - 2i \cosh(b) \sin(a) = 0,$$

or

$$\sinh(b) \cos(a) = 0 \tag{8.42}$$

and

$$\cosh(b) \sin(a) = 0. \tag{8.43}$$

Since $\cosh(b) > 0$ for all values of b, Eq. (8.43) requires that $a = n\pi$, where n is an integer. Further, for this choice of a, $\cos(a) \neq 0$ and Eq. (8.42) reduces to $\sinh(b) = 0$, i.e., $b = 0$. However, if $b = 0$ then we cannot have $a = 0$, because then $\mu = 0$, and we have seen it is not an eigenvalue. Hence, $\mu = n\pi$ where n is a nonzero integer. Thus, the eigenvalues of (8.37), (8.38) are $\lambda_n = \mu^2 = (n\pi)^2$, $n = 1, 2, \cdots$. Further, from (8.40) since $c_2 = -c_1$ for $\lambda_n = (n\pi)^2$ the corresponding nontrivial solutions of the problem (8.37), (8.38) are

$$\phi_n(x) = c_1(e^{in\pi x} - e^{-in\pi x}) = 2ic_1 \sin(n\pi x),$$

or simply $\phi_n(x) = a_n \sin(n\pi x)$.

The trivial solution, $y(x) \equiv 0$, means that the column is straight. It is usually called the unbuckled equilibrium state. For $\lambda < 0$, as we have seen above, it is the only solution of (8.37), (8.38). Since $P < 0$ if $\lambda < 0$, the interpretation of this result is that a stretched column will not buckle.

For $\lambda \neq (n\pi)^2$, the trivial solution $y(x) \equiv 0$ is the only solution of the eigenvalue problem. For $\lambda = (n\pi)^2$, other solutions given by the eigenfunctions $y(x) = \phi_n(x) = a_n \sin(n\pi x)$ are possible states of equilibrium of the rod. They are called buckled states, since the rod is not straight. The eigenfunctions corresponding to the first three eigenvalues are sketched in Fig. 8.3.

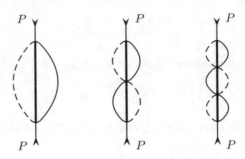

Fig. 8.3

For $\lambda < \pi^2$, the unbuckled state is the only possible equilibrium state of the rod. As P is increased from zero, the column will remain straight for $\lambda < \pi^2$. When λ reaches the lowest eigenvalue π^2, nontrivial solutions occur and the column will buckle. For this reason, $P = \pi^2 EI$ is called the buckling load of the road. The amplitudes a_n of the buckled states are not determined, since eigenfunctions are known only within arbitrary multiplicative constants. In engineering practice, rods usually deform with large amplitudes when the buckling load is reached. If the amplitude is sufficiently large, the material of the column weakens and the column breaks. Consequently, in the actual design and analysis of structures involving columns, engineers must design the columns so that $\lambda = P/(EI) < \pi^2$. For a fixed P, this can be achieved by making I sufficiently large. The moment of inertia is large for cross sections whose area is far from the centroid. This accounts for the widespread use of columns with cross sections resembling the letters I and H.

Example 8.3 Consider a string of length L with constant linear density ρ which is stretched along the x-axis and fixed at $x = 0$ and $x = L$. Suppose the string is then rotated about that axis at a constant angular speed ω. This is similar to two persons holding a jump rope and then twirling it in a synchronous manner. We shall find the differential equation which defines the shape (deflection curve away from its initial position) $y(x)$ of the string. For this, we consider a portion of the string on the interval $[x, x + \Delta x]$, where Δx is small. In what follows, for simplicity, we assume that the magnitude T of the tension $\mathbf{T}$ acting tangential to the string is constant along the string. Now from Fig. 8.4 it is clear that the net vertical force F acting on the string on the interval $[x, x + \Delta x]$ is

$$F = T\sin(\theta_2) - T\sin(\theta_1).$$

If the angles θ_1 and θ_2 (measured in radians) are small, then we have

$$\sin(\theta_2) \simeq \tan(\theta_2) \simeq y'(x + \Delta x) \quad \text{and} \quad \sin(\theta_1) \simeq \tan(\theta_1) \simeq y'(x)$$

and hence

$$F \simeq T[y'(x + \Delta x) - y'(x)]. \tag{8.44}$$

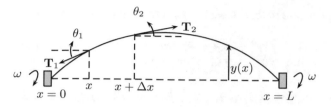

Fig. 8.4

The net force F can also be given by Newton's second law as $F = ma$. Clearly, the mass of the string on the interval $[x, x + \Delta x]$ is $m = \rho \Delta x$, and the centripetal acceleration of a point rotating with angular speed ω in a circle of radius r is $a = r\omega^2$. Since Δx is small, we can assume that $r = y$. Thus another formulation of the net force is

$$F \simeq -(\rho \Delta x)y\omega^2, \tag{8.45}$$

where the minus sign indicates the fact that the acceleration points in the direction opposite to the positive y direction. From (8.44) and (8.45), we get

$$T[y'(x + \Delta x) - y'(x)] \simeq -(\rho \Delta x)y\omega^2,$$

or

$$T \frac{y'(x + \Delta x) - y'(x)}{\Delta x} \simeq -\rho \omega^2 y,$$

which as $\Delta x \to 0$ leads to the differential equation

$$T \frac{d^2 y}{dx^2} = -\rho \omega^2 y. \tag{8.46}$$

Since the string is fixed at the ends, the solution $y(x)$ of (8.46) must also satisfy the boundary conditions $y(0) = 0$, $y(L) = 0$. Thus the shape of the string $y(x)$ can be determined by solving the boundary value problem

$$\frac{d^2 y}{dx^2} + \frac{\rho \omega^2}{T} y = 0, \quad y(0) = y(L) = 0. \tag{8.47}$$

Clearly, the problem (8.47) is exactly the same as the eigenvalue problem (8.37), (8.38).

Finally, we note that if the magnitude T of the tension is not constant throughout the interval $[0, L]$, then the boundary value problem which gives the deflection curve of the string is

$$\frac{d}{dx}\left(T(x)\frac{dy}{dx}\right) + \rho \omega^2 y = 0, \quad y(0) = y(L) = 0. \tag{8.48}$$

As a particular case of (8.48), we consider the singular eigenvalue problem

$$(1 - x^2)y'' - 2xy' + \lambda y = ((1 - x^2)y')' + \lambda y = 0$$
$$\lim_{x \to -1} y(x) < \infty, \quad \lim_{x \to 1} y(x) < \infty.$$

(8.49)

In view of Example 4.4, it is clear that for (8.49) the eigenvalues are $\lambda_n = n(n + 1)$, $n = 0, 1, 2, \cdots$ and the corresponding eigenfunctions are the Legendre polynomials $P_n(x)$.

Example 8.4 Consider the problem of a vertical column of uniform material and cross section, bent by its own weight. Let a long thin rod be set up in a vertical plane so that the lower end is constrained to remain vertical (see, Fig. 8.5). Suppose the rod is of length L and weight W and has the coefficient of flexural rigidity B (> 0). Then, if $p = dy/dx$, the equation describing this system can be written as (see [3, 6])

$$\frac{d^2 p}{dx^2} + \frac{W}{B}\frac{(L - x)}{L}p = 0$$

(8.50)

$$p(0) = 0 = p'(L).$$

(8.51)

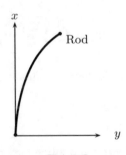

Fig. 8.5

One possibility that is always present is that the rod does not bend at all, which is just another way of saying that the problem has only the trivial solution, i.e., $p(x) \equiv 0$. One would expect that if the rod is short enough (just how short it would need to be depends upon the constants W, and B, of course) the rod *cannot* bend at all, which is to say that the trivial solution is the only solution of the problem, and the problem is accordingly said to be *stable*. However, for all sufficiently large L, the rod can bend and the problem has a nontrivial solution. Clearly, then uniqueness no longer holds for the boundary value problem.

Equation (8.50) can be transformed into Bessel's equation (4.41) by the substitution

$$\xi = \frac{2}{3}\left(\frac{W}{LB}\right)^{1/2}(L - x)^{3/2}, \quad p = \eta(L - x)^{1/2}.$$

(8.52)

In fact, it leads to the equation

$$\frac{d^2\eta}{d\xi^2} + \frac{1}{\xi}\frac{d\eta}{d\xi} + \left(1 - \frac{1}{9\xi^2}\right)\eta = 0, \tag{8.53}$$

whose solution can be written as

$$\eta(\xi) = AJ_{1/3}(\xi) + BJ_{-1/3}(\xi)$$

and hence the solution of (8.50) is

$$p(\xi) = (L - x)^{1/2}[AJ_{1/3}(\xi) + BJ_{-1/3}(\xi)]. \tag{8.54}$$

Now it a simple matter to see that $p'(L) = 0$ only if $A = 0$, and $p(0) = 0$ provided $J_{-1/3}(\xi) = 0$ at $\xi = (2L/3)(W/B)^{1/2}$. Since

$$J_{-1/3}(\xi) = 1 - \frac{1}{3 \cdot 2}\frac{L^2 W}{B} + \frac{1}{3 \cdot 6 \cdot 2 \cdot 5}\left(\frac{L^2 W}{B}\right)^2$$
$$+ \cdots + \frac{(-1)^n}{3 \cdot 6 \cdots (3n) \cdot 2 \cdot 5 \cdots (3n - 1)}\left(\frac{L^2 W}{B}\right)^n + \cdots \tag{8.55}$$

the rod remains stable provided L^2/B is less than the first zero, say, ξ_1 of (8.55). An easy computation shows that $\xi_1 = 7.84 \cdots$, and hence the rod will not bend by its own weight, i.e., remain stable provided

$$L < (2.80 \cdots)\left(\frac{B}{W}\right)^{1/2}. \tag{8.56}$$

In a similar situation, the following problem occurs:

$$\frac{d^2\phi}{ds^2} + \frac{R^2}{AC}(L - s)^2\phi = 0 \tag{8.57}$$

$$\phi(0) = 0 = \phi'(L). \tag{8.58}$$

For this problem, Eqs. (8.52)–(8.56) take the following form:

$$\xi = \frac{1}{2}\frac{R}{\sqrt{AC}}(L - x)^2, \quad \phi = \eta(L - s)^{1/2}, \tag{8.59}$$

$$\frac{d^2\eta}{d\xi^2} + \frac{1}{\xi}\frac{d\eta}{d\xi} + \left(1 - \frac{1}{16\xi^2}\right)\eta = 0, \tag{8.60}$$

$$\phi(\xi) = (L - s)^{1/2}[AJ_{1/4}(\xi) + BJ_{-1/4}(\xi)], \tag{8.61}$$

$$J_{-1/4}(\xi) = 1 - \frac{1}{2.6}\frac{R^2 L^4}{AC} + \frac{1}{2\cdot 4\cdot 6\cdot 14}\left(\frac{R^2 L^4}{AC}\right)^2$$

$$+\cdots+(-1)^n\frac{1}{2\cdot 4\cdots(2n)\cdot 6\cdot 14\cdots(8n-2)}\left(\frac{R^2 L^4}{AC}\right)^n+\cdots$$

$$\text{(8.62)}$$

$$L < \gamma\frac{(AC)^{1/4}}{\sqrt{R}}, \tag{8.63}$$

where γ is a number very close to 2.

Example 8.5 Consider a wedge-shaped canal of uniform depth L that empties into the open sea (see Fig. 8.6). Assume that the water level at the mouth of the canal varies harmonically, i.e., the depth at the mouth of the canal is given by $H\cos(\omega t)$, where H and ω are positive constants. This assumption simulates the motion of the tides. Now the function $h(x, t)$, which gives the depth at a distance x from the inland end of the canal at time t, has the form $h(x, t) = y(x)\cos(\omega t)$, where $y(x)$ satisfies the differential equation

$$x^2 y'' + x y' + k^2 x^2 y = 0, \tag{8.64}$$

where $k > 0$ is a constant.

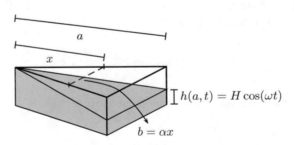

$$h(a, t) = H\cos(\omega t)$$

$$b = \alpha x$$

Fig. 8.6

Comparing (8.64) with (4.83), we find that its solution can be written as

$$y(x) = A J_0(kx) + B J^0(kx). \tag{8.65}$$

Since at $x = 0$, the depth of the water must be finite at all times, $\lim_{x\to 0} y(x)$ has to be finite. However, since $\lim_{x\to 0}|J^0(kx)| = \infty$, (from (4.47) as $x \to 0$, $J^0(x) \simeq \ln|x|$) we need to assume that $B = 0$. Hence, the depth $h(x, t)$ can be written as

$$h(x, t) = A J_0(kx)\cos(\omega t). \tag{8.66}$$

Next using the condition that the depth at the mouth of the canal is $H \cos \omega t$, we have

$$H \cos(\omega t) = h(a, t) = A J_0(ka) \cos(\omega t)$$

and hence $A = H/J_0(ka)$ provided $J_0(ka) \neq 0$. Thus, the depth h appears as

$$h(x, t) = H \frac{J_0(kx)}{J_0(ka)} \cos(\omega t). \tag{8.67}$$

Clearly, from (4.46) we have $J_0(0) = 1$ and hence $J_0(x) \neq 0$ at least for sufficiently small $x > 0$. Thus, the solution (8.67) is meaningful as long as ka is sufficiently small.

Example 8.6 Now in Example 8.5, we assume that the depth of the canal is not uniform, but varies according as $L(x) = \beta x$. Figure 8.7 shows the lengthwise cross section of the canal.

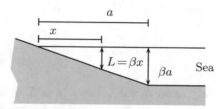

Fig. 8.7

Again we assume that the water level at the mouth of the canal varies harmonically. Then, the depth has the form $h(x, t) = y(x) \cos(\omega t)$, where y satisfies the differential equation

$$x^2 y'' + 2xy' + k^2 xy = 0; \tag{8.68}$$

here, $k > 0$ is a constant. Comparing (8.68) with (4.82), we find that its solution can be written as

$$y(x) = A x^{-1/2} J_1 \left(2kx^{1/2}\right) + B x^{-1/2} J_{-1} \left(2kx^{1/2}\right). \tag{8.69}$$

Since at $x = 0$ the depth of the water must be finite at all times, we need to assume that $B = 0$. At $t = 0$ we have $h(x, 0) = \beta x$. In particular, $\beta a = h(a, 0) = y(a) = a^{-1/2} A J_1(2ka^{1/2})$ and hence $A = \beta a^{3/2}/J_1(2ka^{1/2})$. Thus, the solution appears as

$$h(x, t) = \beta a^{3/2} x^{-1/2} \frac{J_1 \left(2kx^{1/2}\right)}{J_1 \left(2ka^{1/2}\right)} \cos(\omega t),$$

which is meaningful as long as $2ka^{1/2}$ is small.

Example 8.7 We recall that the fundamental principles involved in the problems of heat conduction are (1) Heat flows from a higher temperature to the lower temperature, (2) the quantity of heat in a body is proportional to its mass and temperature, and (3) the rate of heat flow across a surface is proportional to its area and to the rate of change of temperature with respect to its distance normal to the surface. Consider a homogeneous bar of uniform cross section S (cm^2). Suppose that the sides are covered with a material impervious to heat so that the stream lines of heat flow are all parallel and perpendicular to S. Take one end of the bar as the origin and the direction of flow as the positive x-axis (see Fig. 8.8). Let ρ be the density (gr/cm^3), s the specific heat (cal./gr. deg.), and k the thermal conductivity (cal./cm deg. sec.).

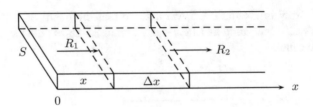

Fig. 8.8

Let $u(x, t)$ be the temperature at a distance x from 0. If Δu is the temperature change in a slab of thickness Δx of the bar then by the principle (2) the quantity of heat in this slab $= s\rho S \Delta x \Delta u$. Hence, the rate of increase of heat in this slab, i.e., $s\rho S \Delta x u_t = R_1 - R_2$, where R_1 and R_2 are, respectively, the rates (cal./sec.) of inflow and outflow of heat. Now since the rate of propagation of heat, (i.e., the quantity of heat passing through a cross-sectional area S with abscissa x in unit time) in view of (3) is given by *Fourier's Law* $q = -ku_x S$, where k is a constant depending upon the material of the body and called as the *thermal conductivity*, it follows that

$$R_1 = -kS \left(\frac{\partial u}{\partial x}\right)_x \quad \text{and} \quad R_2 = -kS \left(\frac{\partial u}{\partial x}\right)_{x+\Delta x} ;$$

here, the negative sign appears as a result of (1). Hence, we have

$$s\rho S \Delta x \frac{\partial u}{\partial t} = -kS \left(\frac{\partial u}{\partial x}\right)_x + kS \left(\frac{\partial u}{\partial x}\right)_{x+\Delta x},$$

which is the same as

$$\frac{\partial u}{\partial t} = \frac{k}{s\rho} \left\{ \frac{\left(\frac{\partial u}{\partial x}\right)_{x+\Delta x} - \left(\frac{\partial u}{\partial x}\right)_x}{\Delta x} \right\}.$$

Denoting the constant $k/(s\rho) = c^2$, known as the *diffusivity* of the substance (cm^2/s), and taking the limit as $\Delta x \to 0$, we obtain the *equation of heat conduction* in a homogeneous rod

$$\frac{\partial u}{\partial t} = c^2 \frac{\partial^2 u}{\partial x^2}, \quad 0 < x < a, \quad t > 0, \quad c > 0. \tag{8.70}$$

This equation also arises in the study of the filtration of liquids and gases in a porous medium, e.g., the filtration of oil and gas in subterranean sandstones, some problems in probability theory, biology to model cell physiology, chemical reactions, nerve impulses, the spread of populations, and so on.

For the solution of (8.70) to be definite, the function $u(x, t)$ must satisfy some initial and boundary conditions corresponding to the physical conditions of the problem. Let initially, i.e., when $t = 0$ a temperature be given in various cross sections of the rod equal to $f(x)$, which gives the initial condition

$$u(x, 0) = f(x), \quad 0 < x < a \tag{8.71}$$

and let for simplicity the ends of the rod, i.e., $x = 0$ and $x = a$, be held at zero temperature all the time, which gives the boundary conditions

$$u(0, t) = 0, \quad t > 0 \tag{8.72}$$

$$u(a, t) = 0, \quad t > 0. \tag{8.73}$$

These boundary conditions are of the *first kind* and known as *Dirichlet conditions*.

Now to solve the initial-boundary value problem (8.70)–(8.73), we shall use the method of separation of variables. For this, we assume a solution of (8.70) of the form $u(x, t) = X(x)T(t) \neq 0$, so that $X(x)T'(t) - c^2 X''(x)T(t) = 0$, or

$$\frac{T'(t)}{T(t)} = c^2 \frac{X''(x)}{X(x)} = \lambda,$$

which leads to the ordinary differential equations

$$X'' - \frac{\lambda}{c^2} X = 0 \tag{8.74}$$

and

$$T' - \lambda T = 0. \tag{8.75}$$

The boundary condition (8.72) demands that $X(0)T(t) = 0$ for all $t \geq 0$, thus $X(0) = 0$. Similarly, the boundary condition (8.73) requires that $X(a)T(t) = 0$ and hence $X(a) = 0$. Thus, the function X has to be a solution of the eigenvalue problem (8.74),

$$X(0) = 0, \quad X(a) = 0. \tag{8.76}$$

From Example 8.2, it is clear that the eigenvalues and eigenfunctions of (8.74), (8.76) are

$$\lambda_n = -\frac{n^2\pi^2c^2}{a^2}, \quad n = 1, 2, \cdots \tag{8.77}$$

$$X_n(x) = \sin\left(\frac{n\pi x}{a}\right), \quad n = 1, 2, \cdots. \tag{8.78}$$

With λ given by (8.77), Eq. (8.75) takes the form

$$T' + \frac{n^2\pi^2c^2}{a^2}T = 0$$

whose general solution appears as

$$T_n(t) = c_n e^{-(n^2\pi^2c^2/a^2)t},$$

where c_n is an arbitrary constant.

We conclude that for each specific value of n $(n = 1, 2, \cdots)$ the function $X_n(x)T_n(t)$ is a solution of (8.70) that satisfies conditions (8.72) and (8.73). Now the condition (8.71), when $u(x, t) = X_n(x)T_n(t)$ with n not specified, but otherwise considered fixed, is satisfied provided $X_n(x)T_n(0) = f(x)$, i.e.,

$$\sin\left(\frac{n\pi x}{a}\right)c_n = f(x).$$

But, the only way this can happen is that $f(x)$ to be restricted to the form $A\sin(n\pi x/a)$, where A is a constant. This places too great a restriction on the permissible forms of f, therefore we consider an alternative approach. Since $X_n(x)T_n(t)$ is a solution of (8.70) for each value of n $(n = 1, 2, \cdots)$ and since (8.70) is a linear partial differential equation, it seems reasonable to expect that $\sum_{n=1}^{\infty} X_n(x)T_n(t)$ is a solution of (8.70). Naturally, the question of whether or not this infinite series converges is always there. We will not investigate this question here, but rather emphasize the method of solution. Thus, we consider

$$u(x, t) = \sum_{n=1}^{\infty} X_n(x)T_n(t) = \sum_{n=1}^{\infty} c_n e^{-(n^2\pi^2c^2/a^2)t} \sin\left(\frac{n\pi x}{a}\right) \tag{8.79}$$

as a solution of (8.70) that satisfies conditions (8.72) and (8.73). Now condition (8.71) is satisfied if and only if

$$\sum_{n=1}^{\infty} c_n \sin\left(\frac{n\pi x}{a}\right) = f(x),$$

i.e., the Fourier sine series for $f(x)$ in the interval $0 \le x \le a$ must be $\sum_{n=1}^{\infty} c_n \sin(n\pi x/a)$. Consequently, c_n is given by

$$c_n = \frac{2}{a} \int_0^a f(x) \sin\left(\frac{n\pi x}{a}\right) dx, \quad n = 1, 2, \cdots. \tag{8.80}$$

Hence, the solution of (8.70)–(8.71) can be written as (8.79) where c_n is given by (8.80).

Next we assume that the ends of the rod, i.e., $x = 0$ and $x = a$, are insulated, which gives the boundary conditions

$$u_x(0, t) = 0, \quad t > 0 \tag{8.81}$$

$$u_x(a, t) = 0, \quad t > 0. \tag{8.82}$$

These boundary conditions are of the *second kind* and known as *Neumann conditions*.

Clearly in this case also, we have the same differential equations (8.74) and (8.75); however, instead of (8.76) the new boundary conditions are

$$X'(0) = 0, \quad X'(a) = 0 \tag{8.83}$$

For (8.74), (8.83), the eigenvalues and eigenfunctions are

$$\lambda_n = -\frac{n^2\pi^2 c^2}{a^2}, \quad n = 0, 1, \cdots \tag{8.84}$$

$$X_n(x) = \cos\left(\frac{n\pi x}{a}\right), \quad n = 0, 1, \cdots \tag{8.85}$$

and correspondingly

$$T_0(t) = \frac{a_0}{2} \quad \text{and} \quad T_n(t) = a_n e^{-(n^2\pi^2 c^2/a^2)t}. \tag{8.86}$$

Therefore,

$$u(x, t) = X_0(x)T_0(t) + \sum_{n=1}^{\infty} X_n(x)T_n(t)$$

$$= \frac{a_0}{2} + \sum_{n=1}^{\infty} a_n e^{-(n^2\pi^2 c^2/a^2)t} \cos\frac{n\pi x}{a}. \tag{8.87}$$

Finally, condition (8.71) implies that

$$f(x) = \frac{a_0}{2} + \sum_{n=1}^{\infty} a_n \cos\left(\frac{n\pi x}{a}\right),$$

which is the Fourier cosine series for $f(x)$, and hence

$$a_n = \frac{2}{a} \int_0^a f(x) \cos\left(\frac{n\pi x}{a}\right) dx, \quad n \geq 0. \tag{8.88}$$

Thus, the solution of (8.70), (8.71), (8.81), (8.82) can be written as (8.87) where a_n is given by (8.88).

Finally, we consider the case when in the rod the material properties vary with position. In this case the partial differential equation that governs the temperature $u(x, t)$ in the rod appears as

$$\frac{\partial}{\partial x}\left(k(x)\frac{\partial u}{\partial x}\right) = \rho(x)c(x)\frac{\partial u}{\partial t}, \quad \alpha < x < \beta, \quad t > 0. \tag{8.89}$$

We shall consider (8.89) with the initial condition

$$u(x, 0) = f(x), \quad \alpha < x < \beta \tag{8.90}$$

and the boundary conditions

$$a_0 u(\alpha, t) - a_1 \frac{\partial u}{\partial x}(\alpha, t) = c_1, \quad t > 0, \quad a_0^2 + a_1^2 > 0 \tag{8.91}$$

$$d_0 u(\beta, t) + d_1 \frac{\partial u}{\partial x}(\beta, t) = c_2, \quad t > 0, \quad d_0^2 + d_1^2 > 0 \tag{8.92}$$

Equations (8.89)–(8.92) make up an initial-boundary value problem. The boundary conditions (8.91) and (8.92) are of the *third kind* and are known as *Robin's conditions* (after Victor Gustave Robin (1855–1897)). These boundary conditions appear when each face loses heat to a surrounding medium according to Newton's law of cooling, which states that a body radiates heat from its surface at a rate proportional to the difference between the skin temperature of the body and the temperature of the surrounding medium.

Clearly, after a long time "under the same conditions" the variation of temperature with time dies away. In terms of the function $u(x, t)$ that represents temperature, we expect that the limit of $u(x, t)$, as t tends to infinity, exists and depends only on x : $\lim_{t\to\infty} u(x, t) = v(x)$ and also that $\lim_{t\to\infty} u_t = 0$. The function $v(x)$, called the *steady-state temperature distribution*, must still satisfy the boundary conditions and the heat equation, which are valid for all $t > 0$. Therefore, $v(x)$ (*steady-state solution*) should be the solution to the problem

$$\frac{d}{dx}\left(k(x)\frac{dv}{dx}\right) = 0, \quad \alpha < x < \beta \tag{8.93}$$

$$\begin{aligned} a_0 v(\alpha) - a_1 v'(\alpha) &= c_1 \\ d_0 v(\beta) + d_1 v'(\beta) &= c_2. \end{aligned} \tag{8.94}$$

Equation (8.93) can be solved to obtain

$$v(x) = A\int_\alpha^x \frac{d\xi}{k(\xi)} + B, \tag{8.95}$$

which satisfies the boundary conditions (8.94), if and only if,

$$\begin{aligned} a_0 B - a_1 \frac{A}{k(\alpha)} &= c_1 \\ d_0\left(A\int_\alpha^\beta \frac{d\xi}{k(\xi)} + B\right) + d_1\frac{A}{k(\beta)} &= c_2. \end{aligned} \tag{8.96}$$

Clearly, we can solve (8.96), if and only if,

$$a_0\left(d_0\int_\alpha^\beta \frac{d\xi}{k(\xi)} + \frac{d_1}{k(\beta)}\right) + \frac{a_1 d_0}{k(\alpha)} \neq 0,$$

or

$$a_0 d_0\int_\alpha^\beta \frac{d\xi}{k(\xi)} + \frac{a_0 d_1}{k(\beta)} + \frac{a_1 d_0}{k(\alpha)} \neq 0. \tag{8.97}$$

Thus, the problem (8.93), (8.94) has a unique solution, if and only if, condition (8.97) is satisfied.

Now we define the function

$$w(x, t) = u(x, t) - v(x), \tag{8.98}$$

where $u(x, t)$ and $v(x)$ are the solutions of (8.89)–(8.92) and (8.93), (8.94) respectively. Clearly,

$$\frac{\partial w(x, t)}{\partial x} = \frac{\partial u(x, t)}{\partial x} - \frac{dv(x)}{dx}$$

and hence

$$k(x)\frac{\partial w(x, t)}{\partial x} = k(x)\frac{\partial u(x, t)}{\partial x} - k(x)\frac{dv(x)}{dx},$$

which gives

$$\frac{\partial}{\partial x}\left(k(x)\frac{\partial w(x, t)}{\partial x}\right) = \frac{\partial}{\partial x}\left(k(x)\frac{\partial u(x, t)}{\partial x}\right) - \frac{d}{dx}\left(k(x)\frac{dv(x)}{dx}\right). \tag{8.99}$$

We also have

$$\rho(x)c(x)\frac{\partial w(x,t)}{\partial t} = \rho(x)c(x)\frac{\partial u(x,t)}{\partial t}. \tag{8.100}$$

Subtraction of (8.100) from (8.99) gives

$$\frac{\partial}{\partial x}\left(k(x)\frac{\partial w}{\partial x}\right) - \rho(x)c(x)\frac{\partial w}{\partial t}$$
$$= \left[\frac{\partial}{\partial x}\left(k(x)\frac{\partial u}{\partial x}\right) - \rho(x)c(x)\frac{\partial u}{\partial t}\right] - \frac{d}{dx}\left(k(x)\frac{dv}{dx}\right) = 0 - 0 = 0.$$

Therefore,

$$\frac{\partial}{\partial x}\left(k(x)\frac{\partial w}{\partial x}\right) = \rho(x)c(x)\frac{\partial w}{\partial t}, \quad \alpha < x < \beta, \quad t > 0. \tag{8.101}$$

We also have

$$a_0 w(\alpha, t) - a_1 \frac{\partial w}{\partial x}(\alpha, t) = a_0(u(\alpha, t) - v(\alpha)) - a_1\left(\frac{\partial u}{\partial x}(\alpha, t) - v'(\alpha)\right)$$
$$= \left[a_0 u(\alpha, t) - a_1 \frac{\partial u}{\partial x}(\alpha, t)\right] - [a_0 v(\alpha) - a_1 v'(\alpha)]$$
$$= c_1 - c_1 = 0,$$

i.e.,

$$a_0 w(\alpha, t) - a_1 \frac{\partial w}{\partial x}(\alpha, t) = 0, \quad t > 0 \tag{8.102}$$

and similarly

$$d_0 w(\beta, t) + d_1 \frac{\partial w}{\partial x}(\beta, t) = 0, \quad t > 0. \tag{8.103}$$

Now we shall solve (8.101)–(8.103) by using the method of separation of variables. We assume that $w(x, t) = X(x)T(t) \neq 0$, to obtain

$$\frac{d}{dx}\left(k(x)\frac{dX(x)}{dx}\right)T(t) = \rho(x)c(x)X(x)T'(t),$$

or

$$\frac{\frac{d}{dx}\left(k(x)\frac{dX}{dx}\right)}{\rho(x)c(x)X(x)} = \frac{T'(t)}{T(t)} = -\lambda,$$

which in view of (8.102) and (8.103) gives

$$\begin{aligned}(k(x)X')' + \lambda\rho(x)c(x)X &= 0 \\ a_0 X(\alpha) - a_1 X'(\alpha) &= 0 \\ d_0 X(\beta) + d_1 X'(\beta) &= 0\end{aligned} \tag{8.104}$$

and

$$T' + \lambda T = 0. \tag{8.105}$$

In the literature the system (8.104) (which is a generalization of (8.37), (8.38)) is called Sturm–Liouville problem for which the following facts are well known:
1. there are infinite number of eigenvalues $0 \le \lambda_1 < \lambda_2 < \cdots$,
2. for each eigenvalue λ_n there exists a unique eigenfunction $X_n(x)$,
3. the set of eigenfunctions $\{X_n(x)\}$ is *orthogonal* with respect to the *weight function* $\rho(x)c(x)$, i.e.,

$$\int_\alpha^\beta \rho(x)c(x)X_n(x)X_m(x)dx = 0, \quad n \ne m.$$

For $\lambda = \lambda_n$ Eq. (8.105) becomes $T_n' + \lambda_n T_n = 0$, and gives

$$T_n(t) = ce^{-\lambda_n t}, \quad n \ge 1.$$

Hence, the solution of (8.101)–(8.103) can be written as

$$w(x, t) = \sum_{n=1}^\infty a_n X_n(x)e^{-\lambda_n t}. \tag{8.106}$$

Finally, we note that condition (8.90) gives

$$w(x, 0) = u(x, 0) - v(x) = f(x) - v(x) = F(x), \quad \text{say.} \tag{8.107}$$

The solution (8.106) satisfies (8.107), if and only if,

$$w(x, 0) = F(x) = \sum_{n=1}^\infty a_n X_n(x),$$

which gives

$$a_n = \frac{\int_\alpha^\beta \rho(x)c(x)X_n(x)F(x)dx}{\int_\alpha^\beta \rho(x)c(x)X_n^2(x)dx}, \quad n \ge 1. \tag{8.108}$$

Therefore, in view of (8.98) and (8.106) the solution of (8.89)–(8.92) appears as

$$u(x, t) = v(x) + \sum_{n=1}^\infty a_n X_n(x)e^{-\lambda_n t}, \tag{8.109}$$

where a_n, $n \ge 1$ are given by (8.108).

From the representation (8.109), the following properties are immediate.
(i) Since each $\lambda_n > 0$, $u(x, t) \to v(x)$ as $t \to \infty$.

(ii) For any $t_1 > 0$ the series for $u(x, t_1)$ converges uniformly in $\alpha \le x \le \beta$ because of the exponential factors; therefore $u(x, t_1)$ is a continuous function of x.
(iii) For large t we can approximate $u(x, t)$ by $\left[v(x) + a_1 X_1(x) e^{-\lambda_1 t} \right]$.

Example 8.8 Consider a tightly stretched elastic string of length a, initially directed along a segment of the x-axis from O to a. We assume that the ends of the string are fixed at the points $x = 0$ and $x = a$. If the string is deflected from its original position and then let loose, or if we give to its points a certain velocity at the initial time, or if we deflect the string and give a velocity to its points, then the points of the string will perform certain motions. In such a stage, we say that the string is set into oscillation, or allowed to vibrate. The problem of interest is then to find the shape of the string at any instant of time.

We assume that the string is subjected to a constant tension T, which is directed along the tangent to its profile. We also assume that T is large compared to the weight of the string so that the effects of gravity are negligible. We further assume that no external forces are acting on the string, and each point of the string makes only small vibrations at right angles to the equilibrium position so that the motion takes place entirely in the xu-plane. Figure 8.9 shows the string in the position $OPQa$ at time t.

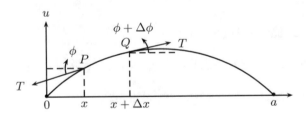

Fig. 8.9

Consider the motion of the element PQ of the string between its points $P(x, u)$ and $Q(x + \Delta x, u + \Delta u)$ where the tangents make angles ϕ and $(\phi + \Delta\phi)$ with the x-axis. Clearly, this element is moving upwards with acceleration $\partial^2 u / \partial t^2$. Also the vertical component of the force acting on this element is

$$
\begin{aligned}
&= T \sin(\phi + \Delta\phi) - T \sin(\phi) \\
&\simeq T[\tan(\phi + \Delta\phi) - \tan(\phi)], \quad \text{since } \phi \text{ is small} \\
&= T\left[\left(\frac{\partial u}{\partial x} \right)_{x+\Delta x} - \left(\frac{\partial u}{\partial x} \right)_{x} \right].
\end{aligned}
$$

If m is the mass per unit length of the string, then by Newton's second law of motion, we have

$$m \Delta x \frac{\partial^2 u}{\partial t^2} = T \left[\left(\frac{\partial u}{\partial x} \right)_{x+\Delta x} - \left(\frac{\partial u}{\partial x} \right)_x \right],$$

which is the same as

$$\frac{\partial^2 u}{\partial t^2} = \frac{T}{m} \left[\frac{\left(\frac{\partial u}{\partial x} \right)_{x+\Delta x} - \left(\frac{\partial u}{\partial x} \right)_x}{\Delta x} \right].$$

Finally, taking the limit as $Q \to P$, i.e., $\Delta x \to 0$, we obtain

$$\frac{\partial^2 u}{\partial t^2} = c^2 \frac{\partial^2 u}{\partial x^2}, \quad c^2 = \frac{T}{m}. \tag{8.110}$$

This partial differential equation gives the transverse vibrations of the string. It is called the one-dimensional *wave equation*.

Equation (8.110) by itself does not describe the motion of the string. The required function $u(x, t)$ must also satisfy the *initial conditions* which describe the state of the string at the initial time $t = 0$ and the *boundary conditions* which indicate to what occurs at the ends of the string, i.e., $x = 0$ and $x = a$. At $t = 0$ the string has a definite shape, that which we gave it. We assume that this shape is defined by the function $f(x)$. This leads to the condition

$$u(x, 0) = f(x), \quad 0 < x < a. \tag{8.111}$$

Further, at $t = 0$ the velocity at each point of the string must be given, we assume that it is defined by the function $g(x)$. Thus, we must also have

$$\left. \frac{\partial u}{\partial t} \right|_{t=0} = u_t(x, 0) = g(x), \quad 0 < x < a. \tag{8.112}$$

Now since we have assumed that the string at $x = 0$ and $x = a$ is fixed, for any t the following conditions must be satisfied

$$u(0, t) = 0, \quad t > 0 \tag{8.113}$$

$$u(a, t) = 0, \quad t > 0. \tag{8.114}$$

The partial differential equation (8.110) together with the initial conditions (8.111), (8.112) and the boundary conditions (8.113), (8.114) constitutes a typical *initial-boundary value problem*.

Now we shall show that the problem of electric oscillations in wires also leads to Eq. (8.110). The electric current in a wire is characterized by the current flow $i(x, t)$ and the voltage $v(x, t)$, which are dependent on the coordinate x of the point of the wire and on the time t. On an element Δx of the wire the drop in voltage is equal to $v(x, t) - v(x + \Delta x, t) \simeq -(\partial v/\partial x)\Delta x$. This voltage drop consists of the ohmic drop which is equal to $iR\Delta x$, and the inductive drop which is the same as $(\partial i/\partial t)L\Delta x$. Thus, we have

$$-\frac{\partial v}{\partial x}\Delta x = iR\Delta x + \frac{\partial i}{\partial t}L\Delta x, \tag{8.115}$$

where R and L are the resistance and the coefficient of self-induction per unit length of wire. In (8.115) the minus sign indicates that the current flow is in a direction opposite to the build-up of v. From (8.115) it follows that

$$\frac{\partial v}{\partial x} + iR + L\frac{\partial i}{\partial t} = 0. \tag{8.116}$$

Further, the difference between the current leaving the element Δx and entering it during the time Δt is

$$i(x,t) - i(x + \Delta x, t) \simeq -\frac{\partial i}{\partial x}\Delta x\Delta t.$$

In charging the element Δx it requires $C\Delta x(\partial v/\partial t)\Delta t$, and in leakage through the lateral surface of the wire due to imperfect insulation we have $Av\Delta x\Delta t$, where A is the leak coefficient and C is the capacitance. Equating these expressions and canceling out $\Delta x\Delta t$, we get the equation

$$\frac{\partial i}{\partial x} + C\frac{\partial v}{\partial t} + Av = 0. \tag{8.117}$$

Equations (8.116) and (8.117) are called *telegraph equations*.

Differentiating equation (8.117) with respect to x, (8.116) with respect to t and multiplying it by C, and subtracting, we obtain

$$\frac{\partial^2 i}{\partial x^2} + A\frac{\partial v}{\partial x} - CR\frac{\partial i}{\partial t} - CL\frac{\partial^2 i}{\partial t^2} = 0.$$

Substituting in this equation the expression $\partial v/\partial x$ from (8.116), we get an equation only in $i(x,t)$,

$$\frac{\partial^2 i}{\partial x^2} = CL\frac{\partial^2 i}{\partial t^2} + (CR + AL)\frac{\partial i}{\partial t} + ARi. \tag{8.118}$$

Similarly, we obtain an equation for determining $v(x,t)$,

$$\frac{\partial^2 v}{\partial x^2} = CL\frac{\partial^2 v}{\partial t^2} + (CR + AL)\frac{\partial v}{\partial t} + ARv. \tag{8.119}$$

If we neglect the leakage through the insulation ($A = 0$) and the resistance ($R = 0$), then Eqs. (8.118) and (8.119) reduce to wave equations

$$c^2 \frac{\partial^2 i}{\partial x^2} = \frac{\partial^2 i}{\partial t^2}, \quad c^2 \frac{\partial^2 v}{\partial x^2} = \frac{\partial^2 v}{\partial t^2}, \tag{8.120}$$

where $c^2 = 1/(CL)$. Again the physical conditions dictate the formulation of the initial and boundary conditions of the problem.

Wave equation (8.110) also occurs in electric oscillations in conductors, the torsional oscillations of shafts, gas vibrations, and so on.

As in Example 8.7 to solve the initial-boundary value problem (8.110)–(8.114) we use the method of separation of variables. For this, we assume a solution of (8.110) to be of the form $u(x, t) = X(x)T(t) \neq 0$, which yields

$$T'' = \lambda T \tag{8.121}$$

and

$$c^2 X'' = \lambda X. \tag{8.122}$$

The boundary conditions (8.113), (8.114) lead to the conditions $X(0) = 0$ and $X(a) = 0$. Thus, in view of (8.122) the function X has to be a solution of the eigenvalue problem

$$X'' - \frac{\lambda}{c^2} X = 0, \quad X(0) = 0, \quad X(a) = 0. \tag{8.123}$$

The eigenvalues and eigenfunctions of (8.123) are

$$\lambda_n = -\frac{n^2 \pi^2 c^2}{a^2}, \quad X_n(x) = \sin\left(\frac{n\pi x}{a}\right), \quad n = 1, 2, \cdots . \tag{8.124}$$

With λ given by (8.124), Eq. (8.121) takes the form

$$T'' + \frac{n^2 \pi^2 c^2}{a^2} T = 0$$

whose solution appears as

$$T_n(t) = a_n \cos\left(\frac{n\pi ct}{a}\right) + b_n \sin\left(\frac{n\pi ct}{a}\right), \quad n = 1, 2, \cdots \tag{8.125}$$

where a_n and b_n are the integration constants in the general solution.

Therefore, it follows that

$$u(x, t) = \sum_{n=1}^{\infty} X_n(x) T_n(t) = \sum_{n=1}^{\infty} \left[a_n \cos\left(\frac{n\pi ct}{a}\right) + b_n \sin\left(\frac{n\pi ct}{a}\right) \right] \sin\left(\frac{n\pi x}{a}\right) \tag{8.126}$$

is a solution of (8.110). Clearly, $u(x, t)$ satisfies conditions (8.113) and (8.114), and it will satisfy (8.111) provided

$$\sum_{n=1}^{\infty} a_n \sin\left(\frac{n\pi x}{a}\right) = f(x),$$ (8.127)

which is the Fourier sine series for $f(x)$. Consequently, a_n is given by

$$a_n = \frac{2}{a} \int_0^a f(x) \sin\left(\frac{n\pi x}{a}\right) dx, \quad n = 1, 2, \cdots.$$ (8.128)

Likewise condition (8.112) will be satisfied provided that

$$\sum_{n=1}^{\infty} \frac{n\pi c}{a} b_n \sin\left(\frac{n\pi x}{a}\right) = g(x)$$ (8.129)

and hence

$$\frac{n\pi c}{a} b_n = \frac{2}{a} \int_0^a g(x) \sin\left(\frac{n\pi x}{a}\right) dx,$$

which gives

$$b_n = \frac{2}{n\pi c} \int_0^a g(x) \sin\left(\frac{n\pi x}{a}\right) dx, \quad n = 1, 2, \cdots.$$ (8.130)

We conclude that the solution of the initial-boundary value problem (8.110)–(8.114) is given by (8.126) where a_n and b_n are as in (8.128) and (8.130), respectively. This solution is due to Daniel Bernoulli.

Now for simplicity we assume that $g(x) \equiv 0$, i.e., the string is initially at rest. We further define $f(x)$ for all x by its Fourier series (8.127). Then, $f(x)$ is an odd function of period $2a$, i.e., $f(-x) = -f(x)$ and $f(x + 2a) = f(x)$. With these assumptions $b_n = 0$, $n \geq 1$ and thus the solution (8.126) by the trigonometric identity

$$\sin\left(\frac{n\pi x}{a}\right) \cos\left(\frac{n\pi ct}{a}\right) = \frac{1}{2}\left[\sin\left(\frac{n\pi}{a}(x + ct)\right) + \sin\left(\frac{n\pi}{a}(x - ct)\right)\right]$$

can be written as

$$u(x, t) = \frac{1}{2} \sum_{n=1}^{\infty} a_n \left[\sin\left(\frac{n\pi}{a}(x + ct)\right) + \sin\left(\frac{n\pi}{a}(x - ct)\right)\right],$$

which in view of (8.127) is the same as

$$u(x, t) = \frac{1}{2}[f(x + ct) + f(x - ct)].$$ (8.131)

This is d'Alembert's solution (Jean le Rond d'Alembert (1717–1783)). It is easy to verify that this indeed satisfies (8.110)–(8.114) with $g(x) \equiv 0$ provided $f(x)$ is twice differentiable. To realize the significance of this solution, consider the term $f(x - ct)$ and evaluate it at two pairs of values (x_1, t_1) and (x_2, t_2), where $t_2 = t_1 + \tau$, and $x_2 = x_1 + c\tau$. Then, $x_1 - ct_1 = x_2 - ct_2$, and $f(x_1 - ct_1) = f(x_2 - ct_2)$, which means that this displacement travels along the string with velocity c. Thus, $f(x - ct)$ represents a wave traveling to the right with velocity c, and similarly, $f(x + ct)$ represents a wave traveling to the left with velocity c. It is for this reason that (8.110) is called the one-dimensional wave equation.

Next, suppose that the vibrating string is subject to a damping force that is proportional at each instance to the velocity at each point. This results in a partial differential equation of the form

$$\frac{\partial^2 u}{\partial x^2} = \frac{1}{c^2}\left(\frac{\partial^2 u}{\partial t^2} + 2k\frac{\partial u}{\partial t}\right), \quad 0 < x < a, \quad t > 0, \quad c > 0. \tag{8.132}$$

We shall consider this equation together with the initial-boundary conditions (8.111)–(8.114). In (8.132) the constant k is small and positive. Clearly, if $k = 0$ the Eq. (8.132) reduces to (8.110).

Again we assume that the solution of (8.132) can be written as $u(x, t) = X(x)T(t) \neq 0$, so that

$$\frac{X''}{X} = \frac{T'' + 2kT'}{c^2 T} = \lambda,$$

which leads to

$$X'' - \lambda X = 0, \quad X(0) = X(a) = 0 \tag{8.133}$$

$$T'' + 2kT' - \lambda c^2 T = 0. \tag{8.134}$$

For (8.133), we have

$$\lambda_n = -\frac{n^2 \pi^2}{a^2}, \quad X_n(x) = \sin\left(\frac{n\pi x}{a}\right).$$

With $\lambda = \lambda_n = -n^2\pi^2/a^2$ Eq. (8.134) takes the form

$$T_n'' + 2kT_n' + \frac{n^2\pi^2 c^2}{a^2}T_n = 0. \tag{8.135}$$

The auxiliary equation for (8.135) is better written as

$$(m + k)^2 = -\left(\frac{n^2\pi^2 c^2}{a^2} - k^2\right)$$

and hence

$$m = -k \pm i\mu_n \quad \text{where} \quad \mu_n = \sqrt{\frac{n^2\pi^2 c^2}{a^2} - k^2}.$$

Recall that $k > 0$ and small, so $\mu_n > 0$, $n \geq 1$.

Thus, the solution of (8.135) appears as

$$T_n(t) = e^{-kt}[a_n \cos(\mu_n t) + b_n \sin(\mu_n t)].$$

Therefore, the solution of (8.132) which satisfies (8.113) and (8.114) can be written as

$$u(x,t) = \sum_{n=1}^{\infty} e^{-kt}[a_n \cos(\mu_n t) + b_n \sin(\mu_n t)] \sin\left(\frac{n\pi x}{a}\right). \tag{8.136}$$

This solution satisfies (8.111) if

$$f(x) = \sum_{n=1}^{\infty} a_n \sin\left(\frac{n\pi x}{a}\right),$$

which gives

$$a_n = \frac{2}{a}\int_0^a f(x) \sin\left(\frac{n\pi x}{a}\right)dx, \quad n = 1, 2, \cdots. \tag{8.137}$$

Finally, condition (8.112) is satisfied if

$$g(x) = \sum_{n=1}^{\infty}(-ka_n + b_n\mu_n) \sin\left(\frac{n\pi x}{a}\right)$$

and hence

$$-ka_n + b_n\mu_n = \frac{2}{a}\int_0^a g(x) \sin\left(\frac{n\pi x}{a}\right)dx,$$

which gives

$$b_n = k\frac{a_n}{\mu_n} + \frac{2}{\mu_n a}\int_0^a g(x) \sin\left(\frac{n\pi x}{a}\right)dx, \quad n = 1, 2, \cdots. \tag{8.138}$$

Now we shall assume that for the vibrating string the ends are free—they are allowed to slide without friction along the vertical lines $x = 0$ and $x = a$. This may seem impossible, but it is a standard mathematically modeled case. This leads to the initial-boundary value problem (8.110)–(8.112), and the Neumann boundary conditions

$$u_x(0, t) = 0, \quad t > 0 \tag{8.139}$$

$$u_x(a, t) = 0, \quad t > 0. \tag{8.140}$$

For the problem (8.110)–(8.112), (8.139), (8.140) if we assume a solution in the form $u(x, t) = X(x)T(t) \neq 0$, then X must satisfy the eigenvalue problem

$$X'' - \frac{\lambda}{c^2}X = 0$$
$$X'(0) = X'(a) = 0$$

for which the eigenvalues and eigenfunctions are

$$\lambda_n = -\frac{n^2\pi^2c^2}{a^2}, \quad X_n(x) = \cos\left(\frac{n\pi x}{a}\right), \quad n = 0, 1, \cdots.$$

For $\lambda_0 = 0$, the equation $T'' - \lambda T = 0$ reduces to $T_0'' = 0$, and hence

$$T_0(t) = b_0t + a_0.$$

For $\lambda_n = -n^2\pi^2c^2/a^2$ the equation $T'' - \lambda T = 0$ is $T_n'' + (n^2\pi^2c^2/a^2)T_n = 0$ and hence the solution is the same as (8.125). Thus, the solution of (8.110) satisfying (8.139), (8.140) can be written as

$$u(x, t) = (b_0t + a_0) + \sum_{n=1}^{\infty}\left[a_n\cos\left(\frac{n\pi ct}{a}\right) + b_n\sin\left(\frac{n\pi ct}{a}\right)\right]\cos\left(\frac{n\pi x}{a}\right).$$

$$(8.141)$$

The solution (8.141) satisfies (8.111) if and only if

$$f(x) = a_0 + \sum_{n=1}^{\infty}a_n\cos\left(\frac{n\pi x}{a}\right)$$

and hence

$$a_0 = \frac{1}{a}\int_0^a f(x)dx$$
$$a_n = \frac{2}{a}\int_0^a f(x)\cos\left(\frac{n\pi x}{a}\right)dx, \quad n = 1, 2, \cdots.$$

$$(8.142)$$

Finally, the solution (8.141) satisfies (8.112) if and only if

$$g(x) = b_0 + \sum_{n=1}^{\infty}b_n\frac{n\pi c}{a}\cos\left(\frac{n\pi x}{a}\right)$$

and hence

$$b_0 = \frac{1}{a}\int_0^a g(x)dx$$
$$b_n = \frac{2}{n\pi c}\int_0^a g(x)\cos\left(\frac{n\pi x}{a}\right)dx, \quad n = 1, 2, \cdots.$$

$$(8.143)$$

Thus, the solution of (8.110)–(8.112), (8.139), (8.140) can be written as (8.141), where the constants a_n, b_n, $n = 0, 1, \cdots$ are given by (8.142) and (8.143) respectively.

Finally, we consider the equation of the general vibrating string

$$\frac{\partial}{\partial x}\left(k(x)\frac{\partial u}{\partial x}\right) = \frac{\rho(x)}{c^2}\frac{\partial^2 u}{\partial t^2}, \quad \alpha < x < \beta, \quad t > 0, \quad c > 0 \tag{8.144}$$

subject to the initial conditions (8.90),

$$u_t(x, 0) = g(x), \quad \alpha < x < \beta \tag{8.145}$$

and Robin's boundary conditions (8.91), (8.92). These boundary conditions describe some type of an elastic, or spring attachment at both ends of the string. Now following as in Example 8.7, although there is no steady-state for the wave equation (8.144), we let $v(x)$ to be the solution of the problem (8.93), (8.94). Again, we define the function $w(x, t)$ as in (8.97), which satisfies the wave equation

$$\frac{\partial}{\partial x}\left(k(x)\frac{\partial w}{\partial x}\right) = \frac{\rho(x)}{c^2}\frac{\partial^2 w}{\partial t^2} \quad \alpha < x < \beta, \quad t > 0 \tag{8.146}$$

the initial conditions (8.107),

$$w_t(x, 0) = g(x), \quad \alpha < x < \beta \tag{8.147}$$

and the boundary conditions (8.102), (8.103). We use the substitution $w(x, t) = X(t)T(t) \neq 0$ which leads to solving

$$\begin{aligned}
(k(x)X')' + \frac{\lambda}{c^2}\rho(x)X &= 0 \\
a_0 X(\alpha) - a_1 X'(\alpha) &= 0 \\
d_0 X(\beta) + d_1 X'(\beta) &= 0
\end{aligned} \tag{8.148}$$

and

$$T'' + \lambda T = 0. \tag{8.149}$$

Thus, the solution of (8.146), (8.102), (8.103) in terms of the eigenvalues $0 \leq \lambda_1 < \lambda_2 < \cdots$ and eigenfunctions $X_n(x)$ of (8.148) appears as

$$w(x, t) = \sum_{n=1}^{\infty}\left[a_n \cos(\sqrt{\lambda_n}t) + b_n \sin(\sqrt{\lambda_n}t)\right]X_n(x). \tag{8.150}$$

This solution satisfies the initial conditions (8.107), (8.147) if and only if

$$a_n = \frac{\int_\alpha^\beta \rho(x)X_n(x)F(x)dx}{\int_\alpha^\beta \rho(x)X_n^2(x)dx}, \quad b_n = \frac{\int_\alpha^\beta \rho(x)X_n(x)g(x)dx}{\sqrt{\lambda_n}\int_\alpha^\beta \rho(x)X_n^2(x)dx}, \quad n \geq 1.$$

Finally, the solution of (8.144), (8.90), (8.145), (8.91), (8.92) is obtained from the relation $u(x,t) = w(x,t) + v(x)$.

Example 8.9 Consider the flow of heat in a metal plate of uniform thickness α (cm), density ρ (gr./cm^3), specific heat s (cal./gr. deg.) and thermal conductivity k (cal./cm. s. deg.). Let XOY plane be taken in one face of the plate. If the temperature at any point is independent of the z-coordinate and depends only on x, y, and time t, (for instance, its two parallel faces are insulated) then the flow is said to be two-dimensional. In this case, the heat flow is in the XY-plane only and is zero along the normal to the XY-plane.

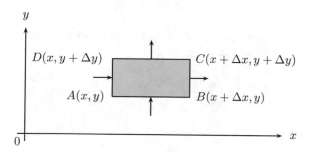

Fig. 8.10

Consider a rectangular element $ABCD$ of the plate with sides Δx and Δy as shown in Fig. 8.10. By Fourier's Law, the amount of heat entering the element in 1 s from the side AB is

$$= -k\alpha\Delta x \left(\frac{\partial u}{\partial y}\right)_y$$

and the amount of heat entering the element in 1 s from the side AD is

$$= -k\alpha\Delta y \left(\frac{\partial u}{\partial x}\right)_x.$$

The quantity of heat flowing out through the side CD in 1 s is

$$= -k\alpha\Delta x \left(\frac{\partial u}{\partial y}\right)_{y+\Delta y}$$

and the quantity of heat flowing out through the side BC in 1 s is

$$= -k\alpha\Delta y \left(\frac{\partial u}{\partial x}\right)_{x+\Delta x}.$$

Hence, the total gain of heat by the rectangular element $ABCD$ in 1 s is

$$= -k\alpha\Delta x \left(\frac{\partial u}{\partial y}\right)_y - k\alpha\Delta y \left(\frac{\partial u}{\partial x}\right)_x + k\alpha\Delta x \left(\frac{\partial u}{\partial y}\right)_{y+\Delta y} + k\alpha\Delta y \left(\frac{\partial u}{\partial x}\right)_{x+\Delta x}$$

$$= k\alpha\Delta x\,\Delta y \left[\frac{\left(\frac{\partial u}{\partial x}\right)_{x+\Delta x} - \left(\frac{\partial u}{\partial x}\right)_x}{\Delta x} + \frac{\left(\frac{\partial u}{\partial y}\right)_{y+\Delta y} - \left(\frac{\partial u}{\partial y}\right)_y}{\Delta y} \right].$$

$$(8.151)$$

Also the rate of gain of heat by the element is

$$= \rho\Delta x\,\Delta y\alpha s\frac{\partial u}{\partial t}. \qquad (8.152)$$

Thus, equating (8.151) and (8.152), dividing both sides by $\alpha\Delta x\,\Delta y$, and taking limits as $\Delta x \to 0$, $\Delta y \to 0$, we get

$$k\left(\frac{\partial^2 u}{\partial x^2} + \frac{\partial^2 u}{\partial y^2}\right) = \rho s\frac{\partial u}{\partial t},$$

which is the same as

$$\frac{\partial u}{\partial t} = c^2\left(\frac{\partial^2 u}{\partial x^2} + \frac{\partial^2 u}{\partial y^2}\right), \qquad (8.153)$$

where $c^2 = k/(\rho s)$ is the *diffusivity coefficient*.

Equation (8.153) gives the temperature distribution of the plate in the *transient state*. In the *steady state*, u is independent of t, so that $u_t = 0$ and the Eq. (8.153) reduces to

$$\Delta_2 u = u_{xx} + u_{yy} = 0, \qquad (8.154)$$

which is the well-known *Laplace's equation* in two dimensions. This equation also occurs in the study of problems dealing with electric and magnetic fields, stationary states, and problems in hydrodynamics. Solutions of (8.154) are often called *potential functions* as well as *harmonic functions*. Since there is no time dependence in (8.154), no initial conditions are required to be satisfied by its solution $u(x, y)$. However, certain boundary conditions on the boundary of the region must be satisfied. Thus, a typical problem associated with Laplace's equation is a boundary value problem. A common way is to specify $u(x, y)$ at each point (x, y) on the boundary, which is known as a *Dirichlet problem*.

Now we shall use the method of separation of variables to solve the Dirichlet problem on the rectangle $R = 0 < x < a$, $0 < y < b$, i.e., find the solution $u(x, y)$ of (8.154) on R satisfying the boundary conditions

$$u(x, 0) = f(x), \quad 0 < x < a \tag{8.155}$$

$$u(x, b) = g(x), \quad 0 < x < a \tag{8.156}$$

$$u(0, y) = 0, \quad 0 < y < b \tag{8.157}$$

$$u(a, y) = 0, \quad 0 < y < b. \tag{8.158}$$

This problem is illustrated in Fig. 8.11.

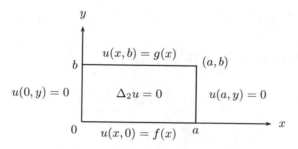

Fig. 8.11

We seek a solution of (8.154) in the form $u(x, y) = X(x)Y(y) \neq 0$, to obtain

$$-\frac{X''(x)}{X(x)} = \frac{Y''(y)}{Y(y)} = \lambda \quad \text{(constant)}.$$

Hence, we have

$$X'' + \lambda X = 0 \tag{8.159}$$

and the conditions (8.157) and (8.158) imply

$$X(0) = 0, \quad X(a) = 0. \tag{8.160}$$

Also Y satisfies the differential equation

$$Y'' - \lambda Y = 0. \tag{8.161}$$

The eigenvalues and eigenfunctions of the problem (8.159), (8.160) are, respectively, given by

$$\lambda_n = \frac{n^2 \pi^2}{a^2}, \quad n = 1, 2, \cdots \tag{8.162}$$

and

$$X_n(x) \;=\; \sin\left(\frac{n\pi x}{a}\right), \quad n = 1, 2, \cdots .\tag{8.163}$$

For λ as given in (8.162) the general solution of the differential equation (8.161) is

$$Y_n(y) \;=\; a_n \cosh\left(\frac{n\pi y}{a}\right) + b_n \sinh\left(\frac{n\pi y}{a}\right).\tag{8.164}$$

Thus, the solution of (8.154) satisfying (8.157) and (8.158) can be written as

$$u(x, y) \;=\; \sum_{n=1}^{\infty}\left[a_n \cosh\left(\frac{n\pi y}{a}\right) + b_n \sinh\left(\frac{n\pi y}{a}\right)\right]\sin\left(\frac{n\pi x}{a}\right).\tag{8.165}$$

Now (8.165) satisfies (8.155), if and only if,

$$f(x) \;=\; \sum_{n=1}^{\infty} a_n \sin\left(\frac{n\pi x}{a}\right),$$

which gives

$$a_n \;=\; \frac{2}{a}\int_0^a f(x)\sin\left(\frac{n\pi x}{a}\right)dx, \quad n = 1, 2, \cdots .\tag{8.166}$$

Finally, (8.165) satisfies (8.156) provided

$$g(x) \;=\; \sum_{n=1}^{\infty}\left[a_n \cosh\left(\frac{n\pi b}{a}\right) + b_n \sinh\left(\frac{n\pi b}{a}\right)\right]\sin\left(\frac{n\pi x}{a}\right),$$

which gives

$$a_n \cosh\left(\frac{n\pi b}{a}\right) + b_n \sinh\left(\frac{n\pi b}{a}\right) \;=\; \frac{2}{a}\int_0^a g(x)\sin\left(\frac{n\pi x}{a}\right)dx$$

and therefore

$$b_n \sinh\left(\frac{n\pi b}{a}\right) \;=\; \frac{2}{a}\int_0^a g(x)\sin\left(\frac{n\pi x}{a}\right)dx - a_n \cosh\left(\frac{n\pi b}{a}\right),$$

which in view of (8.166) gives

$$b_n \;=\; \frac{1}{\sinh\left(\frac{n\pi b}{a}\right)}\left[\frac{2}{a}\int_0^a g(x)\sin\left(\frac{n\pi x}{a}\right)dx \right.$$
$$\left. -\cosh\left(\frac{n\pi b}{a}\right)\frac{2}{a}\int_0^a f(x)\sin\left(\frac{n\pi x}{a}\right)dx\right], \quad n = 1, 2, \cdots$$
$$\tag{8.167}$$

Hence, the solution of the boundary value problem (8.154)–(8.158) is given by (8.165) where a_n and b_n are as in (8.166) and (8.167), respectively.

Next we note that as for the problem (8.154)–(8.158) the solution $u(x, y)$ of the Dirichlet problem (8.154) on the rectangle R satisfying the boundary conditions

$$u(x, 0) = 0, \quad 0 < x < a \tag{8.168}$$

$$u(x, b) = 0, \quad 0 < x < a \tag{8.169}$$

$$u(0, y) = h(y), \quad 0 < y < b \tag{8.170}$$

$$u(a, y) = k(y), \quad 0 < y < b \tag{8.171}$$

(see Fig. 8.12) can be written as

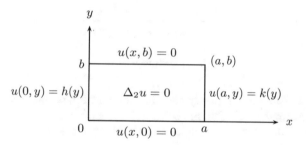

Fig. 8.12

$$u(x, y) = \sum_{n=1}^{\infty} \left[\alpha_n \cosh\left(\frac{n\pi x}{b}\right) + \beta_n \sinh\left(\frac{n\pi x}{b}\right) \right] \sin\left(\frac{n\pi y}{b}\right), \tag{8.172}$$

where

$$\alpha_n = \frac{2}{b} \int_0^b h(y) \sin\left(\frac{n\pi y}{b}\right) dy, \quad n = 1, 2, \cdots \tag{8.173}$$

and

$$\beta_n = \frac{1}{\sinh\left(\frac{n\pi a}{b}\right)} \left[\frac{2}{b} \int_0^b k(y) \sin\left(\frac{n\pi y}{b}\right) dy \right.$$

$$\left. - \cosh\left(\frac{n\pi a}{b}\right) \frac{2}{b} \int_0^b h(y) \sin\left(\frac{n\pi y}{b}\right) dy \right], \quad n = 1, 2, \cdots. \tag{8.174}$$

Finally, from the linearity of the problem as well as by direct substitution it is clear that if $u_1(x, y)$ is the solution of the problem (8.154)–(8.158) and $u_2(x, y)$ is the solution of the problem (8.154), (8.168)–(8.171) then

$$u(x, y) = u_1(x, y) + u_2(x, y) \qquad (8.175)$$

is the solution of the Dirichlet problem (8.154) on the rectangle R satisfying the boundary conditions (8.155), (8.156), (8.170), (8.171) (see Fig. 8.13).

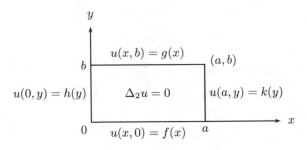

Fig. 8.13

Example 8.10 Consider the steady-state heat conduction problem for a flat plate in the shape of a circular disk with the boundary curve $x^2 + y^2 = a^2$. In what follows we assume that the plate is isotropic, i.e., the flat surfaces are insulated, and that the temperature is known everywhere on the circular boundary. The temperature inside the disk is then a solution of the *Dirichlet problem* (see Fig. 8.14) consisting of Laplace's equation (8.154) in polar coordinates, i.e., $x = r \cos(\theta)$, $y = r \sin(\theta)$,

$$\frac{\partial^2 u}{\partial r^2} + \frac{1}{r}\frac{\partial u}{\partial r} + \frac{1}{r^2}\frac{\partial^2 u}{\partial \theta^2} = 0, \quad 0 < r < a, \quad -\pi < \theta \le \pi \qquad (8.176)$$

and the boundary condition

$$u(a, \theta) = f(\theta), \quad -\pi < \theta \le \pi. \qquad (8.177)$$

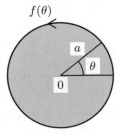

Fig. 8.14

In problem (8.176), (8.177) we notice that $r = 0$ is not a physical boundary, rather we recognize it as a "mathematical boundary", and for a solution $u(r, \theta)$ to be physically meaningful we need to impose at $r = 0$ the implicit boundary condition

$$|u(0, \theta)| < \infty, \tag{8.178}$$

i.e., the solution remains bounded at the origin. We also wish to allow θ to assume any value rather than restricted to the interval $-\pi < \theta \leq \pi$, and hence we assume that $f(\theta)$, and consequently $u(r, \theta)$ to be periodic with period 2π. Thus, we also need the conditions

$$u(r, \pi) = u(r, -\pi), \quad 0 < r < a \tag{8.179}$$

$$\frac{\partial u}{\partial \theta}(r, \pi) = \frac{\partial u}{\partial \theta}(r, -\pi), \quad 0 < r < a \tag{8.180}$$

which are actually continuity requirements along the slit $\theta = \pi$. The problem (8.176)–(8.180) is often called as an interior problem.

To solve (8.176)–(8.180) we assume that $u(r, \theta) = R(r)\Theta(\theta) \neq 0$, which leads to equations

$$\Theta'' + \lambda\Theta = 0, \quad -\pi < \theta \leq \pi \tag{8.181}$$

and

$$r^2 R'' + r R' - \lambda R = 0, \quad 0 < r < a. \tag{8.182}$$

Now (8.179) implies

$$\Theta(-\pi) = \Theta(\pi) \tag{8.183}$$

whereas (8.180) gives

$$\Theta'(-\pi) = \Theta'(\pi). \tag{8.184}$$

For (8.181), (8.183), (8.184) eigenvalues and eigenfunctions are

$$\begin{cases} \lambda_0 = 0, \quad \Theta_0 = 1 \\ \lambda_n = n^2 \ (n \geq 1), \quad \Theta_n = \cos(n\theta) \ \text{and} \ \sin(n\theta) \\ \qquad \text{(two linearly independent eigenfunctions).} \end{cases} \tag{8.185}$$

Next, for $\lambda = 0$, Eq. (8.182) is

$$r^2 R_0'' + r R_0' = 0 \tag{8.186}$$

for which two linearly independent solutions are 1 and $\ln r$. However, in view of (8.178) the solution $\ln r$ is discarded because of its behavior at $r = 0$. Thus, we have

$$R_0(r) = 1. \tag{8.187}$$

For $\lambda = \lambda_n = n^2$, Eq. (8.182) is

$$r^2 R_n'' + r R_n' - n^2 R_n \ = \ 0 \tag{8.188}$$

for which two linearly independent solutions are r^n and r^{-n}. However, since the solution r^{-n} is unbounded as r approaches zero, to fulfill condition (8.178) we need to discard it. Thus, we obtain

$$R_n(r) \ = \ r^n. \tag{8.189}$$

Therefore, the solution $u(r, \theta)$ can be written as

$$u(r, \theta) \ = \ \frac{a_0}{2} + \sum_{n=1}^{\infty} r^n \left[a_n \cos(n\theta) + b_n \sin(n\theta) \right]. \tag{8.190}$$

This solution satisfies (8.177) if

$$u(a, \theta) \ = \ f(\theta) \ = \ \frac{a_0}{2} + \sum_{n=1}^{\infty} a^n \left[a_n \cos(n\theta) + b_n \sin(n\theta) \right]. \tag{8.191}$$

Clearly, (8.191) is a Fourier trigonometric series, and hence

$$\begin{aligned}
a_n &= \frac{1}{\pi a^n} \int_{-\pi}^{\pi} f(\phi) \cos(n\phi) d\phi, \quad n \geq 0 \\
b_n &= \frac{1}{\pi a^n} \int_{-\pi}^{\pi} f(\phi) \sin(n\phi) d\phi, \quad n \geq 1.
\end{aligned} \tag{8.192}$$

In conclusion the solution of (8.176)–(8.180) can be written as (8.190) where a_n and b_n are given in (8.192).

Now in (8.190) we substitute the coefficients a_n, b_n from (8.192), interchange the order of summation and integration, and use some elementary identities, to get

$$\begin{aligned}
u(r, \theta) &= \frac{1}{2\pi} \int_{-\pi}^{\pi} f(\phi) d\phi \\
&\quad + \frac{1}{\pi} \sum_{n=1}^{\infty} \frac{r^n}{a^n} \left[\cos(n\theta) \int_{-\pi}^{\pi} f(\phi) \cos(n\phi) d\phi + \sin(n\theta) \int_{-\pi}^{\pi} f(\phi) \sin(n\phi) d\phi \right] \\
&= \frac{1}{\pi} \int_{-\pi}^{\pi} f(\phi) \left[\frac{1}{2} + \sum_{n=1}^{\infty} \frac{r^n}{a^n} [\cos(n\theta) \cos(n\phi) + \sin(n\theta) \sin(n\phi)] \right] d\phi \\
&= \frac{1}{\pi} \int_{-\pi}^{\pi} f(\phi) \left[\frac{1}{2} + \sum_{n=1}^{\infty} \frac{r^n}{a^n} \cos(n(\theta - \phi)) \right] d\phi \\
&= \frac{1}{\pi} \int_{-\pi}^{\pi} f(\phi) \left[\frac{1}{2} + \sum_{n=1}^{\infty} \frac{r^n}{a^n} \frac{1}{2} \left(e^{n(\theta-\phi)i} + e^{-n(\theta-\phi)i} \right) \right] d\phi
\end{aligned}$$

$$= \frac{1}{2\pi} \int_{-\pi}^{\pi} f(\phi) \left[1 + \sum_{n=1}^{\infty} \left(\left(\frac{r}{a} e^{(\theta-\phi)i} \right)^n + \left(\frac{r}{a} e^{-(\theta-\phi)i} \right)^n \right) \right] d\phi.$$

Now since $|e^{i\psi}| = 1$, for $r < a$ we can sum the geometric series, to obtain

$$u(r, \theta) = \frac{1}{2\pi} \int_{-\pi}^{\pi} f(\phi) \left[1 + \frac{\frac{r}{a} e^{(\theta-\phi)i}}{1 - \frac{r}{a} e^{(\theta-\phi)i}} + \frac{\frac{r}{a} e^{-(\theta-\phi)i}}{1 - \frac{r}{a} e^{-(\theta-\phi)i}} \right] d\phi,$$

which is the same as

$$u(r, \theta) = \frac{a^2 - r^2}{2\pi} \int_{-\pi}^{\pi} \frac{f(\phi)}{a^2 + r^2 - 2ra \cos(\theta - \phi)} d\phi, \quad r < a. \qquad (8.193)$$

This formula is called *Poisson's integral formula*. It shows that the temperature at any interior point (r, θ) of the disk of radius a may be obtained by integrating the boundary temperatures according to the formula (8.193). In particular, if $r = 0$, then the temperature at the center of the disk is

$$u(0, \theta) = \frac{1}{2\pi} \int_{-\pi}^{\pi} f(\phi) d\phi, \qquad (8.194)$$

i.e., the temperature at the center is the integral average of the boundary temperatures. This fact is called the *mean value theorem* and holds for all functions that satisfy Laplace's equation on the disk.

Now we shall find the solution of the Laplace equation (8.176) outside the disk $r = a$ (see Fig. 8.15). For this, again we assume that the conditions (8.177), (8.179), (8.180) are satisfied, but the condition (8.178) has to be replaced by

$$\lim_{r \to \infty} |u(r, \theta)| < \infty. \qquad (8.195)$$

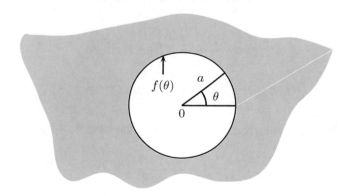

Fig. 8.15

Clearly, for this *exterior problem* also all the steps remain the same as for the case $r < a$, except that the solution of (8.188) which satisfies the condition (8.195) is now r^{-n}. This change leads to the solution

$$u(r, \theta) = \frac{\alpha_0}{2} + \sum_{n=1}^{\infty} r^{-n} [\alpha_n \cos(n\theta) + \beta_n \sin(n\theta)], \qquad (8.196)$$

where

$$\alpha_n = \frac{a^n}{\pi} \int_{-\pi}^{\pi} f(\phi) \cos(n\phi) d\phi, \quad n \geq 0$$

$$\beta_n = \frac{a^n}{\pi} \int_{-\pi}^{\pi} f(\phi) \sin(n\phi) d\phi, \quad n \geq 1. \qquad (8.197)$$

Finally, comparing (8.190), (8.192) with (8.196), (8.197) we see that the only difference between the two sets of formulas is that r and a are replaced by r^{-1} and a^{-1}. Thus, with this change the Poisson's formula for the exterior problem appears as

$$u(r, \theta) = \frac{r^2 - a^2}{2\pi} \int_{-\pi}^{\pi} \frac{f(\phi)}{a^2 + r^2 - 2ra \cos(\theta - \phi)} d\phi, \quad r > a. \qquad (8.198)$$

Example 8.11 Consider the heat flow problem of an infinitely long thin bar insulated on its lateral surface, which is modeled by the following initial-value problem

$$u_t = c^2 u_{xx}, \quad -\infty < x < \infty, \quad t > 0, \quad c > 0$$
$$u \text{ and } u_x \text{ finite as } |x| \to \infty, \quad t > 0 \qquad (8.199)$$
$$u(x, 0) = f(x), \quad -\infty < x < \infty,$$

where the function f is piecewise smooth and absolutely integrable in $(-\infty, \infty)$.

Let $U(\omega, t)$ be the Fourier transform of $u(x, t)$. Thus, from the Fotransfor pair, we have

$$u(x, t) = \frac{1}{\sqrt{2\pi}} \int_{-\infty}^{\infty} U(\omega, t) e^{i\omega x} d\omega$$

$$U(\omega, t) = \frac{1}{\sqrt{2\pi}} \int_{-\infty}^{\infty} u(x, t) e^{-i\omega x} dx.$$

Assuming that the derivatives can be taken under the integral, we get

$$\frac{\partial u}{\partial t} = \frac{1}{\sqrt{2\pi}} \int_{-\infty}^{\infty} \frac{\partial U(\omega, t)}{\partial t} e^{i\omega x} dx$$

$$\frac{\partial u}{\partial x} = \frac{1}{\sqrt{2\pi}} \int_{-\infty}^{\infty} U(\omega, t)(i\omega) e^{i\omega x} d\omega$$

$$\frac{\partial^2 u}{\partial x^2} = \frac{1}{\sqrt{2\pi}} \int_{-\infty}^{\infty} U(\omega, t)(i\omega)^2 e^{i\omega x} d\omega.$$

In order that $u(x, t)$ satisfies the heat equation, we must have

$$0 = \frac{\partial u}{\partial t} - c^2 \frac{\partial^2 u}{\partial x^2} = \frac{1}{\sqrt{2\pi}} \int_{-\infty}^{\infty} \left[\frac{\partial U(\omega, t)}{\partial t} + c^2 \omega^2 U(\omega, t) \right] e^{i\omega x} d\omega.$$

Thus, U must be a solution of the ordinary differential equation

$$\frac{dU}{dt} + c^2 \omega^2 U = 0.$$

The initial condition is determined by

$$U(\omega, 0) = \frac{1}{\sqrt{2\pi}} \int_{-\infty}^{\infty} u(x, 0) e^{-i\omega x} dx$$

$$= \frac{1}{\sqrt{2\pi}} \int_{-\infty}^{\infty} f(x) e^{-i\omega x} dx = F(\omega).$$

Therefore, we have

$$U(\omega, t) = F(\omega) e^{-\omega^2 c^2 t},$$

and hence

$$u(x, t) = \frac{1}{\sqrt{2\pi}} \int_{-\infty}^{\infty} F(\omega) e^{-\omega^2 c^2 t} e^{i\omega x} d\omega. \tag{8.200}$$

Now since from (8.28),

$$\frac{1}{\sqrt{2\pi}} \int_{-\infty}^{\infty} e^{-\omega^2 c^2 t} e^{i\omega x} d\omega = \sqrt{2\pi} \frac{e^{-x^2/(4c^2 t)}}{\sqrt{4\pi c^2 t}},$$

if in (8.27) we denote $F(\omega) = \hat{f}(\omega)$ and $\hat{g}(\omega) = e^{-\omega^2 c^2 t}$, then from (8.200) it follows that

$$u(x, t) = \frac{1}{\sqrt{2\pi}} \int_{-\infty}^{\infty} f(\mu) \sqrt{2\pi} \frac{e^{-(x-\mu)^2/4c^2 t}}{\sqrt{4\pi c^2 t}} d\mu$$

$$= \int_{-\infty}^{\infty} f(\mu) \frac{e^{-(x-\mu)^2/4c^2 t}}{\sqrt{4\pi c^2 t}} d\mu. \tag{8.201}$$

This formula is due to Gauss and Karl Theodor Wilhelm Weierstrass (1815–1897). For each μ, the function $(x, t) \to e^{-(x-\mu)^2/4c^2 t}/\sqrt{4\pi c^2 t}$ is a solution of the heat equation and is called the *fundamental solution*. Thus, (8.201) gives a representation of the solution as a continuous superposition of the fundamental solution.

Example 8.12 Consider the problem

$$
\begin{aligned}
u_t &= c^2 u_{xx}, \quad x > 0, \quad t > 0 \\
u(0, t) &= 0, \quad t > 0 \\
u \text{ and } u_x &\text{ finite as } x \to \infty, \quad t > 0 \\
u(x, 0) &= f(x), \quad x > 0,
\end{aligned}
\tag{8.202}
$$

which appears in heat flow in a semi-infinite region. In (8.202), the function f is piecewise smooth and absolutely integrable in $[0, \infty)$.

We define the odd function

$$
\bar{f}(x) = \begin{cases} f(x), & x > 0 \\ -f(-x), & x < 0. \end{cases}
$$

Then, from (8.201), we have

$$
\begin{aligned}
u(x, t) &= \int_{-\infty}^{\infty} \frac{e^{-(x-\mu)^2/4c^2 t}}{\sqrt{4\pi c^2 t}} \bar{f}(\mu) d\mu \\
&= \int_{-\infty}^{0} \frac{e^{-(x-\mu)^2/4c^2 t}}{\sqrt{4\pi c^2 t}} \bar{f}(\mu) d\mu + \int_{0}^{\infty} \frac{e^{-(x-\mu)^2/4c^2 t}}{\sqrt{4\pi c^2 t}} \bar{f}(\mu) d\mu.
\end{aligned}
$$

In the first integral, we change μ to $(-\mu)$ and use the oddness of $\bar{f}$, to obtain

$$
\int_{-\infty}^{0} \frac{e^{-(x-\mu)^2/4c^2 t}}{\sqrt{4\pi c^2 t}} \bar{f}(\mu) d\mu = -\int_{0}^{\infty} \frac{e^{-(x+\mu)^2/4c^2 t}}{\sqrt{4\pi c^2 t}} f(\mu) d\mu.
$$

Thus, the solution of the problem (8.202) can be written as

$$
u(x, t) = \int_{0}^{\infty} \frac{e^{-(x-\mu)^2/4c^2 t} - e^{-(x+\mu)^2/4c^2 t}}{\sqrt{4\pi c^2 t}} f(\mu) d\mu.
\tag{8.203}
$$

The above procedure to find the solution of (8.202) is called *the method of images*. In an analogous way, it can be shown that the solution of the problem

$$
\begin{aligned}
u_t &= c^2 u_{xx}, \quad x > 0, \quad t > 0 \\
u_x(0, t) &= 0, \quad t > 0 \\
u \text{ and } u_x &\text{ finite as } x \to \infty, \quad t > 0 \\
u(x, 0) &= f(x), \quad x > 0
\end{aligned}
\tag{8.204}
$$

can be written as

$$
u(x, t) = \int_{0}^{\infty} \frac{e^{-(x-\mu)^2/4c^2 t} + e^{-(x+\mu)^2/4c^2 t}}{\sqrt{4\pi c^2 t}} f(\mu) d\mu.
\tag{8.205}
$$

Here, of course, we need to extend $f(x)$ to an even function

$$\bar{f}(x) = \begin{cases} f(x), & x > 0 \\ f(-x), & x < 0. \end{cases}$$

In (8.204), the physical significance of the condition $u_x(0, t) = 0$ is that there is a perfect insulation, i.e., there is no heat flux across the surface.

Example 8.13 Consider the initial-value problem for the wave equation

$$\begin{aligned}
u_{tt} &= c^2 u_{xx}, \quad -\infty < x < \infty, \quad t > 0, \quad c > 0 \\
u \text{ and } & u_x \text{ finite as } |x| \to \infty, \quad t > 0 \\
u(x, 0) &= f_1(x), \quad -\infty < x < \infty \\
u_t(x, 0) &= f_2(x), \quad -\infty < x < \infty
\end{aligned} \tag{8.206}$$

where the functions f_1 and f_2 are piecewise smooth and absolutely integrable in $(-\infty, \infty)$.

To find the solution of this problem, we introduce the Fourier transforms

$$F_j(\omega) = \frac{1}{\sqrt{2\pi}} \int_{-\infty}^{\infty} e^{-i\omega x} f_j(x) dx, \quad j = 1, 2$$

and its inversion formulas

$$f_j(x) = \frac{1}{\sqrt{2\pi}} \int_{-\infty}^{\infty} e^{i\omega x} F_j(\omega) d\omega, \quad j = 1, 2.$$

We also need the Fourier representation of the solution $u(x, t)$,

$$u(x, t) = \frac{1}{\sqrt{2\pi}} \int_{-\infty}^{\infty} U(\omega, t) e^{i\omega x} d\omega,$$

where $U(\omega, t)$ is an unknown function, which we will now determine. For this, we substitute this into the differential equation (8.206), to obtain

$$0 = \frac{1}{\sqrt{2\pi}} \int_{-\infty}^{\infty} \left[\frac{\partial^2 U(\omega, t)}{\partial t^2} + c^2 \omega^2 U(\omega, t) \right] e^{i\omega x} d\omega.$$

Thus, U must be a solution of the ordinary differential equation

$$\frac{d^2 U}{dt^2} + c^2 \omega^2 U = 0,$$

whose solution can be written as

$$U(\omega, t) = c_1(\omega) \cos \omega c t + c_2(\omega) \sin \omega c t.$$

To find $c_1(\omega)$ and $c_2(\omega)$, we note that

$$f_1(x) = u(x, 0) = \frac{1}{\sqrt{2\pi}} \int_{-\infty}^{\infty} c_1(\omega) e^{i\omega x} d\omega$$

$$f_2(x) = \frac{\partial u(x, 0)}{\partial t} = \frac{1}{\sqrt{2\pi}} \int_{-\infty}^{\infty} \omega c c_2(\omega) e^{i\omega x} d\omega$$

and hence $F_1(\omega) = c_1(\omega)$ and $F_2(\omega) = \omega c c_2(\omega)$.

Therefore, it follows that

$$U(\omega, t) = F_1(\omega) \cos(\omega c t) + \frac{F_2(\omega)}{\omega c} \sin(\omega c t)$$

and hence the Fourier representation of the solution is

$$u(x, t) = \frac{1}{\sqrt{2\pi}} \int_{-\infty}^{\infty} \left[F_1(\omega) \cos(\omega c t) + \frac{F_2(\omega)}{\omega c} \sin(\omega c t) \right] e^{i\omega x} d\omega. \qquad (8.207)$$

Now since $\cos\theta = (e^{i\theta} + e^{-i\theta})/2$, $\sin\theta = (e^{i\theta} - e^{-i\theta})/2i$, we have

$$\frac{1}{\sqrt{2\pi}} \int_{-\infty}^{\infty} F_1(\omega) \cos(\omega c t) e^{i\omega x} dx$$

$$= \frac{1}{2} \frac{1}{\sqrt{2\pi}} \int_{-\infty}^{\infty} F_1(\omega) \left(e^{i\omega c t} + e^{-i\omega c t} \right) e^{i\omega x} d\omega$$

$$= \frac{1}{2} \frac{1}{\sqrt{2\pi}} \int_{-\infty}^{\infty} F_1(\omega) \left(e^{i\omega(x+ct)} + e^{i\omega(x-ct)} \right) d\omega$$

$$= \frac{1}{2} [f_1(x + ct) + f_1(x - ct)].$$

Similarly,

$$\frac{1}{\sqrt{2\pi}} \int_{-\infty}^{\infty} F_2(\omega) \frac{\sin(\omega c t)}{\omega c} e^{i\omega x} d\omega$$

$$= \frac{1}{2} \frac{1}{\sqrt{2\pi}} \int_{-\infty}^{\infty} F_2(\omega) \frac{e^{i\omega c t} - e^{-i\omega c t}}{i\omega c} e^{i\omega x} d\omega$$

$$= \frac{1}{2} \frac{1}{\sqrt{2\pi}} \int_{-\infty}^{\infty} F_2(\omega) \frac{e^{i\omega(x+ct)} - e^{i\omega(x-ct)}}{i\omega c} d\omega$$

$$= \frac{1}{2c} \frac{1}{\sqrt{2\pi}} \int_{-\infty}^{\infty} F_2(\omega) \left(\int_{x-ct}^{x+ct} e^{i\omega\xi} d\xi \right) d\omega$$

$$= \frac{1}{2c} \int_{x-ct}^{x+ct} \left[\frac{1}{\sqrt{2\pi}} \int_{-\infty}^{\infty} e^{i\omega\xi} F_2(\omega) d\omega \right] d\xi$$

$$= \frac{1}{2c} \int_{x-ct}^{x+ct} f_2(\xi) d\xi.$$

Putting these together yields d'Alembert's formula

$$u(x, t) = \frac{1}{2}[f_1(x + ct) + f_1(x - ct)] + \frac{1}{2c} \int_{x-ct}^{x+ct} f_2(\xi)d\xi, \qquad (8.208)$$

which is also obtained in Problem 8.36.

Example 8.14 We shall find the solution of the following problem involving the Laplace equation in a half-plane:

$$\begin{aligned} u_{xx} + u_{yy} &= 0, \quad -\infty < x < \infty, \quad y > 0 \\ u(x, 0) &= f(x), \quad -\infty < x < \infty \\ |u(x, y)| &\le M, \quad -\infty < x < \infty, \quad y > 0, \end{aligned} \qquad (8.209)$$

where the function f is piecewise smooth and absolutely integrable in $(-\infty, \infty)$. If $f(x) \to 0$ as $|x| \to \infty$, then we also have the implied boundary conditions $\lim_{|x| \to \infty} u(x, y) = 0$, $\lim_{y \to +\infty} u(x, y) = 0$.

For this, as earlier, we let

$$f(x) = \frac{1}{\sqrt{2\pi}} \int_{-\infty}^{\infty} F(\omega)e^{i\omega x} d\omega, \quad F(\omega) = \frac{1}{\sqrt{2\pi}} \int_{-\infty}^{\infty} f(x)e^{-i\omega x} dx$$

and

$$u(x, y) = \frac{1}{\sqrt{2\pi}} \int_{-\infty}^{\infty} U(\omega, y)e^{i\omega x} d\omega.$$

We find that

$$0 = u_{xx} + u_{yy} = \frac{1}{\sqrt{2\pi}} \int_{-\infty}^{\infty} \left[-\omega^2 U(\omega, y) + \frac{\partial^2 U(\omega, y)}{\partial y^2} \right] e^{i\omega x} d\omega.$$

Thus, U must satisfy the ordinary differential equation

$$\frac{d^2 U}{dy^2} = \omega^2 U$$

and the initial condition $U(\omega, 0) = F(\omega)$ for each ω.

The general solution of this ordinary differential equation is $c_1 e^{\omega y} + c_2 e^{-\omega y}$. If we impose the initial condition and the boundedness condition, the solution becomes

$$U(\omega, y) = \begin{cases} F(\omega)e^{-\omega y}, & \omega \ge 0 \\ F(\omega)e^{\omega y}, & \omega < 0 \end{cases} = F(\omega)e^{-|\omega|y}.$$

Thus, the desired Fourier representation of the solution is

$$u(x, y) = \frac{1}{\sqrt{2\pi}} \int_{-\infty}^{\infty} F(\omega) e^{-|\omega|y} e^{i\omega x} d\omega.$$

To obtain an explicit representation, we insert the formula for $F(\omega)$ and formally interchange the order of integration, to obtain

$$u(x, y) = \frac{1}{2\pi} \int_{-\infty}^{\infty} \left(\int_{-\infty}^{\infty} f(\xi) e^{-i\omega\xi} d\xi \right) e^{-|\omega|y} e^{i\omega x} d\omega$$

$$= \frac{1}{2\pi} \int_{-\infty}^{\infty} \left(\int_{-\infty}^{\infty} e^{i\omega(x-\xi)} e^{-|\omega|y} d\omega \right) f(\xi) d\xi.$$

Now the inner integral is

$$\int_{-\infty}^{\infty} e^{i\omega(x-\xi)} e^{-|\omega|y} d\omega = 2Re\left(\int_{0}^{\infty} e^{i\omega(x-\xi)} e^{-\omega y} d\omega \right)$$

$$= 2Re\left(\frac{1}{y - i(x-\xi)} \right) = \frac{2y}{y^2 + (x-\xi)^2}.$$

Therefore, the solution $u(x, y)$ can be explicitly written as

$$u(x, y) = \frac{1}{\pi} \int_{-\infty}^{\infty} \frac{y}{y^2 + (x-\xi)^2} f(\xi) d\xi. \tag{8.210}$$

This representation is known as *Poisson's integral formula*.

Example 8.15 We shall employ the Fourier sine transform to find the solution of the following problem involving the Laplace equation in a semi-infinite strip

$$\begin{aligned}
u_{xx} + u_{yy} &= 0, & 0 < x < \infty, & \quad 0 < y < b \\
u(x, 0) &= f(x), & 0 < x < \infty \\
u(0, y) &= 0, & 0 < y < b \\
u(x, b) &= 0, & 0 < x < \infty,
\end{aligned} \tag{8.211}$$

where the function f is piecewise smooth and absolutely integrable in $[0, \infty)$. We shall also need the boundary conditions $\lim_{x \to \infty} u(x, y) = 0$ and $\lim_{x \to \infty} u_x (x, y) = 0$.

For this, we let

$$f(x) = \sqrt{\frac{2}{\pi}} \int_0^{\infty} F_s(\omega) \sin(\omega x) d\omega, \quad F_s(\omega) = \sqrt{\frac{2}{\pi}} \int_0^{\infty} f(x) \sin(\omega x) dx$$

and

$$u(x, y) = \sqrt{\frac{2}{\pi}} \int_0^{\infty} U_s(\omega, y) \sin(\omega x) d\omega.$$

This as in Example 8.14 leads to the same ordinary d.e. $U_s'' = \omega^2 U_s$, and hence

$$U_s(\omega, y) = c_1(\omega)\cosh(\omega y) + c_2(\omega)\sinh(\omega y).$$

Now the boundary condition $U_s(\omega, b) = 0$ yields

$$c_1(\omega) = -c_2(\omega)\frac{\sinh(\omega b)}{\cosh(\omega b)}.$$

Thus, we have

$$U_s(\omega, y) = -c_2(\omega)\frac{\sinh(\omega b)}{\cosh(\omega b)}\cosh(\omega y) + c_2(\omega)\sinh(\omega y) = c_2(\omega)\frac{\sinh(\omega(y - b))}{\cosh(\omega b)}.$$

Now since $U_s(\omega, 0) = F_s(\omega)$, we find

$$c_2(\omega) = -F_s(\omega)\frac{\cosh(\omega b)}{\sinh(\omega b)},$$

and therefore

$$U_s(\omega, y) = F_s(\omega)\frac{\sinh(\omega(b - y))}{\sinh(\omega b)}.$$

This gives the solution

$$u(x, y) = \sqrt{\frac{2}{\pi}}\int_0^\infty F_s(\omega)\frac{\sinh(\omega(b - y))}{\sinh(\omega b)}\sin(\omega x)d\omega$$

$$= \frac{2}{\pi}\int_0^\infty\int_0^\infty f(t)\sin(\omega t)\frac{\sinh(\omega(b - y))}{\sinh(\omega b)}\sin(\omega x)dtd\omega.$$

Problems

8.1 A telephone cable stretched tightly with constant tension T between supports at $x = 0$ and $x = 1$ hangs at rest under its own weight. For small displacements y, the equation of equilibrium and the boundary conditions are

$$y'' = -\frac{mg}{T}, \quad 0 < x < 1, \quad y(0) = 0 = y(1), \tag{8.212}$$

where m is the mass per unit length of the cable, and g is the gravitational constant. Show that the solution of (8.212) can be written as $y(x) = mgx(1 - x)/(2T)$, i.e., the telephone cable hangs in a parabolic arc.

8.2 A beam resting on two supports near its ends is called a *simple beam*. Let its length be L and let W be its weight per foot. Taking the origin $O = (0, 0)$ at the left support and considering the portion of the beam to the left of any section at point $P = (x, y)$ we find that the external forces are the reaction $WL/2$ and the load $(-wx)$. Their moments about the section are $WLx/2$ and $(-Wx^2/2)$, respectively. The algebraic sum of these moments is $M(x) = WLx/2 - Wx^2/2$. Now a result from the theory of the strength of materials states that $M(x) = EIy''$. Equating these two expressions, we obtain the boundary value problem

$$EIy'' = \frac{WLx}{2} - \frac{Wx^2}{2}, \quad y(0) = 0 = y(L).$$

Show that $y(x) = -W(x^4 - 2Lx^3 + L^3x)/(24EI)$ and $y'(L/2) = 0$ (which naturally follows from the symmetry), also $|y|_{max} = |y(L/2)| = 5WL^4/(384EI)$ feet. If we consider the portion of the beam to the right of the section, then it leads to the boundary value problem

$$EIy'' = \frac{WL(L - x)}{2} - \frac{W(L - x)^2}{2}, \quad y(0) = 0 = y(L).$$

Verify that it gives the same solution.

8.3 A beam is said to have a *fixed* or *clamped* end if the tangent to the elastic curve at that end always remains horizontal. Consider a beam of length L, weight W pound per foot, and fixed at both ends. For the portion to the left of any section x, the forces are the same as in Problem 8.2, but there is an unknown moment M at the left support because of the fixing of the beam. This leads to the boundary value problem

$$EIy'' = \frac{WLx}{2} - \frac{Wx^2}{2} + M, \quad y(0) = y'(0) = 0 = y(L).$$

Show that $y(x) = -W(x^4 - 2Lx^3 + L^2x^2)/(24EI)$ (the three boundary conditions determine the constant M) and $|y|_{max} = |y(L/2)| = WL^4/(384EI)$ feet.

8.4 A *cantilever* is a long projecting beam or girder fixed at one end and the other end free, used chiefly in bridge construction. Such a beam leads to the boundary value problem

$$EIy'' = -\frac{Wx^2}{2}, \quad y(0) = 0 = y'(L).$$

Show that $y(x) = -W(x^4 - 4L^3x)/(24EI)$ and $|y|_{max} = |y(L)| = WL^4/(8EI)$ feet.

8.5 A simple beam carries a concentrated load P at a point s feet from the left support. Letting $L - s = r$, we find that the reactions are Pr/L at the left support and Ps/L at the right support, assuming the weight of the beam can be neglected. This leads to the differential equations

$$EIy_1'' = Prx/L, \quad 0 \le x \le s, \quad y_1(0) = 0$$
$$EIy_2'' = Prx/L - P(x - s), \quad s \le x \le L, \quad y_2(L) = 0.$$

The elastic curve also have common ordinate and tangent under the load, i.e., $y_1(s) = y_2(s)$ and $y_1'(s) = y_2'(s)$. Show that

$$y_1(x) = \frac{Pr}{6LEI}[x^3 - s(r + L)x], \quad 0 \le x \le s$$

$$y_2(x) = \frac{P}{6LEI}[rx^3 - L(x - s)^3 - rs(r + L)x], \quad s \le x \le L.$$

Note that if $s = L/2$, then $y_1'(s) = y_2'(s) = 0$ and $y_1(s) = y_2(s) = -PL^3/(48EI)$.

8.6 The equation of equilibrium of a tightly stretched and initially straight elastic string embedded in an elastic foundation of modulus $k > 0$ is given by

$$y'' - (k/T)y = 0,$$

where y is the deflection of the string. Here the weight of the string is neglected, the deflections are assumed to be small, and the tension T is considered as a constant. The end $x = 0$ of the string is fixed, i.e., $y(0) = 0$, and at the end $x = L$ there is a displacement given by $y(L) = \beta > 0$. Show that $y(x) = \beta \sinh(\sqrt{k/T}x)/\sinh(\sqrt{k/T}L)$ is the solution of this boundary value problem, and $\max_{0 \le x \le L} y(x) = y(L) = \beta$.

8.7 A cantilever of length L, weighing W per foot, has a horizontal compressive force P acting at the free end. Taking the origin at the left free end, leads to the following boundary value problem

$$EIy'' = -Py - \frac{Wx^2}{2}, \quad y(0) = 0 = y'(L).$$

Show that

$$y(x) = A\cos(mx) + B\sin(mx) - \frac{W}{2Qm^2}\left(x^2 - \frac{2}{m^2}\right),$$

where $Q = EI$, $m^2 = P/Q$, $A = -W/(Qm^4)$, $B = W[\theta \sec\theta - \tan\theta)]/(Qm^4)$, and $\theta = mL$. Further, show that

$$|y|_{max} = |y(L)| = \frac{WQ}{P^2}\left(1 - \frac{\theta^2}{2} - \sec(\theta) + \theta\tan(\theta)\right).$$

Note that if $P \to 0$, then $m \to 0$ and $\theta \to 0$, and then $|y|_{max} \to WL^4/(8Q)$ as it should (see Problem 8.4).

8.8 The deflection of a beam of length L with the end $x = 0$ fixed and the other end $x = L$ simply supported and subjected to end force P satisfies the differential equation

$$\frac{d^2y}{dx^2} + \omega^2 y = \frac{\omega^2 R}{P}(L - x).$$

Show that

$$y(x) = \frac{R}{P}\left(\frac{\sin(\omega x)}{\omega} - L\cos(\omega x) + L - x\right),$$

where $\omega L = \tan(\omega L)$.

8.9 A horizontal beam of length L, weighing W pounds per foot, is fixed at both ends. In addition, there is a horizontal pull P at each end so that the beam acts as a tie. If the origin is taken at the left end, then it leads to the following boundary value: problem

$$EIy'' = M + \frac{WLx}{2} - \frac{Wx^2}{2} + Py, \quad y(0) = y'(0) = 0 = y(L) = y'(L),$$

where M is the unknown moment at the left support due to the clamping, and $WL/2$ is the reaction. Show that

$$y(x) = \frac{WL}{2mP}\left[[1 - \cosh(mx)]\coth\left(\frac{mL}{2}\right) + \sinh(mx) - mx\right] + \frac{Wx^2}{2P},$$

where $m = P/(EI)$. Further, show that

$$|y|_{max} = |y(L/2)| = \frac{WL}{2mP}\left(\tanh\left(\frac{mL}{4}\right) - \frac{mL}{4}\right).$$

8.10 A gas diffuses into a liquid in a narrow pipe. Let $y(x)$ denote the concentration of the gas at the distance x in the pipe. The gas is absorbed by the liquid at a rate proportional to $y'(x)$, and the gas reacts chemically with the liquid and as a result disappears at a rate proportional to $y(x)$. This leads to the balance equation

$$y'' - (k/D)y = 0,$$

where k is the reaction rate and D is the diffusion coefficient. If the initial concentration is α, i.e., $y(0) = \alpha$ and at $x = L$ the gas is completely absorbed by the liquid, i.e., $y(L) = 0$, show that $y(x) = \alpha \sinh[\sqrt{k/D}(L - x)]/\sinh[\sqrt{k/D}L]$.

8.11 The pressure $p(x)$ in the lubricant under a plane pad bearing satisfies the boundary value problem

$$\frac{d}{dx}\left(x^3\frac{dp}{dx}\right) = -K, \quad p(a) = 0 = p(b), \quad 0 < a < b.$$

Show that

$$p(x) = K\left(\frac{1}{x} - \frac{1}{a}\right) + \frac{Kb}{a+b}\left(\frac{1}{a} - \frac{a}{x^2}\right).$$

8.12 A long river flows through a populated region with uniform velocity u. Sewage continuously enters at a constant rate at the beginning of the river $x = 0$. The sewage is convected down the river by the flow and it is simultaneously decomposed by bacteria and other biological activities. Assume that the river is sufficiently narrow so that the concentration y of sewage is uniform over the cross section and that the polluting has been going on for a long time, so that y is a function only of the distance x downstream from the sewage plant. If the rate of decomposition at x is proportional to the concentration $y(x)$ and k is the proportionality constant, then y satisfies the differential equation

$$y'' - \beta y' - \alpha^2 y = 0,$$

where $\beta = u/D$ and $\alpha^2 = k/(AD)$; A is the cross-sectional area of the river and $D > 0$ is a constant. If the concentrations at $x = 0$ and $x = L$ are known to be

$$y(0) = y_0, \quad y(L) = y_1 \ (< y_0)$$

then show that the concentration in the stream for $0 \le x \le L$ is

$$y(x) = e^{\beta x/2}\left[y_0\cosh(\theta x) + \frac{y_1 e^{-L/2} - y_0\cosh(\theta L)}{\sinh(\theta L)}\sinh(\theta x)\right],$$

where $\theta = \sqrt{\beta^2 + 4\alpha^2}/2$.

8.13 For the telephone cable considered in Problem 8.1, the large displacements y are governed by the equation and boundary conditions

$$y'' = -\frac{mg}{T}\sqrt{1 + (y')^2}, \quad 0 < x < 1, \quad y(0) = 0 = y(1). \tag{8.213}$$

Show that the solution of (8.213) can be written as

$$y(x) = \frac{T}{mg}\left[\cosh\left(\frac{mg}{2T}\right) - \cosh\left(\frac{mg}{T}\left(x - \frac{1}{2}\right)\right)\right],$$

i.e., the telephone cable hangs in a catenary.

8.14 Suppose a hollow spherical shell has an inner radius $r = \alpha$ and outer radius $r = \beta$, and the temperature at the inner and outer surfaces are u_α and u_β, respectively. The temperature u at a distance r from the center ($\alpha \le r \le \beta$) is determined by the boundary value problem

$$r\frac{d^2u}{dr^2} + 2\frac{du}{dr} = 0, \quad u(\alpha) = u_\alpha, \quad u(\beta) = u_\beta.$$

Show that

$$u(r) = \frac{u_\beta\alpha^{-1} - u_\alpha\beta^{-1}}{\alpha^{-1} - \beta^{-1}} + \left(\frac{u_\alpha - u_\beta}{\alpha^{-1} - \beta^{-1}}\right)r^{-1}.$$

8.15 A steam pipe has temperature u_α at its inner surface $r = \alpha$ and temperature u_β at its outer surface $r = \beta$. The temperature u at a distance r from the center $(\alpha \le r \le \beta)$ is determined by the boundary value problem

$$r\frac{d^2u}{dr^2} + \frac{du}{dr} = 0, \quad u(\alpha) = u_\alpha, \quad u(\beta) = u_\beta.$$

Show that

$$u(r) = \frac{u_\alpha \ln(r/\beta) - u_\beta \ln(r/\alpha)}{\ln(\alpha/\beta)}.$$

8.16 In the kidney dialysis process the following boundary value problem: occurs (see [4])

$$Q_B u' = -k(u - v)$$
$$-Q_D v' = k(u - v)$$
$$u(0) = u_0, \quad v(L) = 0$$

where $u(x)$ represents the concentration of wastes in the blood, $v(x)$ is the concentration of wastes in the dialysis, x is the distance along the dialyser, Q_D is the flow rate of the dialysate through the machine, Q_B is the flow rate of the blood through the machine, k is the proportionality constant, L is the length of the dialyser, and u_0 is the initial concentration of wastes in the blood. Show that

(1) $u(x) = u_0\dfrac{Q_B e^{\alpha x} - Q_D e^{\alpha L}}{e^{\alpha x}(Q_B - Q_D e^{\alpha L})}, \quad v(x) = u_0\dfrac{Q_B(e^{\alpha x} - e^{\alpha L})}{e^{\alpha x}(Q_B - Q_D e^{\alpha L})},$ where $\alpha = (k/Q_B) - (k/Q_D),$

(2) the amount of waste removed, i.e., $\int_0^L k[u(x) - v(x)]dx$ satisfies the relation

$$\int_0^L k[u(x) - v(x)]dx = Q_B[u_0 - u(L)],$$

(3) the clearance of a dialyser, i.e., $Q_B(u_0 - u(L))/u_0$ satisfies the relation

$$\frac{Q_B}{u_0}(u_0 - u(L)) = Q_B\frac{1 - e^{-\alpha L}}{1 - \frac{Q_B}{Q_D}e^{-\alpha L}}.$$

8.17 The temperature $y(x)$ in a cooling fin satisfies the following boundary value problem

$$y'' = \frac{hC}{\kappa A}(y - T), \quad y(0) = T_0, \quad -\kappa y'(a) = h[y(a) - T].$$

This means that the temperature at the left end is kept at $T_0 > T$, while the surface of the rod and its right end exchange heat with a surrounding medium at temperature T. Show that $y(x) = T + c_1 \cosh(\lambda x) + c_2 \sinh(\lambda x)$, where $\lambda = \sqrt{hC/(\kappa A)}$ and $c_1 = T_0 - T$, $c_2 = -[\kappa \lambda \sinh(\lambda a) + h \cosh(\lambda a)][T_0 - T]/[\kappa \lambda \cosh(\lambda a) + h \sinh(\lambda a)]$.

8.18 [7] A collection of nuclear fuel rods is housed in a pressure vessel that is shaped roughly like a cylinder with flat or hemispherical ends. The temperature in the thick steel wall of the vessel affects its strength and thus must be studied for design and safety. Assuming the vessel as a long cylinder, so that we can ignore the effects of the ends, for the temperature $u(r)$ in the wall (in cylindrical coordinates) the following differential equation occurs

$$\frac{1}{r}\frac{d}{dr}\left(r\frac{du}{dr}\right) = 0, \quad \alpha < r < \beta$$

(same as in Problem 8.15), where α and β are the inner and outer radii. The boundary conditions involve convention, with hot pressurized at the inner radius and with air at the outer radius, i.e.,

$$-\kappa u'(\alpha) = h_0[T_w - u(\alpha)], \quad \kappa u'(\beta) = h_1[T_a - u(\beta)].$$

Show that $u(r) = c_1 \ln(r/\alpha) + c_2$, where $c_1 = h_0 h_1(T_a - T_w)/D$, $c_2 = [h_0(\kappa/\beta + h_1 \ln(\beta/\alpha))T_w + (\kappa/\alpha)h_1 T_a]/D$, and $D = h_1\kappa/\alpha + h_0\kappa/\beta + h_0 h_1 \ln(\beta/\alpha)$.

8.19 Imposing some ideal conditions sandwich beam analysis leads to a third-order linear differential equation

$$y''' - k^2 y' + r = 0 \tag{8.214}$$

together with three-point boundary conditions

$$y'(0) = y(1/2) = y'(1) = 0, \tag{8.215}$$

where k and r are some physical constants, for details see [5]. Show that the problem (8.214), (8.215) has a unique solution and can be expressed in terms of elementary functions (see [1])

$$y(x) = \frac{r}{k^3}\left[\sinh\left(\frac{k}{2}\right) - \sinh(kx) + k\left(x - \frac{1}{2}\right) + \tanh\left(\frac{k}{2}\left[\cosh(kx) - \cosh\left(\frac{k}{2}\right)\right]\right)\right].$$

8.20 Consider the differential equation (4.65) with $n = 4$, i.e.,

$$x^4 y'' + k^2 y = 0. \tag{8.216}$$

(1) Verify that the general solution of (8.216) is

$$y(x) = x \left[A \cos\left(\frac{k}{x}\right) + B \sin\left(\frac{k}{x}\right) \right].$$

(2) Find the eigenvalues and eigenfunctions of the problem (8.216), $y(\alpha) = y(\beta) = 0, 0 < \alpha < \beta.$ $\left(k_n = \frac{n\pi\alpha\beta}{\beta-\alpha}, \ \phi_n(x) = x \sin\left(\frac{n\pi\beta(x-\alpha)}{x(\beta-\alpha)}\right) \right).$

8.21 In many beam problems, the forces or loads are moving or changing with time. This causes the beam to vibrate. For example, some of the loads acting on a monorail track are its own weight, the weight of the moving cars, and thermal forces. The weight of the rail itself is a constant load, but the latter two loads are time-dependent. In the analysis of the effect of time-dependent loads on the deflection of a simply supported beam, the following eigenvalue problem occurs:

$$y^{(4)} - \lambda y = 0 \tag{8.217}$$

$$y(0) = 0, \quad y(L) = 0, \quad y''(0) = 0, \quad y''(L) = 0. \tag{8.218}$$

The quantity λ in Eq. (8.217) is given by

$$\lambda = \frac{Am}{EI}\omega^2, \tag{8.219}$$

where A is the cross-sectional area of the beam, m is the mass per unit volume of the beam's material, and ω is the frequency of free vibration of the beam. Find the eigenvalues and eigenfunctions of the problem (8.217), (8.218). For each eigenvalue ω obtained from (8.219) gives a corresponding frequency of free vibration of the beam, and the eigenfunction $y(x)$ gives the shape of the beam as it oscillates at the corresponding natural frequency. $\left(\lambda_n = \frac{n^4\pi^4}{L^4}, \ \phi_n(x) = \sin\left(\frac{n\pi x}{L}\right) \right)$

8.22 Find the eigenvalues and eigenfunctions of the differential equation (8.217) with the boundary conditions

(1) $\qquad\qquad\qquad y(0) = y'(0) = y''(L) = y'''(L) = 0. \tag{8.220}$

$\Big(\lambda_n = (\mu_n/L)^4$ where μ_n is the nth root of the equation $\cosh(\mu)\cos(\mu) + 1 = 0,$
$\phi_n(x) = \left[\frac{\cosh(\mu_n x/L)-\cos(\mu_n x/L)}{\cosh(\mu_n)+\cos(\mu_n)} - \frac{\sinh(\mu_n x/L)-\sin(\mu_n x/L)}{\sinh(\mu_n)+\sin(\mu_n)} \right] \Big)$

(2) $\qquad\qquad\qquad y(0) = y'(0) = y(L) = y'(L) = 0. \tag{8.221}$

$\Big(\lambda_n = (\mu_n/L)^4$ where μ_n is the n–th root of the equation $\cosh(\mu)\cos(\mu) - 1 = 0,$
$\phi_n(x) = \left[\frac{\cos(\mu_n x/L)-\cosh(\mu_n x/L)}{\cos(\mu_n)-\cosh(\mu_n)} - \frac{\sin(\mu_n x/L)-\sinh(\mu_n x/L)}{\sin(\mu_n)-\sinh(\mu_n)} \right] \Big)$

8.23 Consider a long column or rod of length L that is supporting an axial compressive force P applied at the top (see Fig. 8.2). We want to determine the magnitude of the compressive force P that can be applied to the column without the occurrence of buckling, and if buckling does occur, we then want to know the possible modes of lateral deflection of the column from the equilibrium position ($y = 0$). From the principles of beam theory, the governing equation for this problem is the differential equation

$$\frac{d^2}{dx^2}\left[EIy''\right] + Py'' = 0, \quad 0 < x < L.$$

where E is Young's modulus of the beam material and I is the moment of inertia of the beam's cross section. The same differential equation occurs in studying the deflection of beams in various structures that are likewise subject to compressive forces. In many cases of interest the cross section of the beam is uniform throughout the beam. Hence, in this case both E and I are constant and then the above differential equation is expressed in the form

$$y^{(4)} + \lambda y'' = 0, \quad 0 < x < L \tag{8.222}$$

where $\lambda = P/(EI)$. Find the eigenvalues and eigenfunctions of problem (8.222), (8.218). Further, show that (8.222), (8.220) is not an eigenvalue problem. $\left(\lambda_n = \frac{n^2\pi^2}{L^2}, \ \phi_n(x) = \sin\left(\frac{n\pi x}{L}\right)\right)$

8.24 The differential equation for the displacement y of a whirling shaft when the weight of the shaft is taken into account is

$$EIy^{(4)} - \frac{W\omega^2}{g}y = W.$$

Taking the shaft of length $2L$ with the origin at the center and simply supported at both ends, show that

$$y(x) = \frac{g}{2\omega^2}\left[\frac{\cosh(ax)}{\cosh(aL)} + \frac{\cos(ax)}{\cos(aL)} - 2\right], \quad a^4 = \frac{W\omega^2}{EIg}$$

and hence the maximum deflection of the shaft is

$$\max_{-L \le x \le L} y(x) = y(0) = \frac{g}{2\omega^2}[\operatorname{sech}(aL) + \sec(aL) - 2].$$

8.25 Consider the differential equation in (8.49) together with the boundary conditions

$$y'(0) = 0, \quad \lim_{x \to 1} y(x) < \infty.$$

Show that the eigenvalues of this problem are $\lambda_n = 2n(2n + 1)$, $n = 0, 1, 2, \cdots$ and the corresponding eigenfunctions are the even Legendre polynomials $P_{2n}(x)$.

8.26 Consider the differential equation in (8.49) together with the boundary conditions

$$y(0) = 0, \quad \lim_{x \to 1} y(x) < \infty.$$

Show that the eigenvalues of this problem are $\lambda_n = (2n + 1)(2n + 2)$, $n = 0, 1, 2, \cdots$ and the corresponding eigenfunctions are the odd Legendre polynomials $P_{2n+1}(x)$.

8.27 Consider the singular eigenvalue problem

$$y'' - 2xy' + \lambda y = 0 = \left(e^{-x^2} y'\right)' + \lambda e^{-x^2} y$$

$$\lim_{x \to -\infty} \frac{y(x)}{|x|^k} < \infty, \quad \lim_{x \to \infty} \frac{y(x)}{x^k} < \infty \quad \text{for some positive integer } k.$$

Show that the eigenvalues of this problem are $\lambda_n = 2n$, $n = 0, 1, 2, \cdots$ and the corresponding eigenfunctions are the Hermite polynomials $H_n(x)$.

8.28 Consider the singular eigenvalue problem

$$xy'' + (1 - x)y' + \lambda y = 0 = \left(xe^{-x} y'\right)' + \lambda e^{-x} y$$

$$\lim_{x \to 0} |y(x)| < \infty, \quad \lim_{x \to \infty} \frac{y(x)}{x^k} < \infty \quad \text{for some positive integer } k.$$

Show that the eigenvalues of this problem are $\lambda_n = n$, $n = 0, 1, 2, \cdots$ and the corresponding eigenfunctions are the Laguerre polynomials $L_n(x)$.

8.29 Let $a \geq 0$ be fixed, and b_n, $n = 0, 1, 2, \cdots$ be the zeros of the Bessel function $J_a(x)$ (see Problem 4.22). Show that the singular eigenvalue problem

$$x^2 y'' + xy' + (\lambda x^2 - a^2)y = 0 = (xy')' + \left(\lambda x - \frac{a^2}{x}\right) y$$

$$\lim_{x \to 0} y(x) < \infty, \quad y(1) = 0$$

has the eigenvalues $\lambda_n = b_n^2$, $n = 0, 1, 2, \cdots$ and the corresponding eigenfunctions are $J_a(b_n x)$.

8.30 Show that for the heat equation (8.70) with $c = 1$ the function

$$\phi(x, t) = A + B \int_0^{x/\sqrt{t}} e^{-s^2/4} ds$$

as well as its partial derivatives $\phi_x(x, t)$ and $\phi_t(x, t)$ are solutions; here, A and B are arbitrary constants.

8.31 Let $\psi(\tau)$ be an arbitrary function such that

$$\phi(x, t) = \int_0^\infty \psi(\tau) \frac{1}{(t + \tau)^{1/2}} \exp\left(-\frac{x^2}{4(t + \tau)}\right) d\tau$$

is convergent. Show that $\phi(x, t)$ is a solution of (8.70) with $c = 1$.

8.32 In Example 8.7, suppose that the ends of the rod, i.e., $x = 0$ and $x = a$, are kept at the fixed temperatures A and B, respectively. This means that we have the boundary conditions

$$u(0, t) = A, \quad t > 0 \tag{8.223}$$

$$u(a, t) = B, \quad t > 0. \tag{8.224}$$

Let $u(x, t)$ be a solution of (8.70), (8.71), (8.223), (8.224). Show that the function

$$v(x, t) = u(x, t) + \left(\frac{x - a}{a}\right) A - \frac{x}{a} B \tag{8.225}$$

is a solution of the initial-boundary value problem

$$\begin{aligned}
v_t - c^2 v_{xx} &= 0, \quad 0 < x < a, \quad t > 0, \quad c > 0 \\
v(0, t) &= 0, \quad t > 0 \\
v(a, t) &= 0, \quad t > 0 \\
v(x, 0) &= f(x) + \left(\frac{x - a}{a}\right) A - \frac{x}{a} B, \quad 0 < x < a,
\end{aligned} \tag{8.226}$$

which is of the type (8.70)–(8.73), and thus we can find its solution $v(x, t)$. The solution $u(x, t)$ is then obtained by the relation (8.225).

8.33 Heat conduction in a thin circular ring (consider it as a rod, bent into the shape of a circular ring by tightly joining the two ends) of length $2a$, labeled from $-a$ to a leads to the equation $u_t = c^2 u_{xx}$, $-a < x < a$, $t > 0$, $c > 0$ with the initial condition $u(x, 0) = f(x)$, $-a < x < a$ and the *periodic boundary conditions*

$$u(-a, t) = u(a, t)$$
$$u_x(-a, t) = u_x(a, t), \quad t > 0.$$

Find the solution of this initial-boundary value problem.

8.34 If the lateral surface of the rod is not insulated, there is a heat exchange by convection into the surrounding medium. If the surrounding medium has constant temperature T_0, the rate at which heat is lost from the rod is proportional to the difference $u - T_0$. The governing partial differential equation in this situation is

$$c^2 u_{xx} = u_t + b(u - T_0), \quad 0 < x < a, \quad b > 0.$$

Show that the change of variable $u(x, t) = T_0 + v(x, t)e^{-bt}$ leads to the heat equation (8.70) in v.

8.35 Consider the particular case of the telegraph equation (8.119),

$$v_{tt} + 2av_t + a^2 v = c^2 v_{xx}$$

with the initial conditions $v(x, 0) = \phi(x)$, $v_t(x, 0) = 0$ and the boundary conditions $v(0, t) = v(a, 0) = 0$. Show that the change of variable $v(x, t) = e^{-at} u(x, t)$ transforms this initial-boundary value problem to (8.110)–(8.114) with $f(x) = \phi(x)$ and $g(x) = a\phi(x)$.

8.36 Show that the solution (8.126) of (8.110)–(8.114) can be written as

$$u(x, t) = \frac{1}{2}[f(x + ct) + f(x - ct)] + \frac{1}{2c} \int_{x-ct}^{x+ct} g(z)dz.$$

This is *d'Alembert's solution*. Thus, to find the solution $u(x, t)$ we need to know only the initial displacement $f(x)$ and the initial velocity $g(x)$. This makes d'Alembert's solution easy to apply as compared to the infinite series (32.24).

8.37 The partial differential equation which describes the small displacement $w = w(x, t)$ of a heavy flexible chain of length a from equilibrium is

$$\frac{\partial^2 w}{\partial t^2} = -g\frac{\partial w}{\partial x} + g(a - x)\frac{\partial^2 w}{\partial x^2},$$

where g is the gravitational constant. This equation was studied extensively by Daniel Bernoulli around 1732 and later by Euler in 1781.
(1) Set $y = a - x$, $u(y, t) = w(a - x, t)$ to transform the above partial differential equation to

$$\frac{\partial^2 u}{\partial t^2} = g\frac{\partial u}{\partial y} + gy\frac{\partial^2 u}{\partial y^2}. \tag{8.227}$$

(2) Use separation of variables to show that the solution of (8.227) which is bounded for $0 \le y \le a$ and satisfies $u(a, t) = 0$ is

$$\sum_{n=1}^{\infty} J_0\left(2\lambda_n\sqrt{\frac{y}{g}}\right)[a_n \cos(\lambda_n t) + b_n \sin(\lambda_n t)],$$

where $\lambda_n = (1/2)b_{0,n}\sqrt{g/a}$ and $b_{0,n}$ is a positive root of $J_0(x)$.

8.38 A bar has length a, density δ, cross-sectional area A, Young's modulus E, and total mass $M = \delta Aa$. Its end $x = 0$ is fixed and a mass m is attached to its free end.

The bar initially is stretched linearly by moving m a distance $d = ba$ to the right, and at time $t = 0$ the system is released from rest. Find the subsequent vibrations of the bar by solving the initial-boundary value problem (8.110), $u(x, 0) = bx$, $u_t(x, 0) = 0$, (8.113) and $mu_{tt}(a, t) = -AEu_x(a, t)$.

8.39 Small transverse vibrations of a beam are governed by the partial differential equation

$$\frac{\partial^2 u}{\partial t^2} + c^2 \frac{\partial^4 u}{\partial x^4} = 0, \quad 0 < x < a, \quad t > 0,$$

where $c^2 = EI/(A\mu)$, and E is the modulus of elasticity, I is the moment of inertia of any cross section about the x-axis, A is the area of cross section, and μ is the mass per unit length. Boundary conditions at the ends of the beam are usually of the following type:

(1) a *fixed end* also known as *built-in* or a *clamped end* has the displacement and slope equal to zero, (see Fig. 8.16a)

$$u(a, t) = \frac{\partial u}{\partial x}(a, t) = 0$$

(2) a *simply supported end* has displacement and moment equal to zero, (see Fig. 8.16b)

$$u(a, t) = \frac{\partial^2 u}{\partial x^2}(a, t) = 0$$

(3) a *free end* has zero moment and zero shear, (see Fig. 8.16c)

$$\frac{\partial^2 u}{\partial x^2}(a, t) = \frac{\partial^3 u}{\partial x^3}(a, t) = 0.$$

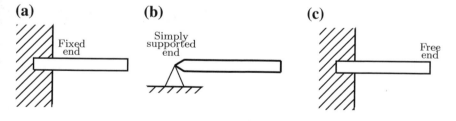

(a) Fixed end

(b) Simply supported end

(c) Free end

Fig. 8.16

Find the solution for the vibration of a beam that has simply supported ends at $x = 0$ and $x = a$ with the initial conditions $u(x, 0) = f(x)$, $u_t(x, 0) = g(x)$, $0 < x < a$.

8.40 Show that Neumann boundary value problem $\Delta_2 u = 0$, $u_y(x, 0) = f(x)$, $u_y(x, b) = g(x)$, $u_x(0, y) = 0 = u_x(a, y)$ has an infinite number of solutions. (If u is a solution, then $(u + K)$ is also a solution)

8.41 A rectangular plate with insulated surface is a cm. wide and so long compared to its width that it may be considered infinite in length without introducing an appreciable error. If the temperature of the short edge $y = 0$ is given by $f(x)$, $0 < x < a$ and the two edges $x = 0$, $x = a$ are kept at $0\,°C$, determine the temperature at any point of the plate in the steady state. $\left(u(x, t) = \sum_{n=1}^{\infty} e^{-n\pi y/a} \sin\left(\frac{n\pi x}{a}\right), \; a_n = \frac{2}{a} \int_0^a f(x) \sin\left(\frac{n\pi x}{a}\right) dx \right)$

8.42 A circular plate of unit radius, whose faces are insulated, has upper half of its boundary kept at constant temperature T_1 and the lower half at constant temperature T_2. Find the steady-state temperature of the plate. $\left(\frac{T_1+T_2}{2} + \frac{2}{\pi}(T_1 - T_2) \sum_{n=1}^{\infty} \frac{1}{2n-1} r^{2n-1} \sin\left((2n-1)\theta\right) = \frac{T_1+T_2}{2} + \frac{T_1-T_2}{\pi} \tan^{-1}\left(\frac{2r \sin\theta}{1-r^2}\right) \right)$

8.43 Show that a necessary condition for the existence of a solution to the Neumann problem (8.176),

$$\frac{\partial u}{\partial r}(a, \theta) = f(\theta), \quad -\pi < \theta \le \pi \tag{8.228}$$

is that

$$\int_{-\pi}^{\pi} f(\phi)d\phi = 0,$$

i.e., the mean value of the normal derivative on the boundary is zero. (Use Green's theorem $\int\int_S (f\Delta_2 g - g\Delta_2 f)dS = \int_\Gamma \left(f\frac{\partial g}{\partial n} - g\frac{\partial f}{\partial n} \right) ds$)

8.44 Solve the Laplace equation (8.176) in the wedge with three sides $\theta = 0$, $\theta = \beta$, and $r = a$ (see Fig. 8.17) and the boundary conditions $u(r, 0) = 0 = u(r, \beta)$, $0 < r < a$, and (8.177) for $0 < \theta < \beta$. $\left(u(r, \theta) = \sum_{n=1}^{\infty} A_n r^{n\pi/\beta} \sin\left(\frac{n\pi\theta}{\beta}\right), \; A_n = \frac{2}{\beta} a^{-n\pi/\beta} \int_0^\beta f(\phi) \sin\left(\frac{n\pi\phi}{\beta}\right) d\phi \right)$

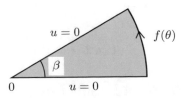

Fig. 8.17

8.45 Solve the same problem as in Problem 8.44 with condition (8.177) replaced by the Neumann condition (8.228) for $0 < \theta < \beta$. $(u(r, \theta) = \sum_{n=1}^{\infty} A_n r^{n\pi/\beta}$
$\sin\left(\frac{n\pi\theta}{\beta}\right)$, $A_n = \frac{2}{n\pi} a^{1-n\pi/\beta} \int_0^{\beta} f(\phi) \sin\left(\frac{n\pi\phi}{\beta}\right) d\phi)$

8.46 The diameter of a semi-circular plate of radius a is kept at $0°$ C and the temperature at the semi-circular boundary at T °C. Show that the steady-state temperature in the plate is given by

$$u(r, \theta) = \frac{4T}{\pi} \sum_{n=1}^{\infty} \frac{1}{2n-1} \left(\frac{r}{a}\right)^{2n-1} \sin\left((2n-1)\theta\right).$$

8.47 A semi-circular plate of radius a has its circumference kept at temperature $k\theta(\pi - \theta)$ while the boundary diameter is kept at zero temperature. Find the steady state temperature distribution $u(r, \theta)$ of the plate assuming the lateral surfaces of the plate to be insulated. $(\frac{8k}{\pi} \sum_{n=1}^{\infty} \frac{1}{(2n-1)^3} \left(\frac{r}{a}\right)^{2n-1} \times \sin\left((2n-1)\theta\right))$

8.48 Solve the Laplace equation (8.176) in the annulus $0 < a^2 < x^2 + y^2 < b^2$ (see Fig. 8.18) with the Dirichlet conditions $u(a, \theta) = f(\theta)$, $u(b, \theta) = g(\theta)$, $-\pi < \theta < \pi$. $(u(r, \theta) = \frac{1}{2}(c_0 + d_0 \ln(r)) + \sum_{n=1}^{\infty} (a_n r^n + d_n r^{-n})[A_n \cos(n\theta) + B_n \sin(n\theta)]$, where the unknowns are determined by using the boundary conditions)

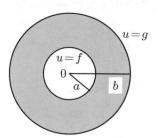

Fig. 8.18

8.49 The velocity potential function $u(r, \theta)$ for steady flow of an ideal fluid around a cylinder of radius $r = a$ satisfies (8.176) for $r > a$ with the boundary conditions

$$u_r(a, \theta) = 0, \quad u(r, \theta) = u(r, -\theta)$$
$$\lim_{r \to \infty} [u(r, \theta) - U_0 r \cos(\theta)] = 0.$$

Find its solution and the components of the velocity. $(\frac{U_0}{r}(r^2 + a^2) \cos\theta, u_x = \frac{U_0}{r^2}[r^2 - a^2 \cos(2\theta)], u_y = -\frac{U_0}{r^2} a^2 \sin(2\theta))$

8.50 From the real part of the solution of Laplace's equation in two independent variables

$$f(x + iy) = \frac{ae^{i\phi} + x + iy}{ae^{i\phi} - (x + iy)}$$

show that *Poisson's integral*

$$\frac{a^2 - r^2}{2\pi} \int_0^{2\pi} \frac{V(\phi)}{a^2 + r^2 - 2ar\cos(\theta - \phi)} d\phi,$$

where $x = r\cos\theta$, $y = r\sin\theta$ and V is an arbitrary function, is a solution.

8.51 In elasticity certain problems in plane stress can be solved with the aid of *Airy's stress function* ϕ, which satisfies the partial differential equation

$$\phi_{xxxx} + 2\phi_{xxyy} + \phi_{yyyy} = 0. \tag{8.229}$$

This equation is called *biharmonic equation* and also occurs in the study of hydrodynamics. Show that
(1) $\phi(x, y) = f_1(y - ix) + xf_2(y - ix) + f_3(y + ix) + xf_4(y + ix)$ is a solution of (8.229)
(2) if $u(x, y)$ and $v(x, y)$ are any two harmonic functions, then $\phi(x, y) = u(x, y) + xv(x, y)$ is a solution of (8.229).

8.52 Show that the solution (8.201) can be written as

$$u(x, t) = \frac{1}{\sqrt{\pi}} \int_{-\infty}^{\infty} e^{-w^2} f(x + 2c\sqrt{t}\, w) dw.$$

(Use the substitution $\mu = x + 2c\sqrt{t}\, w$)

8.53 Show that the solution (8.203) can be written as

$$u(x, t) = \frac{2}{\pi} \int_0^{\infty} \int_0^{\infty} f(\mu) e^{-c^2\omega^2 t} \sin(\omega\mu) \sin(\omega x) d\omega d\mu.$$

(Use $e^{-ax^2} = \frac{1}{\sqrt{\pi a}} \int_0^{\infty} e^{-\omega^2/4a} \cos(\omega x) d\omega$)

8.54 Show that the solution (8.205) can be written as

$$u(x, t) = \frac{2}{\pi} \int_0^{\infty} \int_0^{\infty} f(\mu) e^{-c^2\omega^2 t} \cos(\omega\mu) \cos(\omega x) d\omega d\mu.$$

8.55 Use the Fourier cosine transform to solve the following problem:

$$u_t = u_{xx}, \quad x > 0, \quad t > 0$$
$$u(x, 0) = 0, \quad x > 0$$
$$u(x, t) \to 0 \quad \text{as} \quad x \to \infty, \quad t > 0$$
$$u_x(0, t) = f(t), \quad t > 0$$

where the function f is piecewise smooth and absolutely integrable in $[0, \infty)$.

$(u(x, t) = -\frac{1}{\sqrt{\pi}} \int_0^t \frac{1}{\sqrt{t-\mu}} f(\mu) \exp\left(-\frac{x^2}{4(t-\mu)}\right) d\mu)$

8.56 Use the Fourier transform to solve the following problem for a heat equation with transport term:

$$u_t = c^2 u_{xx} + k u_x, \quad -\infty < x < \infty, \quad t > 0, \quad c > 0, \quad k > 0$$
$$u(x, 0) = f(x), \quad -\infty < x < \infty$$
$$u(x, t) \text{ and } u_x(x, t) \text{ finite as } |x| \to \infty, \quad t > 0,$$

where the function f is piecewise smooth and absolutely integrable in $(-\infty, \infty)$.

$(u(x, t) = \int_{-\infty}^{\infty} f(\mu) \frac{e^{-(x-\mu+kt)^2/(4c^2 t)}}{\sqrt{4\pi c^2 t}} d\mu)$

8.57 Use the Fourier transform to solve the following nonhomogeneous problem:

$$u_t = c^2 u_{xx} + q(x, t), \quad -\infty < x < \infty, \quad t > 0$$
$$u(x, 0) = f(x), \quad -\infty < x < \infty$$
$$u(x, t) \to 0, \quad u_x(x, t) \to 0 \quad \text{as} \quad |x| \to \infty, \quad t > 0,$$

where the function f is piecewise smooth and absolutely integrable in $(-\infty, \infty)$.

$(\int_{-\infty}^{\infty} f(\mu) \frac{e^{-(x-\mu)^2/(4c^2 t)}}{\sqrt{4\pi c^2 t}} d\mu + \int_0^t \int_{-\infty}^{\infty} q(\mu, \tau) \frac{e^{-(x-\mu)^2/(4c^2(t-\tau))}}{\sqrt{4\pi c^2(t-\tau)}} d\mu d\tau)$

8.58 Use the Fourier sine transform to solve the following problem:

$$u_{tt} = u_{xx}, \quad x > 0, \quad t > 0$$
$$u(x, 0) = 0, \quad x > 0$$
$$u_t(x, 0) = 0, \quad x > 0$$
$$u(0, t) = f(t), \quad t > 0$$
$$u(x, t) \text{ and } u_x(x, t) \to 0 \quad \text{as} \quad x \to \infty, \quad t > 0$$

where the function f is piecewise smooth and absolutely integrable in $[0, \infty)$.

$(u(x, t) = \frac{2}{\pi} \int_0^{\infty} \int_0^t f(\mu) \sin(\omega(t - \mu)) \sin(\omega x) d\mu d\omega)$

8.59 Find the solution of the wave equation

$$u_{tt} = c^2 u_{xx} - ku, \quad -\infty < x < \infty, \quad t > 0, \quad c > 0, \quad k > 0$$

satisfying the same conditions as in (8.206). $\quad (\frac{1}{\sqrt{2\pi}} \int_{-\infty}^{\infty} [F_1(\omega) \cos(t\sqrt{k+\omega^2 c^2})$

$+ F_2(\omega) \frac{\sin(t\sqrt{k+\omega^2 c^2})}{\sqrt{k+\omega^2 c^2}}] e^{i\omega x} d\omega)$

8.60   Show that the solution of the following Neumann problem:

$$u_{xx} + u_{yy} = 0, \quad -\infty < x < \infty, \quad y > 0$$
$$u_y(x, 0) = f(x), \quad -\infty < x < \infty$$
$$u(x, y) \text{ and } u_y(x, y) \to 0 \text{ as } (x^2 + y^2) \to \infty,$$

where the function f is piecewise smooth and absolutely integrable in $(-\infty, \infty)$, can be written as

$$u(x, y) = c + \frac{1}{2\pi} \int_{-\infty}^{\infty} f(\xi) \ln[y^2 + (x - \xi)^2] d\xi,$$

where c is an arbitrary constant.

8.61   Find the solution of the following problem:

$$u_{xx} + u_{yy} = 0, \quad 0 < x < \infty, \quad 0 < y < b$$
$$u(x, 0) = f(x), \quad 0 < x < \infty$$
$$u_x(0, y) = 0, \quad 0 < y < b$$
$$u_y(x, b) = 0, \quad 0 < x < \infty$$

where the function f is piecewise smooth and absolutely integrable in $[0, \infty)$.
($u(x, y) = \frac{2}{\pi} \int_0^\infty \int_0^\infty f(t) \cos(\omega t) \frac{\cosh(\omega(b-y))}{\cosh(\omega b)} \cos(\omega x) dt d\omega$)

8.62   Find the bounded solution of the following problem

$$u_{xx} + u_{yy} = 0, \quad 0 < x < c, \quad 0 < y < \infty$$
$$u_y(x, 0) = 0, \quad 0 < x < c$$
$$u(0, y) = 0, \quad 0 < y < \infty$$
$$u_x(c, y) = f(y), \quad 0 < y < \infty$$

where the function f is piecewise smooth and absolutely integrable in $[0, \infty)$.
($u(x, y) = \frac{2}{\pi} \int_0^\infty \int_0^\infty f(t) \cos(\omega t) \frac{\sinh(\omega x)}{\omega \cosh(\omega c)} \cos(\omega y) dt d\omega$)

8.63   The nonhomogeneous Laplace equation known as *Poisson's equation*

$$\Delta_2 u = u_{xx} + u_{yy} = q(x, y), \quad 0 < x < a, \quad 0 < y < b \qquad (8.230)$$

appears in many problems in engineering and physics, in particular, in electrostatics theory. Show that solution of the boundary value problem (8.230), (8.155)–(8.158) can be written as

$$u(x, y) = \sum_{n=1}^{\infty} \frac{1}{\sinh \frac{n\pi b}{a}} \left(a_n \sinh\left(\frac{n\pi(b - y)}{a}\right) + b_n \sinh\left(\frac{n\pi y}{a}\right) \right.$$
$$\left. - \frac{a}{n\pi} \left[\sinh\left(\frac{n\pi(b - y)}{a} Vr\right) \int_0^y S_n(\tau) \sinh\left(\frac{n\pi\tau}{a}\right) d\tau \right. \right.$$

$$+ \sinh\left(\frac{n\pi y}{a}\right) \int_y^b S_n(\tau) \sinh\left(\frac{n\pi(b-\tau)}{a}\right) d\tau \bigg] \sin\left(\frac{n\pi x}{a}\right),$$

where

$$a_n = \frac{2}{a} \int_0^a f(x) \sin\left(\frac{n\pi x}{a}\right) dx, \quad b_n = \frac{2}{a} \int_0^a g(x) \sin\left(\frac{n\pi x}{a}\right) dx$$

and

$$S_n(y) = \frac{2}{a} \int_0^a q(x, y) \sin\left(\frac{n\pi x}{a}\right) dx.$$

8.64 The boundary value problem

$$\frac{d^2 U}{dR^2} + \frac{1}{R}\frac{dU}{dR} = -P + \epsilon\theta$$

$$\frac{d^2\theta}{dR^2} + \frac{1}{R}\frac{d\theta}{dR} = -U$$

$$U'(0) = \theta'(0) = 0, \quad U(1) = \theta(1) = 0$$

occurs in a flow of a fluid. Show that

$$U(R) = -\frac{P}{4}(R^2 - 1) + \frac{\epsilon P}{2304}(R^6 - 9R^4 + 27R^2 - 19)$$

$$-\frac{\epsilon^2 P}{14745600}(R^{10} - 25R^8 + 300R^6 - 1900R^4 + 5275R^2 - 3651) + O(\epsilon^3)$$

$$\theta(R) = \frac{P}{64}(R^4 - 4R^2 + 3) + \frac{\epsilon P}{147456}(R^8 + 16R^6 - 108R^4 + 304R^2 - 211)$$

$$+\frac{\epsilon^2 P}{2123366400}(R^{12} - 36R^{10} + 675R^8 - 7600R^6 + 47475R^4 - 131436R^2$$

$$+90921) + O(\epsilon^3).$$

References

1. R.P. Agarwal, J. Comput. Appl. Math. **17**, 271–289 (1987)
2. Agarwal, D. O'Regan, *Ordinary and Partial Differential Equations with Special Functions. Fourier Series and Boundary Value Problems* (Springer, New York, 2009)
3. P. Bailey, L. Shampine, P. Waltman *Nonlinear Two Point Boundary Value Problems* (Academic Press, New York, 1968)
4. D.M. Burley, Math. Spectr. **8**, 69–75 (1975/76)
5. D. Krajcinvic, J. Appl. Mech. **39**, 773–778 (1972)
6. A.H. Love, *A Treatise on the Mathematical Theory of Elasticity* (Dover, New York, 1944)
7. D.L. Powers, *Boundary Value Problems*, 4th edn. (Harcourt Academic Press, San Diego, 1972)

Chapter 9
Nonlinear Boundary Value Problems

In this chapter, mainly we shall show the importance of the method of upper and lower solutions [6, 13, 22, 24] to nonlinear boundary value problems which appear in real-world phenomena. The main advantage of this method is that besides proving the existence of the solutions it also provides upper and lower bounds on the solutions. These bounds are of immense value in numerical computations of the solutions [99–101].

Example 9.1 (*Electrically Conducting Solids*) The boundary value problem

$$y'' = \lambda e^{\mu y}, \quad y(0) = y(1) = 0 \tag{9.1}$$

arises in applications involving the diffusion of heat generated by positive temperature-dependent sources. For instance, if $\mu = 1$, it arises in the analysis of Joule losses in electrically conducting solids, with λ representing the square of the constant current and e^y the temperature-dependent resistance, or in frictional heating with λ representing the square of the constant shear stress and e^y the temperature-dependent fluidity, see [3, 33]. The problem (9.1) was first studied by Gheorghe Bratu (1881–1941), and now in the literature known as Bratu's problem. This problem has been discussed by several numerical analysts, e.g., see [119] and the references therein.

If $\mu\lambda = 0$, the problem (9.1) has a unique solution
(i) if $\lambda = 0$, then $y(t) \equiv 0$,
(ii) if $\mu = 0$, then $y(t) = \lambda t(t - 1)/2$.
If $\mu\lambda < 0$, the problem (9.1) has as many solutions as the number of roots of the equation $c = \sqrt{2|\mu\lambda|} \cosh(c/4)$, also for each such c_i,

$$y_i(t) = -\frac{2}{\mu}\left[\ln\left(\cosh\left(\frac{c_i}{2}\left(t - \frac{1}{2}\right)\right)\right) - \ln\left(\cosh\left(\frac{c_i}{4}\right)\right)\right].$$

© Springer Nature Switzerland AG 2019
R. P. Agarwal et al., *500 Examples and Problems of Applied Differential Equations*, Problem Books in Mathematics,
https://doi.org/10.1007/978-3-030-26384-3_9

From the equation $c = \sqrt{2|\mu\lambda|} \cosh(c/4)$, it follows that

$$
\sqrt{\frac{|\mu\lambda|}{8}} \min_{c \geq 0} \frac{\cosh(c/4)}{(c/4)}
\begin{cases}
< 1, & (9.1) \text{ has two solutions} \\
= 1, & (9.1) \text{ has one solution} \\
> 1, & (9.1) \text{ has no solution.}
\end{cases}
$$

Note that $\min_{c \geq 0} \cosh(c/4)/(c/4) \simeq 1.50887956$.

If $\mu\lambda > 0$, the problem (9.1) has a unique solution and it can be written as

$$
y(t) = \frac{2}{\mu} \left[\ln\left(\cos\left(\frac{c}{4} \right) \right) - \ln\left(\cos\left(\frac{c}{4}\left(t - \frac{1}{2} \right) \right) \right) \right], \quad c \in (-2\pi, 2\pi)
$$

where c is the unique root of the equation $c = \sqrt{2\mu\lambda} \cos(c/4)$.

Thus, in particular if $\lambda = 1$ and $\mu = -1$ the boundary value problem (9.1) has two solutions $y_1(t)$ and $y_2(t)$ of parabolic form, concave to the x-axis and pass through the points $x = 0$ and $x = 1$. Solution $y_1(t)$ drops below up to $(-0.14050941\cdots)$ and $y_2(t)$ up to $(-4.0916146\cdots)$.

Motivated by problem (9.1), we present a general existence result for the two-point boundary value problem

$$
\begin{aligned}
y'' &= f(t, y), \quad t \in [0, a] \\
y(0) &= c_0, \quad y(a) = c_1.
\end{aligned}
\tag{9.2}
$$

We say $\beta \in C^2[0, a]$ is an upper solution to (9.2) if $\beta'' \leq f(t, \beta)$, $t \in [0, a]$, $\beta(0) \geq c_0$, $\beta(a) \geq c_1$, and $\alpha \in C^2[0, a]$ is a lower solution to (9.2) if $\alpha'' \geq f(t, \alpha)$, $t \in [0, a]$, $\alpha(0) \leq c_0$, $\alpha(a) \leq c_1$.

Theorem 9.1 ([3]) *Suppose that*
(C1) $f : [0, a] \times \mathbb{R} \to \mathbb{R}$ *is continuous, and*
(C2) *there exist* α, β, *respectively, lower and upper solutions of (9.2) with* $\alpha(t) \leq \beta(t)$ *for* $t \in [0, a]$.
Then, (9.2) has a solution $y \in C^2[0, a]$ *with* $\alpha(t) \leq y(t) \leq \beta(t)$ *for* $t \in [0, a]$.

To show that (9.1) for $\lambda = 1$, $\mu = -1$ has a solution we will apply Theorem 9.1 with $f(t, y) = e^{-y}$, $c_0 = c_1 = 0$, $a = 1$. Clearly (C1) holds. We claim that $\beta = 0$ is an upper solution of (9.1). For this notice that

$$
\beta(0) = 0, \quad \beta(1) = 0 \quad \text{and} \quad \beta'' - f(t, \beta) = -1 \leq 0 \quad \text{for } t \in [0, 1].
$$

Next we shall find $k > 0$ so that $\alpha = -kt(1 - t)$ is a lower solution of (9.1). Since

$$
\alpha(0) = \alpha(1) = 0 \quad \text{and} \quad \alpha'' - f(t, \alpha) = 2k - e^{kt(1-t)}, \quad t \in [0, 1]
$$

this α is a lower solution provided $2k - e^{k/4} \geq 0$, i.e., $0.577685412 \leq k \leq 13.04674274$. Note that the lower solution $\alpha = -0.577685412\, t(1 - t)$ drops below up to (-0.144421353), and hence this choice of the lower solution is reasonably good.

Explicit solutions of (9.1) when λ and/or μ are functions of t have been obtained in [45, 63, 82].

Example 9.2 (*Electrical Potential Theory*) In 1927, Thomas [108] and Fermi [58] independently derived a boundary value problem for determining the electrical potential in an atom. Their analysis leads to the nonlinear second-order differential equation

$$y'' = t^{-1/2} y^{3/2}. \tag{9.3}$$

The boundary conditions in investigating
(1) the ionized atom are given by

$$y(0) = 1, \quad y(a) = 0; \tag{9.4}$$

(2) the neutral atom with Niels Henrik David Bohr (1885–1962) radius a are given by

$$y(0) = 1, \quad ay'(a) = y(a); \tag{9.5}$$

(3) the isolated neutral atom are given by

$$y(0) = 1, \quad \lim_{t \to \infty} y(t) = 0. \tag{9.6}$$

Motivated by the ionized atom Thomas–Fermi problem (9.3), (9.4) we present a general existence result for the two point boundary value problem

$$\begin{aligned} y'' &= qf(t, y), \quad 0 < t < a, \\ y(0) &= c_0, \quad y(a) = c_1. \end{aligned} \tag{9.7}$$

By an upper solution β to (9.7), we mean a function $\beta \in C^1[0, a] \cap C^2(0, a)$ with $\beta'' \leq qf(t, \beta)$, $0 < t < a$, $\beta(0) \geq c_0$, $\beta(a) \geq c_1$, and by a lower solution α to (9.7) we mean a function $\alpha \in C^1[0, a] \cap C^2(0, a)$ with $\alpha'' \geq qf(t, \alpha)$, $0 < t < a$, $\alpha(0) \leq c_0$, $\alpha(a) \leq c_1$.

Theorem 9.2 ([9, 19]) *Suppose that*
(C3) $q \in C(0, a) \cap L^1[0, a]$ *with* $q > 0$ *on* $(0, a)$,
(C4) $f : [0, a] \times \mathbb{R} \to \mathbb{R}$ *is continuous, and*
(C5) *there exist* α, β, *respectively, lower and upper solutions of (9.7) with* $\alpha(t) \leq \beta(t)$ *for* $t \in [0, a]$.
Then, (9.7) has a solution $y \in C^1[0, a] \cap C^2(0, a)$ *with* $\alpha(t) \leq y(t) \leq \beta(t)$ *for* $t \in [0, a]$.

To show that (9.3), (9.4) has a solution we will apply Theorem 9.2 to the boundary value problem

$$y'' = t^{-1/2}|y|^{3/2}, \quad 0 < t < a$$
$$y(0) = 1, \quad y(a) = 0 \tag{9.8}$$

with $c_0 = 1$, $c_1 = 0$, $q(t) = t^{-1/2}$ and $f(t, y) = |y|^{3/2}$. Clearly (C3) and (C4) hold. We claim that $\alpha = 0$ is a lower solution of (9.8). For this notice that

$$\alpha(0) = 0 \leq 1, \quad \alpha(a) = 0 \quad \text{and} \quad \alpha'' - qf(t, \alpha) = 0 \quad \text{for } t \in (0, a).$$

Next we shall show that $\beta(t) = (a - t)/a$ is an upper solution of (9.8). Indeed, we have

$$\beta(0) = 1, \quad \beta(a) = 0 \quad \text{and} \quad \beta'' - qf(t, \beta) = -t^{-1/2}\left|\frac{a-t}{a}\right|^{3/2} < 0, \quad 0 < t < a.$$

Theorem 9.2 thus guarantees that there exists a solution $y \in C^1[0, a] \cap C^2(0, a)$ to (9.8) with

$$0 \leq y(t) \leq \frac{a-t}{a} \quad \text{for } t \in [0, a].$$

Since $y(t) \geq 0$ for $t \in [0, a]$ we have that y is a solution of (9.3), (9.4).

Now we shall show that the Bohr radius Thomas–Fermi problem (9.3), (9.5) has a solution. Thus, we shall deduce from a general existence theorem for the boundary value problem

$$y'' = qf(t, y), \quad 0 < t < a$$
$$y(0) = c_0 \tag{9.9}$$
$$ky'(a) = y(a), \quad k \geq a$$

where $a > 0$ is fixed. By an upper solution β to (9.9) we mean a function $\beta \in C^1[0, a] \cap C^2(0, a)$ with $\beta'' \leq qf(t, \beta)$, $0 < t < a$, $\beta(0) \geq c_0$, $k\beta'(a) \geq \beta(a)$, and by a lower solution α to (9.9) we mean a function $\alpha \in C^1[0, a] \cap C^2(0, a)$ with $\alpha'' \geq qf(t, \alpha)$, $0 < t < a$, $\alpha(0) \leq c_0$, $k\alpha'(a) \leq \alpha(a)$.

Theorem 9.3 ([9, 19]) *Suppose that in addition to (C3) and (C4) the following holds:*
(C6) there exist α, β respectively lower and upper solutions of (9.9) with $\alpha(t) \leq \beta(t)$ for $t \in [0, a]$.
Then, (9.9) has a solution $y \in C^1[0, a] \cap C^2(0, a)$ with $\alpha(t) \leq y(t) \leq \beta(t)$ for $t \in [0, a]$.

To show that (9.3), (9.5) with $a^3 \geq 9/4$ has a solution, we will apply Theorem 9.3 to the boundary value problem

$$y'' = t^{-1/2}|y|^{3/2}, \quad 0 < t < a$$
$$y(0) = 1, \quad ay'(a) = y(a), \tag{9.10}$$

with $k = a$, $c_0 = 1$, $q(t) = t^{-1/2}$ and $f(t, y) = |y|^{3/2}$. Clearly (C3) and (C4) hold and $\alpha = 0$ is a lower solution of (9.10). Next we shall show that

$$\beta(t) = \frac{4}{3}t^{3/2} + \frac{2}{3}a^{3/2}$$

is an upper solution of (9.10). Notice that

$$\beta(0) = \frac{2}{3}a^{3/2} \geq \frac{2}{3}\frac{3}{2} = 1 \quad \text{since} \quad a^3 \geq \frac{9}{4},$$

and since $\beta'(t) = 2t^{1/2}$ we have

$$a\,\beta'(a) = 2a^{3/2} \quad \text{and} \quad \beta(a) = \frac{4}{3}a^{3/2} + \frac{2}{3}a^{3/2} = 2a^{3/2}, \quad \text{so} \quad a\,\beta'(a) = \beta(a).$$

Now since $\beta''(t) = t^{-1/2}$, we have

$$\beta'' - qf(t, \beta) = t^{-1/2} - t^{-1/2}[\beta]^{3/2} = t^{-1/2}\left(1 - [\beta]^{3/2}\right) \leq 0$$

for $t \in (0, a)$, since $\beta(t) \geq (2/3)a^{3/2} \geq 1$ for $t \in (0, a)$. Thus β is an upper solution of (9.10) so (C6) holds. Theorem 9.3 now guarantees that there exists a solution $y \in C^1[0, a] \cap C^2(0, a)$ to (9.10) with

$$0 \leq y(t) \leq \frac{4}{3}t^{3/2} + \frac{2}{3}a^{3/2} \quad \text{for} \quad t \in [0, a].$$

Again since $y(t) \geq 0$ for $t \in [0, a]$ we have that y is a solution of (9.3), (9.5).

For the problem (9.3), (9.6) Fermi used graphical method to obtain the following approximation for values of t in the neighborhood of the origin:

$$y(t) = 1 - Bt + \frac{4}{3}t^{3/2}, \quad B \sim 1.58.$$

This approximation was extended and improved by Edward B. Baker in 1930 to

$$y(t) = 1 - Bt + \frac{1}{3}t^3 - \frac{2}{15}Bt^4 + \cdots$$
$$+ t^{3/2}\left[\frac{4}{3} - \frac{2}{5}Bt + \frac{3}{70}B^2t^2 + \frac{4}{63}\left(\frac{2}{3} + \frac{B^3}{16}\right)t^3 + \cdots\right],$$

where $B \sim 1.588588$. Arnold Johannes Wilhelm Sommerfeld (1868–1951) observed that $y_1(t) = 144/t^3$ is a particular solution of (9.3), which satisfies the second condition of (9.6) but not the first, he gave the following interesting approximation:

$$y(t) = y_1(t)\left\{1 + [y_1(t)]^{\lambda_1/3}\right\}^{\lambda_2/2},$$

where $\lambda_1 = 0.772$ is the positive root and $\lambda_2 = -7.772$ is the negative root of the equation $\lambda^2 + 7\lambda - 6 = 0$.

Finally, we shall show that the generalized Emden–Fowler (Sir Ralph Howard Fowler 1889–1944) problem which includes isolated neutral atom problem (9.3), (9.6) as a special case

$$(t^j y')' = d_0 t^\theta y^\eta, \quad 0 < t < \infty$$
$$y(0) = c \geq 0, \quad \lim_{t \to \infty} y(t) = 0, \tag{9.11}$$

where $0 \leq j < 1$, $d_0 > 0$, $\eta \geq 0$, $\theta > -1$ has a nonnegative solution.

For this, we state an existence result for the general boundary value problem

$$\frac{1}{p}(py')' = qf(t, y), \quad 0 < t < \infty$$
$$-a_0 y(0) + b_0 \lim_{t \to 0^+} p(t)y'(t) = c_0, \quad a_0 > 0, \ b_0 \geq 0, \ c_0 \leq 0 \tag{9.12}$$
$$\lim_{t \to \infty} y(t) = 0.$$

Theorem 9.4 ([9, 19]) *Assume that*
(C7) $f : [0, \infty) \times \mathbb{R} \to \mathbb{R}$ *is continuous,*
(C8) $q \in C(0, \infty)$ *with* $q > 0$ *on* $(0, \infty)$,
(C9) $p \in C[0, \infty) \cap C^1(0, \infty)$ *with* $p > 0$ *on* $(0, \infty)$,
(C10) $\int_0^\mu ds/p(s) < \infty$ *and* $\int_0^\mu p(s)q(s)ds < \infty$ *for any* $\mu > 0$,
(C11) $f(t, 0) \leq 0$ *for* $t \in (0, \infty)$,
(C12) *there exists* $r_0 \geq -c_0/a_0$ *with* $f(t, r_0) \geq 0$ *for* $t \in (0, \infty)$,
(C13) *there exists* $M > 0$ *with* $|f(t, u)| \leq M$ *for* $t \in [0, \infty)$ *and* $u \in [0, r_0]$,
(C14) *for any constant* $A \in (0, r_0]$ *there exists a constant* $K > 0$ *(which may depend on* A*) and a constant* $c_2 > 0$ *(which may depend on* A*) with* $f(t, u) \geq K$ *for* $A \leq u \leq r_0$ *and* $t \geq c_2$,
(C15) $f(t, u) \geq 0$ *for* $0 \leq u \leq r_0$ *and* $t \in [0, \infty)$, *and*
(C16) $\lim_{t \to \infty} \left(B_0 \int_\mu^t ds/p(s) + C_0 \int_\mu^t (1/p(s)) \int_\mu^s p(c12)q(c12)dxds\right) = \infty$ *for any constants* B_0, $C_0 > 0$ *and* $\mu > 0$.
Then, (9.12) has a solution $y \in C[0, \infty) \cap C^2(0, \infty)$ *with* $py' \in C[0, \infty)$ *and* $0 \leq y(t) \leq r_0$ *for* $t \in [0, \infty)$.

To see that (9.11) has a solution, we apply Theorem 9.4 to the boundary value problem

$$(t^j y')' = d_0 t^\theta |y|^\eta, \quad 0 < t < \infty$$
$$y(0) = c \geq 0, \quad \lim_{t \to \infty} y(t) = 0. \tag{9.13}$$

We let $p(t) = t^j$, $q(t) = t^\theta$, $f(t, y) = d_0|y|^\eta$, $a_0 = 1$, $b_0 = 0$ and $c_0 = -c$. Clearly, (C7)–(C11), (C12) with $r_0 = c$, (C13), (C14) with $K = d_0 A^\eta$, (C15) and (C16) hold. Now Theorem 9.4 implies that (9.13) has a solution y with $y \geq 0$ on $m0, \infty)$. This y is a solution of (9.11).

More detailed results on these problems are available in [27, 48, 64, 73, 84, 87, 88, 98].

Example 9.3 (Circular Membrane Theory) The equation for a circular membrane subjected to a normal uniform pressure can be reduced to (see [52, 55, 102])

$$y'' + \frac{k}{y^2} + \frac{3}{x}y' = 0, \quad 0 < x < 1; \tag{9.14}$$

here $k > 0$ is a constant, x is the radial coordinate and $y(x)$ the radial stress. At the edge $(x = 1)$, we have the condition

$$y(1) = \lambda > 0, \tag{9.15}$$

or

$$a_0 y(1) + y'(1) = 0, \quad a_0 > 0 \tag{9.16}$$

and at the center (for symmetry)

$$y'(0) = 0. \tag{9.17}$$

Making the change of variable $x = 1/t$, the problem (9.14), (9.15) is transformed to

$$y'' + \frac{k}{t^4}\frac{1}{y^2} - \frac{1}{t}y' = 0, \quad 1 < t < \infty$$
$$y(1) = \lambda, \tag{9.18}$$

which is a second-order problem on the infinite interval.

First, we shall prove the existence of solutions of (9.14), (9.15), (9.17) and (9.14), (9.16), (9.17) by applying the following general existence theorem for the boundary value problem:

$$\frac{1}{p}(py')' = qf(t, y), \quad 0 < t < 1$$
$$\lim_{t \to 0^+} p(t)y'(t) = 0 \tag{9.19}$$
$$a_0 y(1) + b_0 \lim_{t \to 1^-} p(t)y'(t) = c_0, \quad a_0 > 0, \ b_0 \geq 0.$$

Recall that by an upper solution β to (9.19), we mean a function $\beta \in C[0, 1] \cap C^2(0, 1)$ with $p\beta' \in C[0, 1]$ that satisfies $(1/p)(p\beta')' \leq qf(t, \beta)$, $0 < t < 1$, $\lim_{t \to 0^+} p(t)\beta'(t) \leq 0$, $a_0\beta(1) + b_0 \lim_{t \to 1^-} p(t)\beta'(t) \geq c_0$. Similarly, by a lower solution α to (9.19), we mean a function $\alpha \in C[0, 1] \cap C^2(0, 1)$ with $p\alpha' \in C[0, 1]$ that satisfies $(1/p)(p\alpha')' \geq qf(t, \alpha)$, $0 < t < 1$, $\lim_{t \to 0^+} p(t)\alpha'(t) \geq 0$, $a_0\alpha(1) + b_0 \lim_{t \to 1^-} p(t)\alpha'(t) \leq c_0$.

Theorem 9.5 ([7, 17]) *Assume that*
(C17) $p \in C[0, 1] \cap C^1(0, 1)$ with $p > 0$ on $(0, 1)$,
(C18) $q \in C(0, 1)$, $q > 0$ on $(0, 1)$ and $\int_0^1 p(s)q(s)ds < \infty$,
(C19) $f : [0, 1] \times \mathbb{R} \to \mathbb{R}$ is continuous,
(C20) $\int_0^1 (1/p(t)) \int_0^t p(s)q(s)dsdt < \infty$, and
(C21) there exist α, β respectively lower and upper solutions of (9.19) with $\alpha \leq \beta$.
Then, (9.19) has a solution $y \in C[0, 1] \cap C^2(0, 1)$ with $py' \in C[0, 1]$ and $\alpha(t) \leq y(t) \leq \beta(t)$ for $t \in [0, 1]$. In addition, if $p(0) \neq 0$, or $p(0) = 0$ and $\lim_{t \to 0^+}[p(t) q(t)/p'(t)]$ exists, then

$$\lim_{t \to 0^+} \frac{y(t) - y(0)}{t} = \begin{cases} 0 & \text{if } p(0) \neq 0 \\ f(0, y(0)) \lim_{t \to 0^+} \dfrac{p(t)q(t)}{p'(t)} & \text{if } p(0) = 0 \end{cases}$$

and $y \in C[0, 1] \cap C^1[0, 1) \cap C^2(0, 1)$.

Now with (9.14), (9.15), (9.17) in mind, we consider the problem

$$y'' + \frac{3}{t}y' = -\frac{q(t)}{y^2}, \quad 0 < t < 1$$

$$y'(0) = 0, \quad y(1) = \lambda > 0,$$
(9.20)

where $q : [0, 1] \to [0, \infty)$ continuous, $q > 0$ on $(0, 1)$ and $0 \leq q(t) \leq M$ for $t \in [0, 1]$. Existence of a solution to (9.20) will be established using Theorem 9.5. For this we first look at (9.19) with $p(t) = t^3$, $a_0 = 1$, $b_0 = 0$, $c_0 = \lambda$ and

$$f(t, y) = \begin{cases} -1/y^2, & y \geq \lambda \\ -1/\lambda^2, & y < \lambda. \end{cases}$$

Clearly (C17)–(C20) hold. It remains to show (C21). Let $\alpha(t) = \lambda$ for $t \in [0, 1]$. Notice that $\alpha(1) = 1$, $\lim_{t \to 0^+} t^3\alpha'(t) = 0$ and

$$\frac{1}{t^3}(t^3\alpha')' - qf(t, \alpha) = 0 + \frac{q(t)}{\lambda^2} \geq 0 \quad \text{for } t \in (0, 1),$$

so α is a lower solution of (9.19). Next, let

$$\beta(t) = \frac{M}{8\lambda^2}(1 - t^2) + \lambda \quad \text{for } t \in [0, 1].$$

Clearly $\beta(t) \geq \alpha(t)$ for $t \in [0, 1]$. Notice $\beta(1) = 1$, $\lim_{t \to 0^+} t^3\beta'(t) = 0$ and

$$\frac{1}{t^3}(t^3\beta')' - qf(t,\beta) = -\frac{M}{\lambda^2} + \frac{q(t)}{\beta^2(t)} \leq -\frac{M}{\lambda^2} + \frac{M}{\lambda^2} = 0 \text{ for } t \in (0,1).$$

So β is an upper solution of (9.19). Now Theorem 9.5 guarantees that (9.19) has a solution $y \in C[0,1] \cap C^2(0,1)$ with $t^3 y' \in C[0,1]$ and

$$\lambda \leq y(t) \leq \frac{M}{8\lambda^2}(1 - t^2) + \lambda \text{ for } t \in [0,1].$$

Finally, since

$$\lim_{t \to 0^+} \frac{p(t)q(t)}{p'(t)} = \frac{1}{3} \lim_{t \to 0^+} tq(t) = 0,$$

we have $y'(0) = 0$. As a result, $y \in C^1[0,1]$ and y is a solution of (9.20).

Now with (9.14), (9.16), (9.17) in mind, we consider the problem

$$y'' + \frac{3}{t} y' = -\frac{q(t)}{y^2}, \quad 0 < t < 1$$

$$y'(0) = 0, \quad y'(1) + (1-v)y(1) = 0, \quad 1 - v > 0, \tag{9.21}$$

where $q : [0,1] \to [0,\infty)$ continuous and $0 < K \leq q(t) \leq M$ for $t \in [0,1]$. Existence of a solution of (9.21) will be established by using Theorem 9.5. For this, we first look at (9.19) with $p(t) = t^3$, $b_0 = 1$, $a_0 = 1 - v$, $c_0 = 0$ and

$$f(t,y) = \begin{cases} -1/\beta^2(t), & y > \beta(t) \\ -1/y^2, & \alpha(t) \leq y \leq \beta(t) \\ -1/\alpha^2(t), & y < \alpha(t) \end{cases}$$

with

$$\alpha(t) = \frac{1}{2}\left[\frac{K(1-v)^2}{(3-v)^2}\right]^{1/3}\left[\frac{3-v}{1-v} - t^2\right] \tag{9.22}$$

and

$$\beta(t) = \left[\frac{M(1-v)^2}{32}\right]^{1/3}\left[\frac{3-v}{1-v} - t^2\right] \tag{9.23}$$

for $t \in [0,1]$. Clearly (C17)–(C19) (note $\alpha \neq 0$ and $\beta \neq 0$ since $(3-v)/(1-v) = 1 + [2/(1-v)] > 1$) and (C20) hold. So it remains to show (C21). Notice

$$\alpha'(t) = -\left[\frac{K(1-v)^2}{(3-v)^2}\right]^{1/3} t,$$

so $\lim_{t \to 0^+} t^3 \alpha'(t) = 0$ and

$$(1 - v)\alpha(1) + \lim_{t \to 1^-} t^3 \alpha'(t) = \frac{1 - v}{2} \left[\frac{K(1 - v)^2}{(3 - v)^2} \right]^{1/3} \frac{2}{1 - v}$$

$$- \left[\frac{K(1 - v)^2}{(3 - v)^2} \right]^{1/3} = 0.$$

Also for $t \in (0, 1)$, we have

$$\frac{1}{t^3} (t^3 \alpha')' - q f(t, \alpha) = -4 \left[\frac{K(1 - v)^2}{(3 - v)^2} \right]^{1/3} + \frac{q(t)}{\alpha^2(t)}$$

$$\geq -4 \left[\frac{K(1 - v)^2}{(3 - v)^2} \right]^{1/3} + \frac{K}{\left\{ \frac{1}{2} \left[\frac{K(1-v)^2}{(3-v)^2} \right]^{1/3} \frac{3-v}{1-v} \right\}^2}$$

$$= -4 \left[\frac{K(1 - v)^2}{(3 - v)^2} \right]^{1/3} + 4 \left[\frac{K(1 - v)^2}{(3 - v)^2} \right]^{1/3} = 0,$$

so α is a lower solution of (9.19). Notice $\alpha(t) \leq \beta(t)$ for $t \in [0, 1]$ since we note that

$$\frac{M}{4} (3 - v)^2 \geq \frac{M}{4} 4 = M \geq K,$$

and so

$$\frac{K(1 - v)^2}{8(3 - v)^2} \leq \frac{M(1 - v)^2}{32}.$$

Also notice that

$$\beta'(t) = -2 \left[\frac{M(1 - v)^2}{32} \right]^{1/3} t,$$

so $\lim_{t \to 0^+} t^3 \beta'(t) = 0$ and

$$(1 - v)\beta(1) + \lim_{t \to 1^-} t^3 \beta'(t) = (1 - v) \left[\frac{M(1 - v)^2}{32} \right]^{1/3} \frac{2}{1 - v}$$

$$- 2 \left[\frac{M(1 - v)^2}{32} \right]^{1/3} = 0.$$

Also for $t \in (0, 1)$ we have

$$\frac{1}{t^3}(t^3\beta')' - qf(t,\beta) = -8\left[\frac{M(1-v)^2}{32}\right]^{1/3} + \frac{q(t)}{\beta^2(t)}$$

$$\leq -8\left[\frac{M(1-v)^2}{32}\right]^{1/3} + \frac{M}{\left\{\left[\frac{M(1-v)^2}{32}\right]^{1/3}\frac{2}{1-v}\right\}^2}$$

$$= -8\left[\frac{M(1-v)^2}{32}\right]^{1/3} + \frac{32}{4}\left[\frac{M(1-v)^2}{32}\right]^{1/3} = 0,$$

so β is an upper solution of (9.19). Now Theorem 9.5 guarantees that (9.19) has a solution $y \in C[0,1] \cap C^2(0,1)$ with $t^3 y' \in C[0,1]$ and $\alpha(t) \leq y(t) \leq \beta(t)$ for $t \in [0,1]$ where α (respectively β) is given in (9.22) (respectively (9.23)). Finally, since $\lim_{t\to 0^+}[p(t)q(t)/p'(t)] = 0$ we have $y'(0) = 0$. As a result, $y \in C^1[0,1]$ and y is a solution of (9.21).

Now we shall establish the existence of a solution of (9.18) by applying the following general existence theorem for the infinite interval boundary value problem

$$y'' = qf(t,y,y'), \quad a < t < \infty$$
$$-a_0 y(a) + b_0 y'(a) = c_0, \quad a_0 > 0, \quad b_0 \geq 0 \qquad (9.24)$$
$$y(t) \text{ bounded on } [a,\infty);$$

here $a \geq 0$ is fixed. We recall that by an upper solution β to (9.24) we mean a function $\beta \in BC[a,\infty) \cap C^2(a,\infty)$, $\beta' \in BC[a,\infty)$ with $\beta'' \leq qf(t,\beta,\beta')$, $a < t < \infty$, $-a_0\beta(a) + b_0\beta'(a) \leq c_0$, $\beta(t)$ bounded on $[a,\infty)$; here $BC[a,\infty)$ denotes the space of continuous, bounded functions from $[a,\infty)$ to $\mathbb{R}$. Similarly, by a lower solution α to (9.24) we mean a function $\alpha \in BC[a,\infty) \cap C^2(a,\infty)$, $\alpha' \in BC[a,\infty)$ with $\alpha'' \geq qf(t,\alpha,\alpha')$, $a < t < \infty$, $-a_0\alpha(a) + b_0\alpha'(a) \geq c_0$, $\alpha(t)$ bounded on $[a,\infty)$.

Theorem 9.6 ([7, 17]) *Suppose that*
(C22) $f : [a,\infty) \times \mathbb{R}^2 \to \mathbb{R}$ is continuous,
(C23) $q \in C(a,\infty)$ with $q > 0$ on (a,∞),
(C24) $\int_a^\mu q(s)ds < \infty$ for any $\mu > a$,
(C25) there exists α, β respectively lower and upper solutions of (9.24) with $\alpha(t) \leq \beta(t)$ for $t \in [a,\infty)$,
(C26) there exists a continuous $\psi : [0,\infty) \to (0,\infty)$ with $|f(t,u,v)| \leq \psi(|v|)$ for $(t,u) \in [a,\infty) \times [\alpha(t),\beta(t)]$, and
(C27) q is bounded on $[a,\infty)$ with $A_0 \sup_{t\in[a,\infty)} q(t) < \int_{d_0}^\infty [u/\psi(u)]du$; here

$$A_0 = \max\{\beta(t) - \alpha(x) : t, x \in [a,\infty)\}$$

with

$$d_0 = \frac{|c_0| + |a_0| \max\{|\alpha(a)|, |\beta(a)|\}}{b_0} \quad \text{if } b_0 > 0$$

whereas

$$d_0 = \left[\max\{|\alpha(a+1)|, \; |\beta(a+1)|\} + \frac{|c_0|}{a_0} \right] \quad \text{if} \;\; b_0 = 0.$$

Then, (9.24) has a solution $y \in BC[a, \infty) \cap C^2(a, \infty)$, $y' \in BC[a, \infty)$ *with*

$$\alpha(t) \leq y(t) \leq \beta(t) \;\; \text{and} \;\; |y'(t)| \leq J^{-1}\left(A_0 \sup_{t \in [a, \infty)} q(t) \right) \;\; \text{for} \;\; t \in [a, \infty);$$

here $J : [d_0, \infty) \to [0, \infty)$ *is given by* $J(z) = \int_{d_0}^{z} [u/\psi(u)] du.$

Now with Theorem 9.6 in mind for (9.18) we consider the boundary value problem

$$y'' + f(t, y, y') = 0, \quad 1 < t < \infty$$
$$y(1) = \lambda \qquad\qquad\qquad\qquad (9.25)$$
$$y(t) \;\; \text{bounded on} \;\; [1, \infty)$$

with

$$f(t, y, v) = \begin{cases} -\left[\dfrac{k}{t^4 y^2} - \dfrac{v}{t} \right], & y \geq \lambda \\[3mm] -\left[\dfrac{k}{t^4 \lambda^2} - \dfrac{v}{t} \right], & y \leq \lambda. \end{cases}$$

We will apply Theorem 9.6 with

$$a = 1, \quad q = 1, \quad \alpha(t) = \lambda \;\; \text{and} \;\; \beta(t) = \lambda + \frac{k}{8\lambda^2}\left(1 - \frac{1}{t^2}\right).$$

Clearly (C22)–(C24) hold. In addition, α is a lower solution of (9.25) since $\alpha(1) = \lambda$ and

$$\alpha'' - q(t)f(t, \alpha, \alpha') = \frac{k}{t^4 \lambda^2} \geq 0 \;\; \text{for} \;\; 1 < t < \infty.$$

We next show β is an upper solution of (9.25). Note $\beta(1) = \lambda$ and for $1 < t < \infty$ we have

$$\beta'' - q(t)f(t, \beta, \beta') = \frac{-3k}{4\lambda^2}\frac{1}{t^4} + \left(\frac{k}{t^4}\frac{1}{[\beta(t)]^2} - \frac{k}{4\lambda^2 t^4} \right)$$
$$\leq \frac{-3k}{4\lambda^2}\frac{1}{t^4} + \left(\frac{k}{t^4}\frac{1}{\lambda^2} - \frac{k}{4\lambda^2 t^4} \right)$$
$$= 0.$$

Thus (C25) holds. Also since

$$|f(t, y, v)| \leq \frac{k}{\lambda^2} + |v| \;\; \text{for} \;\; (t, y) \in [1, \infty) \times [\lambda, \beta(t)],$$

clearly (C26) and (C27) hold. Theorem 9.6 now guarantees that (9.25) has a solution $y \in BC[1, \infty) \cap C^2(1, \infty)$, $y' \in BC[1, \infty)$ with

$$\lambda \leq y(t) \leq \lambda + \frac{k}{8\lambda^2}\left(1 - \frac{1}{t^2}\right) \quad \text{for } t \in [1, \infty).$$

As a result (from the transformation $x = 1/t$), we find that $y(x)$ is a solution of (9.14), (9.15) with

$$\lambda \leq y(x) \leq \lambda + \frac{k}{8\lambda^2}\left(1 - x^2\right) \quad \text{for } 0 \leq x \leq 1. \tag{9.26}$$

From (9.26), it is clear that this solution $y(x)$ remains bounded as $x \to 0^+$. Now we will now show $y'(0) = 0$ and $y \in C[0, 1]$ (if these are true then $y \in C[0, 1]$ is a solution of (9.14), (9.15), (9.17)). Certainly $y \in C(0, 1]$ with

$$x^3 y'(x) = y'(1) + k \int_x^1 \frac{s^3}{y^2(s)} ds,$$

so

$$y(x) = \lambda + \frac{y'(1)}{2}\left(1 - \frac{1}{x^2}\right) - k \int_x^1 \frac{1}{t^3} \int_t^1 \frac{s^3}{y^2(s)} ds dt.$$

Thus

$$x^2 y(x) = \lambda x^2 + \frac{y'(1)}{2}\left(x^2 - 1\right) - kx^2 \int_x^1 \frac{1}{t^3} \int_t^1 \frac{s^3}{y^2(s)} ds dt.$$

Now (9.26) implies

$$\int_0^1 \frac{1}{t^3} \int_t^1 \frac{s^3}{y^2(s)} ds dt \geq \frac{1}{\left(\lambda + \frac{k}{8\lambda^2}\right)^2} \int_0^1 \frac{1}{t^3} \int_t^1 s^3 ds dt = \infty,$$

so L'Hopital's rule gives

$$\frac{y'(1)}{2} = -\frac{k}{2} \int_0^1 \frac{s^3}{y^2(s)} ds.$$

Consequently,

$$y(x) = \lambda - \frac{k}{2} \int_0^1 \frac{s^3}{y^2(s)} ds \left(1 - \frac{1}{x^2}\right) - k \int_x^1 \frac{1}{t^3} \int_t^1 \frac{s^3}{y^2(s)} ds dt$$

$$= \lambda - \frac{k}{2} \int_0^1 \frac{s^3}{y^2(s)} ds \left(1 - \frac{1}{x^2}\right)$$

$$+k\left[\frac{1}{2}\int_x^1 \frac{s}{y^2(s)}ds - \frac{1}{2x^2}\int_x^1 \frac{s^3}{y^2(s)}ds\right]$$

$$= \lambda - \frac{k}{2}\int_0^1 \frac{s^3}{y^2(s)}ds + \frac{k}{2x^2}\int_0^x \frac{s^3}{y^2(s)}ds + \frac{k}{2}\int_x^1 \frac{s}{y^2(s)}ds$$

with

$$y'(x) = -\frac{k}{x^3}\int_0^x \frac{s^3}{y^2(s)}ds.$$

Note that (9.26) implies

$$\int_0^1 \frac{s^3}{y^2(s)}ds \leq \frac{1}{\lambda^2}\int_0^1 s^3 ds.$$

Now L'Hopital's rule gives

$$y'(0) = -k\lim_{x\to 0^+} \frac{\int_0^x \frac{s^3}{y^2(s)}ds}{x^3} = -\frac{1}{3}k\lim_{x\to 0^+} \frac{x}{y^2(x)} = 0.$$

It is also immediate that $y \in C[0, 1]$ since (L'Hopital's rule)

$$y(0) = \lambda - \frac{k}{2}\int_0^1 \frac{s^3}{y^2(s)}ds + \frac{k}{2}\int_0^1 \frac{s}{y^2(s)}ds.$$

Example 9.4 (Kinetics and Heat Transfer) Let Q be the monomolecular heat of reaction, k the thermal conductivity, W the reaction velocity, c the concentration, a the frequency factor, E the energy of activation, R the gas constant, and T the temperature. Then, an appropriate equation for the thermal balance between the heat generated by a chemical reaction and that conducted away can be written as

$$k\nabla^2 T = -QW, \qquad (9.27)$$

where ∇^2 represents the Laplace operator, and the expression for W is the Svante August Arrhenius (1859–1927) relation

$$W = ca\exp(-E/(RT)). \qquad (9.28)$$

In the case of small temperature ranges, we can write

$$-\frac{E}{RT} = -\frac{E}{RT_0}\left[\frac{1}{1 + (T - T_0)/T_0}\right]$$

$$= -\frac{E}{RT_0}\left[1 - \frac{T - T_0}{T_0} + \left(\frac{T - T_0}{T_0}\right)^2 - \cdots\right]$$

$$\simeq -\frac{E}{RT_0}\left[1 - \frac{T - T_0}{T_0}\right].$$

Using this approximation in (9.28), we find the equation for dimensionless temperature, $\theta = E(T - T_0)/(RT_0^2)$ to be

$$\nabla^2\theta + Ae^\theta = 0, \tag{9.29}$$

where

$$A = \frac{QEca}{kRT_0^2}\exp\left(-\frac{E}{RT_0}\right).$$

Equation (9.29) also appears in certain problems of two- dimensional vortex motion of incompressible fluids, in the theory of the space charge of electricity around a glowing wire, and in the nebular theory for the distribution of mass of gaseous interstellar material under the influence of its own gravitational field, see [35, 78, 83, 114].

In one dimension, Eq. (9.29) by using Cartesian, circular cylindrical, and spherical coordinates becomes

$$\frac{d^2\theta}{dx^2} + \frac{n}{x}\frac{d\theta}{dx} = -Ae^\theta \tag{9.30}$$

with the boundary conditions

$$\theta'(0) = 0, \quad \theta(1) = 0 \tag{9.31}$$

where, respectively, $n = 0, 1, 2$. In the literature, (9.30) is known as the radial Liouville–Gelfand (Israel Moiseevich Gelfand 1913–2009) equation.

An explicit solution for the case $n = 0$ can be obtained and appears as

$$\theta(x) = 2\left[\ln(\cosh(c)) - \ln(\cosh(cx))\right], \tag{9.32}$$

where c is a root of the equation $c = \sqrt{A/2}\cosh(c)$. Thus, for the problem (9.30) with $n = 0$, (9.31) if

$$\sqrt{\frac{A}{2}}\min_{c\geq 0}\frac{\cosh c}{c} \begin{cases} < 1 & \text{there are two solutions} \\ = 1 & \text{there is a unique solution} \\ > 1 & \text{there is no solution} \end{cases}.$$

We note that $\min_{c\geq 0}[\cosh(c)/c] \simeq 1.50887956$.

For $n = 1$, we shall follow the ingenious method of Chambré (see [47]). For $n = 1$ Eq. (9.30) is the same as

$$\frac{d}{dx}\left(x\frac{d\theta}{dx}\right) = -Axe^{\theta}. \tag{9.33}$$

We use the transformation

$$\omega = x\frac{d\theta}{dx}, \quad u = x^{p}e^{\theta} \tag{9.34}$$

where p will be chosen later. Clearly,

$$\frac{d\omega}{dx} = \frac{d\omega}{du}\frac{du}{dx} = \frac{d\omega}{du}\left[px^{p-1}e^{\theta} + x^{p}e^{\theta}\frac{d\theta}{dx}\right] = u\frac{d\omega}{du}\left[px^{-1} + \frac{d\theta}{dx}\right]$$

and hence

$$x\frac{d\omega}{dx} = u\frac{d\omega}{du}[p + \omega],$$

which in view of (9.33) is the same as

$$-Ax^{2}e^{\theta} = u\frac{d\omega}{du}[p + \omega]. \tag{9.35}$$

We can integrate (9.35) provided its left-hand side is a function of u and ω. We note that for $p = 2$ the left-hand side is $(-Au)$, and therefore

$$\frac{d\omega}{du} = -\frac{A}{2 + \omega},$$

which immediately gives

$$\omega^{2} + 4\omega = -2Au + c_{1}. \tag{9.36}$$

Clearly, at $x = 0$, $u = 0$ and since $\omega = xd\theta/dx$, $\omega = 0$. Thus, from (9.36), it follows that $c_{1} = 0$. Using (9.34) in (9.36), we obtain the relation in terms of the original variables

$$x^{2}\left(\frac{d\theta}{dx}\right)^{2} + 4x\frac{d\theta}{dx} = -2Ax^{2}e^{\theta}. \tag{9.37}$$

Now multiplying (9.30) for $n = 1$ by $2x^{2}$, we get

$$2x^{2}\frac{d^{2}\theta}{dx^{2}} + 2x\frac{d\theta}{dx} = -2Ax^{2}e^{\theta}. \tag{9.38}$$

Equating (9.37) and (9.38), we find a simpler differential equation

$$\frac{d^{2}\theta}{dx^{2}} - \frac{1}{x}\frac{d\theta}{dx} - \frac{1}{2}\left(\frac{d\theta}{dx}\right)^{2} = 0. \tag{9.39}$$

Since (9.39) is a Bernoulli equation in $d\theta/dx$, we can integrate to obtain

$$\frac{d\theta}{dx} = -\frac{4Bx}{Bx^2 + 1},$$ (9.40)

where B is an integration constant. Now an integration of (9.40) gives

$$\theta(x) = C - 2\ln(Bx^2 + 1),$$ (9.41)

where C is another integration constant. We note that to get (9.41) the condition $\theta'(0) = 0$ has already been used, and the condition $\theta(1) = 0$ immediately gives $C = 2\ln(B + 1)$ and hence, we have

$$\theta(x) = 2\ln(B+1) - 2\ln(Bx^2 + 1) = 2\ln\left(\frac{B+1}{Bx^2+1}\right).$$ (9.42)

Finally, the constant B is computed by requiring that (9.42) is a solution of (9.30) for $n = 1$. This leads to the equation $8B = A(B+1)^2$, and hence

$$B = \frac{8 - 2A \pm [(8 - 2A)^2 - 4A^2]^{1/2}}{2A},$$ (9.43)

which is valid only if $0 < A \leq 2$. Thus, the problem (9.30) with $n = 1$, (9.31) has two solutions, one solution, and no solution according as $0 < A < 2$, $A = 2$, $A > 2$.

For $n = 2$, there does not seem to be a way to find an explicit solution of (9.30), (9.31). However, we can use Theorem 9.5 to prove the existence of a solution for each $n \geq 1$ (not necessarily an integer). We note that for $n \geq 1$ Eq. (9.30) is the same as

$$\frac{1}{x^n}\frac{d}{dx}\left(x^n\frac{d\theta}{dx}\right) = -Ae^\theta.$$ (9.44)

We consider the boundary value problem

$$\frac{1}{t^n}(t^n y')' = -Ae^y$$
$$\lim_{t\to 0^+} t^n y'(t) = 0, \quad y(1) = 0.$$ (9.45)

Comparing (9.45) with (9.19), we find $p(t) = t^n$, $q(t) = 1$, $f(t, y) = -Ae^y$, $b_0 = c_0 = 0$, $a_0 = 1$. Clearly, conditions (C17)–(C20) are satisfied. Now we shall show that $\alpha(t) = 0$ is a lower solution of (9.45). For this, it suffices to note that $(1/t^n)(t^n\alpha'(t))' = 0 \geq -Ae^{\alpha(t)} = -A$, $0 < t < 1$. Finally, we shall compute B so that

$$\beta(t) = 2\ln(B+1) - 2\ln(Bt^2 + 1)$$

is an upper solution of (9.45). For this, all we need is to satisfy the inequality

$$\frac{1}{t^n}\frac{d}{dt}\left(t^n\frac{-4Bt}{Bt^2+1}\right) \le -A\left(\frac{B+1}{Bt^2+1}\right)^2, \quad 0 < t < 1$$

which is the same as

$$A(B+1)^2 \le 4(n+1)B + 4(n-1)B^2t^2, \quad 0 < t < 1.$$

This inequality certainly holds if $A(B+1)^2 \le 4(n+1)B$, or

$$\frac{[4(n+1)-2A]-\sqrt{[4(n+1)-2A]^2-4A^2}}{2A}$$

$$\le B \le \frac{[4(n+1)-2A]+\sqrt{[4(n+1)-2A]^2-4A^2}}{2A}, \qquad (9.46)$$

which is valid only if $0 < A \le (n+1)$.

Thus, in view of Theorem 9.5, the problem (9.45) has a solution $y \in C[0,1] \cap C^2(0,1)$ with $t^n y' \in C[0,1]$ and $0 \le y(t) \le 2\ln(B+1) - 2\ln(Bt^2+1)$ for $t \in [0,1]$. Further, since $p(0) = 0$ and $\lim_{t\to 0^+}[p(t)q(t)/p'(t)]$ exists,

$$\lim_{t\to 0^+}\frac{y(t)-y(0)}{t} = f(0,y(0))\lim_{t\to 0^+}\frac{p(t)q(t)}{p'(t)} = -Ae^{y(0)}\lim_{t\to 0^+}\frac{t^n}{nt^{n-1}} = 0,$$

i.e., $y'(0) = 0$, and $y \in C[0,1] \cap C^1[0,1) \cap C^2(0,1)$.

Thus, in conclusion the problem (9.44), (9.31) for each $n \ge 1$ has a solution $\theta \in C[0,1] \cap C^1[0,1) \cap C^2(0,1)$ with

$$0 \le \theta(x) \le 2\ln(B+1) - 2\ln(Bx^2+1), \qquad (9.47)$$

where B satisfies the inequality (9.46).

For the boundary value problem (9.30), (9.31) one of the best possible results is given in Problem 9.9.

Example 9.5 (*Emission of Electricity from Hot Bodies*) A counterpart of (9.30) is the following equation:

$$\frac{d^2\theta}{dx^2} + \frac{n}{x}\frac{d\theta}{dx} = -Ae^{-\theta}, \qquad (9.48)$$

which for $n = 2$ appears in Richardson's [93] (also see [54]) theory of thermionic currents when one seeks to determine the density and electric force of an electron gas in the neighborhood of a hot body in thermal equilibrium (also see Problem 4.19). According to Richardson, the general condition for this equilibrium at constant temperature is that the force on the electron in any element of volume arising from the electric field should balance the force on the same element of volume arising from the pressure gradient. This is equivalent to having the electric intensity E satisfy the following nonlinear partial differential equation

$$E \operatorname{div}(E) + \frac{KT}{e_0} \nabla^2 E = 0$$

with the additional condition

$$\operatorname{grad}(\ln(n)) = \frac{e_0}{\kappa T} E,$$

where κ is Boltzmann's constant, T the constant temperature, e_0 the charge on the electron, and n the number of electrons per unit volume.

Let v be the volume of unit mass. To find the differential equation for v, we first need to compute the work done when unit mass is moved from a point A in the field to a second point B against a pressure p. Denoting this work by w, we have

$$w = \int_A^B p \, dv. \tag{9.49}$$

But since the gas is in equilibrium, the work done on the electron by the electric force is equal to w. Denoting this work by w', we get

$$w' = -\int_A^B N_0 e_0 \frac{dV}{ds} ds, \tag{9.50}$$

where V is the potential, N_0 the number of electrons per unit mass, and ds is the element of the path from A to B.

Equating (9.49) and (9.50), we obtain

$$\int_A^B p \, dv + \int_A^B N_0 e_0 \frac{dV}{ds} ds = 0. \tag{9.51}$$

However, since $p = RT/v$ Eq. (9.51) can be written as

$$\int_A^B \frac{RT}{v} \frac{dv}{ds} ds + \int_A^B N_0 e_0 \frac{dV}{ds} ds = 0,$$

which leads to the equation

$$\frac{RT}{v} \frac{dv}{ds} + N_0 e_0 \frac{dV}{ds} = 0. \tag{9.52}$$

Now the potential V satisfies Poisson's equation

$$\nabla^2 V = 4\pi\rho = 4\pi N_0 e_0 / v. \tag{9.53}$$

If the thermionic emission is from a flat plate of infinite extent, then $ds = dx$ and (9.53) is the same as

$$\frac{d^2V}{dx^2} = 4\pi\rho. \tag{9.54}$$

Eliminating V from (9.52) and (9.54), we get the equation of Richardson for thermionic distribution in the neighborhood of flat surfaces

$$\frac{d^2v}{dx^2} - \frac{1}{v}\left(\frac{dv}{dx}\right)^2 + C = 0, \quad C = 4\pi N_0^2 e_0^2/(RT). \tag{9.55}$$

If the emission is from a spherical surface with radial symmetry, then Eq. (9.53) must be expressed in polar coordinates, and hence

$$\frac{1}{r^2}\frac{d}{dr}\left(r^2\frac{dV}{dr}\right) = 4\pi\rho. \tag{9.56}$$

Using (9.56), we eliminate V from (9.52) to get the following nonlinear differential equation:

$$\frac{d^2v}{dr^2} - \frac{1}{v}\left(\frac{dv}{dr}\right)^2 + \frac{2}{r}\frac{dv}{dr} + C = 0. \tag{9.57}$$

Finally, the substitution $v = e^\theta$, $r = \alpha x$ transforms (9.57) to the equation

$$\frac{d^2\theta}{dx^2} + \frac{2}{x}\frac{d\theta}{dx} = -C\alpha^2 e^{-\theta}. \tag{9.58}$$

Now, we shall discuss the existence and uniqueness of the solutions of the boundary value problem (9.48), (9.31). For the case $n = 0$, there exists a unique solution and it can be written as

$$\theta(x) = 2\left[\ln(\cos(cx)) - \ln(\cos(c))\right], \quad c \in (-\pi/2, \pi/2) \tag{9.59}$$

where c is the unique root of the equation $c = \sqrt{A/2}\cos(c)$.

Next, we shall use Theorem 9.5 to prove the existence of a solution for each $n > 0$ (not necessarily an integer). We note that for $n > 0$ Eq. (9.48) is the same as

$$\frac{1}{x^n}\frac{d}{dx}\left(x^n\frac{d\theta}{dx}\right) = -Ae^{-\theta}. \tag{9.60}$$

We consider the boundary value problem

$$\frac{1}{t^n}(t^n y')' = -Ae^{-y}$$
$$\lim_{t\to 0^+} t^n y'(t) = 0, \quad y(1) = 0 \tag{9.61}$$

and compare it with (9.19), so that $p(t) = t^n$, $q(t) = 1$, $f(t, y) = -Ae^{-y}$, $b_0 = c_0 = 0$, $a_0 = 1$. Clearly, conditions (C17)–(C20) are satisfied, and $\alpha(t) = 0$ is a lower solution of (9.61). Now we shall that

$$\beta(t) = \frac{A}{2(n+1)}(1 - t^2)$$

is an upper solution of (9.61). For this, all we need is to satisfy the inequality

$$\frac{1}{t^n}\frac{d}{dt}\left(t^n \frac{-2At}{2(n+1)}\right) \leq -A\exp\left(-\frac{A}{2(n+1)}(1 - t^2)\right), \quad 0 < t < 1$$

which is the same as

$$1 \geq \exp\left(-\frac{A}{2(n+1)}(1 - t^2)\right), \quad 0 < t < 1.$$

Finally, following exactly the same as for (9.45) we find that the problem (9.60), (9.31) has a solution $\theta \in C[0, 1] \cap C^1[0, 1) \cap C^2(0, 1)$ with

$$0 \leq \theta(x) \leq \frac{A}{2(n+1)}(1 - x^2). \tag{9.62}$$

Finally, we shall show that the problem (9.60), (9.31) for all $n \geq 0$ has at most one solution. For this, let $\theta_1(x)$ and $\theta_2(x)$ be two solutions, and without loss of generality assume that there exists a point $x_0 \in [0, 1)$ where the function $\phi(x) = \theta_1(x) - \theta_2(x)$ assumes a positive maximum, i.e., $\phi(x_0) > 0$. If $x_0 \in (0, 1)$, then we must have $\phi'(x_0) = 0$, $\phi''(x_0) \leq 0$. However, from Eq. (9.48) we have

$$\phi''(x_0) = -A\left(e^{-\theta_1(x_0)} - e^{-\theta_2(x_0)}\right) > 0,$$

which is a contradiction. If $x_0 = 0$, then there exists an interval $[0, \delta)$, $\delta < 1$ in which $\phi(x) > 0$. Thus for all $0 < \epsilon < x < \delta$, from Eq. (9.60) we have

$$\frac{1}{x^n}\frac{d}{dx}\left(x^n \frac{d\phi}{dx}\right) = -A\left(e^{-\theta_1(x)} - e^{-\theta_2(x)}\right) > 0,$$

and hence $x^n \phi'(x)$ is increasing in (ϵ, δ). However, since $\phi'(0) = 0$ there must be a $\mu > 0$ so that $\phi'(x) > 0$, $x \in (0, \mu)$. But, this contradicts the fact that $\phi(x)$ attains its maximum at $x_0 = 0$.

Example 9.6 (Non-Newtonian Fluid Flows) The Augustin–Louis Cauchy (1789–1857) stress **T** in an incompressible homogeneous fluid of third grade has the form

$$\mathbf{T} = -\rho \mathbf{I} + \mu \mathbf{A}_1 + \alpha_1 \mathbf{A}_2 + \alpha_2 \mathbf{A}_1^2 + \beta_1 \mathbf{A}_3 + \beta_2[\mathbf{A}_1 \mathbf{A}_2 + \mathbf{A}_2 \mathbf{A}_1] + \beta_3 Tr(\mathbf{A}_1^2)\mathbf{A}_2, \tag{9.63}$$

where $-\rho\mathbf{I}$ is the spherical stress due to the constraint of incompressibility, μ, α_1, α_2, β_1, β_2, β_3 are material moduli, and $\mathbf{A}_1$, $\mathbf{A}_2$, $\mathbf{A}_3$ are the first three Rivlin–Ericksen (after R. S. Rivlin and J. L. V. Ericksen 1955) tensors given by

$$\mathbf{A}_1 = \mathbf{L} + \mathbf{L}^T, \quad \mathbf{A}_2 = \frac{d\mathbf{A}_1}{dt} + \mathbf{L}^T\mathbf{A}_1 + \mathbf{A}_1\mathbf{L}$$

and

$$\mathbf{A}_3 = \frac{d\mathbf{A}_2}{dt} + \mathbf{L}^T\mathbf{A}_2 + \mathbf{A}_2\mathbf{L};$$

here $\mathbf{L}$ represents the spatial gradient of velocity and d/dt the material time derivative. Now consider the flow of a third grade fluid, obeying (9.63), maintained at a cylinder (of radius R) by its angular velocity (Ω). The steady-state equation for this fluid is

$$0 = \mu\left[\frac{d^2\tilde{v}}{d\tilde{r}^2} + \frac{1}{\tilde{r}}\frac{d\tilde{v}}{d\tilde{r}} - \frac{\tilde{v}}{\tilde{r}^2}\right] + \beta\left(\frac{d\tilde{v}}{d\tilde{r}} - \frac{\tilde{v}}{\tilde{r}}\right)^2\left[6\frac{d^2\tilde{v}}{d\tilde{r}^2} - \frac{2}{\tilde{r}}\frac{d\tilde{v}}{d\tilde{r}} + \frac{2\tilde{v}}{\tilde{r}^2}\right]$$

with boundary conditions

$$\tilde{v} = R\Omega \quad \text{at} \quad \tilde{r} = R, \quad \text{and} \quad \tilde{v} \to 0 \quad \text{as} \quad \tilde{r} \to \infty;$$

here $\tilde{v}$ is the nonzero velocity in polar coordinates and μ and β are material constants. Making the change of variables

$$r = \frac{\tilde{r}}{R} \quad \text{and} \quad v = \frac{\tilde{v}}{R\Omega}$$

our problem is transformed to

$$\frac{d^2v}{dr^2} + \frac{1}{r}\frac{dv}{dr} - \frac{v}{r^2} + \epsilon\left(\frac{dv}{dr} - \frac{v}{r}\right)^2\left[6\frac{d^2v}{dr^2} - \frac{2}{r}\frac{dv}{dr} + \frac{2v}{r^2}\right] = 0 \qquad (9.64)$$

for $1 < r < \infty$, with boundary conditions

$$v = 1 \quad \text{if} \quad r = 1, \quad v \to 0 \quad \text{as} \quad r \to \infty; \qquad (9.65)$$

here $\epsilon = \Omega^2\beta/\mu$. As a result, our non-Newtonian fluid problem reduces to a second-order boundary value problem on the infinite interval, see [60, 113].

It is easily seen that (9.64), (9.65) can be written as

$$v'' + \frac{1}{r}\frac{\left(1 - 2\epsilon\left(v' - \frac{v}{r}\right)^2\right)}{\left(1 + 6\epsilon\left(v' - \frac{v}{r}\right)^2\right)}\left(v' - \frac{v}{r}\right) = 0, \quad 1 < r < \infty$$

$$v(1) = 1, \quad \lim_{r\to\infty} v(r) = 0. \tag{9.66}$$

To show that (9.66) with $\epsilon < 1/2$ has a solution $v \in BC[1, \infty) \cap C^2(1, \infty)$ we will apply Theorem 9.6. With this in mind we consider the boundary value problem, see [15]

$$v'' + \frac{1}{r} \frac{\left(1 - 2\epsilon\left(v' - \frac{v}{r}\right)^2\right)}{\left(1 + 6\epsilon\left(v' - \frac{v}{r}\right)^2\right)} \left(v' - \frac{v}{r}\right) = 0, \quad 1 < r < \infty \tag{9.67}$$

$$v(1) = 1, \quad v(r) \text{ bounded on } [1, \infty).$$

Since $\epsilon < 1/2$ there exists $\delta > 0$ such that $1/\sqrt{2\epsilon} - \delta \geq 1$. Now choose $\theta \in (0, 1)$ with

$$\sqrt{\frac{1 - \theta}{2\epsilon(3\theta + 1)(\theta + 1)^2}} \geq \frac{1}{\sqrt{2\epsilon}} - \delta, \tag{9.68}$$

and let

$$\tau = \sqrt{\frac{1 - \theta}{2\epsilon(3\theta + 1)(\theta + 1)^2}}. \tag{9.69}$$

We will apply Theorem 9.6 with

$$a = 1, \quad q(r) = \frac{1}{r}, \quad f(r, y, y') = -\frac{\left(1 - 2\epsilon\left(y' - \frac{y}{r}\right)^2\right)}{\left(1 + 6\epsilon\left(y' - \frac{y}{r}\right)^2\right)} \left(y' - \frac{y}{r}\right),$$

$$\alpha = 0 \quad \text{and} \quad \beta(r) = \frac{\tau}{r^\theta}.$$

Note that if

$$g(w) = \frac{1 - 2w}{1 + 6w}, \quad w > 0$$

then

$$g'(w) = \frac{-8}{(1 + 6w)^2} < 0 \quad \text{with} \quad \lim_{w \to 0} g(w) = 1 \quad \text{and} \quad \lim_{w \to \infty} g(w) = -\frac{1}{3}. \tag{9.70}$$

As a result

$$-\frac{1}{3} \leq g(w) \leq 1 \quad \text{for} \quad w \geq 0. \tag{9.71}$$

Clearly (C22)–(C24) of Theorem 9.6 hold. In addition α is a lower solution of (9.67) since $\alpha(1) = 0 \leq 1$ and

$$\alpha'' - qf(r, \alpha, \alpha') = \alpha'' + \frac{1}{r} g\left(\epsilon\left(\alpha' - \frac{\alpha}{r}\right)^2\right) \left(\alpha' - \frac{\alpha}{r}\right) = 0 \text{ for } 1 < r < \infty.$$

Next we show β is an upper solution of (9.67). First note $\beta(1) = \tau \geq (1/\sqrt{2\epsilon}) - \delta \geq 1$. Also for $1 < r < \infty$ we have

$$\beta'' - qf(r, \beta, \beta') = \beta'' + \frac{1}{r}g\left(\epsilon\left(\beta' - \frac{\beta}{r}\right)^2\right)\left(\beta' - \frac{\beta}{r}\right)$$

$$= \frac{\tau\theta(\theta + 1)}{r^{\theta+2}} + \frac{1}{r}g\left(\epsilon\left(\beta' - \frac{\beta}{r}\right)^2\right)\left[-\frac{\tau(\theta + 1)}{r^{\theta+1}}\right]$$

$$= \frac{\tau(\theta + 1)}{r^{\theta+2}}\left\{\theta - g\left(\epsilon\left(\beta' - \frac{\beta}{r}\right)^2\right)\right\}.$$

Now since

$$\epsilon\left(\beta' - \frac{\beta}{r}\right)^2 = \frac{\epsilon\tau^2(\theta + 1)^2}{r^{2\theta+2}} \leq \epsilon\tau^2(\theta + 1)^2$$

we have from (9.70) that

$$g\left(\epsilon\left(\beta' - \frac{\beta}{r}\right)^2\right) \geq g\left(\epsilon\tau^2(\theta + 1)^2\right) \quad \text{for} \quad r > 1,$$

and this together with the above inequality yields

$$\beta'' - qf(r, \beta, \beta') \leq \frac{\tau(\theta + 1)}{r^{\theta+2}}\left\{\theta - g\left(\epsilon\tau^2(\theta + 1)^2\right)\right\} = 0 \quad \text{for} \quad 1 < r < \infty$$

since from (9.69) we have

$$\theta - g\left(\epsilon\tau^2(\theta + 1)^2\right) = \theta - \frac{1 - 2\epsilon\tau^2(\theta + 1)^2}{1 + 6\epsilon\tau^2(\theta + 1)^2}$$

$$= \theta - \left\{\frac{1 - \frac{1-\theta}{3\theta+1}}{1 + \frac{3(1-\theta)}{3\theta+1}}\right\}$$

$$= \theta - \theta = 0.$$

Thus (C25) of Theorem 9.6 holds. Also notice that (9.70) implies

$$|f(r, u, v)| \leq |v| + \tau \quad \text{for} \quad (r, u) \in [1, \infty) \times \left[0, \frac{\tau}{r^\theta}\right],$$

so (C26) of Theorem 9.6 is satisfied with $\psi(z) = |z| + \tau$. Finally notice that (C27) of Theorem 9.6 holds since

$$\int_{d_0}^\infty \frac{u}{\psi(u)}du = \int_{d_0}^\infty \frac{u}{u + \tau}du = \infty \quad \text{and} \quad A_0 \sup_{r\in[1,\infty)} q(r) < \infty.$$

Thus, Theorem 9.6 guarantees that (9.67) has a solution $v \in BC[1, \infty) \cap C^2(1, \infty)$, $v' \in BC[1, \infty)$ with

$$0 \leq v(r) \leq \frac{\tau}{r^\theta} \quad \text{for} \quad r \in [1, \infty).$$

Now $\lim_{r \to \infty} (\tau/r^\theta) = 0$, so v is a solution of (9.64), (9.65).

Example 9.7 (*Membrane Response of a Spherical Cap*) The boundary value problem

$$y'' + \left(\frac{t^2}{32 y^2} - \frac{\lambda^2}{8} \right) = 0, \quad 0 < t < 1 \tag{9.72}$$

$$y(0) = 0, \quad 2y'(1) - (1+v)y(1) = 0, \quad 0 < v < 1 \text{ and } \lambda > 0 \tag{9.73}$$

arises in nonlinear mechanics. The problem models the large deflection membrane response of a spherical cap, see [38]. Here $S_r = y/t$ is the radial stress at points on the membrane, $d(\rho S_r)/d\rho$ is the circumferential stress ($\rho = t^2$), λ is a load geometry parameter and v is the Poisson ratio.

Motivated by the above problem, we present two existence results for the differential equation

$$y'' + qf(t, y) = 0, \quad 0 < t < 1. \tag{9.74}$$

The first result is for the problem (9.74), (9.73), whereas the second is for (9.74) together with the more general conditions

$$y(0) = 0, \quad y'(1) + \psi(y(1)) = 0. \tag{9.75}$$

Theorem 9.7 ([4]) *Suppose that*
(C28) there exists $\beta \in C[0, 1] \cap C^2(0, 1)$ with $\beta' \in AC[0, 1]$, $\beta(0) \geq 0$, $q(t)f(t, \beta(t)) + \beta''(t) \leq 0$ for $t \in (0, 1)$, and $2\beta'(1) - (1+v)\beta(1) \geq 0$,
(C29) there exists $\alpha \in C[0, 1] \cap C^2(0, 1)$ with $\alpha' \in AC[0, 1]$, $\alpha(t) \leq \beta(t)$ on $[0, 1]$, $\alpha(0) \leq 0$, $q(t)f(t, \alpha(t)) + \alpha''(t) \geq 0$ for $t \in (0, 1)$, and $2\alpha'(1) - (1+v)\alpha(1) \leq 0$,
(C30) $q \in C(0, 1) \cap L^1[0, 1]$ with $q > 0$ on $(0, 1)$ and for each $t \in [0, 1]$, $f(t, u) \in \mathbb{R}$ for $u \in [\alpha(t), \beta(t)]$, and
(C31) the function $f^* : [0, 1] \times \mathbb{R} \to \mathbb{R}$ defined by

$$f^*(t, y) = \begin{cases} f(t, \beta(t)) + r(\beta(t) - y), & y \geq \beta(t) \\ f(t, y), & \alpha(t) < y < \beta(t) \\ f(t, \alpha(t)) + r(\alpha(t) - y), & y \leq \alpha(t) \end{cases}$$

where $r : \mathbb{R} \to [-1, 1]$ is the radial retraction defined by

$$r(x) = \begin{cases} x, & |x| \le 1 \\ x/|x|, & |x| > 1 \end{cases}$$

is continuous.

Then, (9.74), (9.73) has a solution $y \in C[0, 1] \cap C^2(0, 1)$ with $y' \in AC[0, 1]$ and $\alpha(t) \le y(t) \le \beta(t)$ for $t \in [0, 1]$.

We shall apply Theorem 9.7 to show that (9.72), (9.73) has a solution $y \in C[0, 1] \cap C^2(0, 1)$ with $y' \in AC[0, 1]$ and

$$t\left(\frac{a}{2\lambda} - bt\right) \le y(t) \le \frac{t}{2\lambda} \quad \text{for } t \in [0, 1]; \tag{9.76}$$

here

$$b = \frac{a}{2\lambda}\left(\frac{1-v}{3-v}\right) \quad \text{and} \quad a = \sqrt{\frac{\lambda^2}{\lambda^2 + (8/\lambda)}}. \tag{9.77}$$

Notice first that $(a/2\lambda) > b$ since $0 < (1-v)/(3-v) < 1$ and so $a/(2\lambda) - bt > 0$ for $t \in [0, 1]$. In Theorem 9.7 we let $q = 1$, $f(t, y) = (t^2/32y^2) - \lambda^2/8$ and

$$\alpha(t) = t\left(\frac{a}{2\lambda} - bt\right) \quad \text{and} \quad \beta(t) = \frac{t}{2\lambda}. \tag{9.78}$$

Let

$$f^*(t, y) = \begin{cases} r\left(\frac{t}{2\lambda} - y\right), & y \ge \frac{t}{2\lambda} \\ \dfrac{t^2}{32y^2} - \dfrac{\lambda^2}{8}, & t\left(\dfrac{a}{2\lambda} - bt\right) < y < \dfrac{t}{2\lambda} \\ \dfrac{1}{32\left(\frac{a}{2\lambda} - bt\right)^2} - \dfrac{\lambda^2}{8} + r\left(t\left[\dfrac{a}{2\lambda} - bt\right] - y\right), & y \le t\left(\dfrac{a}{2\lambda} - bt\right). \end{cases}$$

Notice (C30) and (C31) are satisfied since $a/(2\lambda) - bt > 0$ for $t \in [0, 1]$. To see (C28) notice that $\beta(0) = 0$, $2\beta'(1) - (1+v)\beta(1) = [1 - (1+v)/2]/\lambda \ge 0$ and $q(t)f(t, \beta(t)) + \beta''(t) = 0$ for $t \in (0, 1)$. Finally, to show (C29) we notice that $\alpha(0) = 0$ and

$$2\alpha'(1) - (1+v)\alpha(1) = 2\left(\frac{a}{2\lambda} - 2b\right) - (1+v)\left(\frac{a}{2\lambda} - b\right)$$
$$= b(-3 + v) + \frac{a}{2\lambda}(1-v) = 0.$$

Now since $a < 1$ in view of $8 \ge 8a(1-v)/(3-v)$ it follows that $8/\lambda \ge 16b$ and hence

$$a^2 = \frac{\lambda^2}{\lambda^2 + (8/\lambda)} \le \frac{\lambda^2}{\lambda^2 + 16b}$$

and so for $t \in (0, 1)$,

$$
\begin{aligned}
q(t)f(t, \alpha(t)) + \alpha''(t) &= \frac{t^2}{32[at/(2\lambda) - bt^2]^2} - \frac{\lambda^2}{8} - 2b \\
&\geq \frac{1}{32[a/(2\lambda)]^2} - \frac{\lambda^2}{8} - 2b \\
&= \frac{\lambda^2}{8}\left(\frac{1}{a^2} - 1\right) - 2b \\
&\geq \frac{\lambda^2}{8}\left(\frac{\lambda^2 + 16b}{\lambda^2} - 1\right) - 2b = 0.
\end{aligned}
$$

Now Theorem 9.7 guarantees that (9.72), (9.73) has the desired solution.

Theorem 9.8 ([11]) *Let $n_0 \in \{3, 4, \cdots\}$ be fixed and suppose that*
(C32) $f : [0, 1] \times (0, \infty) \to \mathbb{R}$ *is continuous,*
(C33) $q : C(0, 1)$ *with $q > 0$ on $(0, 1)$ and $tq \in L^1[0, 1]$,*
(C34) $\psi : \mathbb{R} \to \mathbb{R}$ *is continuous,*
(C35) *let $n \in \{n_0, n_0 + 1, \cdots\} = \mathbb{N}_0$ and associated with each n there exists a constant ρ_n such that $\{\rho_n\}$ is a nonincreasing sequence with $\lim_{n\to\infty} \rho_n = 0$ and such that for $1/n \leq t < 1$, $q(t)f(t, \rho_n) \geq 0$,*
(C36) *there exists $\alpha \in C[0, 1] \cap C^1(0, 1] \cap C^2(0, 1)$ with $\alpha(0) = 0$, $\alpha'(1) + \psi(\alpha(1)) \leq 0$, $\alpha > 0$ on $(0, 1]$ such that for each $n \in \mathbb{N}_0$, $q(t)f(t, \alpha(t)) + \alpha''(t) \geq 0$ for $t \in [1/n, 1)$ and $q(t)f(1/n, \alpha(t)) + \alpha''(t) \geq 0$ for $t \in (0, 1/n)$,*
(C37) for each $n \in \mathbb{N}_0$ there exists $\beta_n \in C[0, 1] \cap C^1(0, 1] \cap C^2(0, 1)$ with $\beta_n(t) \geq \alpha(t)$ and $\beta_n(t) \geq \rho_n$ for $t \in [0, 1]$, $\beta_n'(1) + \psi(\beta_n(1)) \geq 0$, and $q(t)f(t, \beta_n(t)) + \beta_n''(t) \leq 0$ for $t \in [1/n, 1)$ with $q(t)f(1/n, \beta_n(t)) + \beta_n''(t) \leq 0$ for $t \in (0, 1/n)$,
(C38) $a_0 \equiv \max\{\sup_{t\in[0,1]} \beta_n(t) : n \in \mathbb{N}_0\} < \infty$,
(C39) $|f(t, y)| \leq g(y)$ *on $[0, 1] \times (0, a_0]$ with $g > 0$ continuous and nonincreasing on $(0, \infty)$, and*
(C40) for any $R > 0$, $1/g$ is differentiable on $(0, R]$ with $g' < 0$ a.e. on $(0, R]$, $g'/g^2 \in L^1[0, R]$ and $\int_0^\infty (|g'(t)|^{1/2}/g(t))dt = \infty$.
Then, (9.74), (9.75) has a solution $y \in C[0, 1] \cap C^1(0, 1] \cap C^2(0, 1)$ with $y(t) \geq \alpha(t)$ for $t \in [0, 1]$.

Corollary 9.1 *If in Theorem 9.8 conditions (C35)–(C37) are replaced by*
(C41) for each $n \in \mathbb{N}_0$ there exists $\beta_n \in C[0, 1] \cap C^1(0, 1] \cap C^2(0, 1)$ with $\beta_n(t) \geq \rho_n$ for $t \in [0, 1]$, $\beta_n'(1) + \psi(\beta_n(1)) \geq 0$ and $q(t)f(t, \beta_n(t)) + \beta_n''(t) \leq 0$ for $t \in [1/n, 1)$ with $q(t)f(1/n, \beta_n(t)) + \beta_n''(t) \leq 0$ for $t \in (0, 1/n)$,
(C42) if $x > 0$, $y > 0$ with $x - y \leq \psi(y) - \psi(x)$ then $x - y \leq 0$, (if $\psi(u) = -b_0 u$ for $b_0 < 1$ then this condition is obviously satisfied),
(C43) $\psi(u) = -b_0 u$, $0 \leq b_0 < 1$ *and there exists $\tau \in (0, 1)$ with $f(t, y) > 0$ for $t \in [\tau, 1)$ and $0 < y \leq \rho_{n_0}[1 - b_0(1 - \tau)]^{-1}$,*
(C44) let $n \in \mathbb{N}_0$ and associated with each n there exist a constant ρ_n such that $\{\rho_n\}$ is a decreasing sequence with $\lim_{n\to\infty} \rho_n = 0$ and there exists a constant $k_0 > 0$ such that for $1/n \leq t \leq 1$ and $0 < y \leq \rho_n$, $q(t)f(t, y) \geq k_0$, and

(C45) $f(\cdot, y)$ *is nondecreasing on* $(0, 1/3)$ *for each fixed* $y \in (0, \infty)$.
Then, (9.74), (9.75) *has a solution* $y \in C[0, 1] \cap C^1(0, 1] \cap C^2(0, 1)$ *with* $y(t) > 0$
for $t \in (0, 1]$.

First, we shall apply Corollary 9.1 to show that (9.72), (9.73) has a solution $y \in C[0, 1] \cap C^1(0, 1] \cap C^2(0, 1)$ with $y(t) > 0$ for $t \in (0, 1]$. For this, we let $q \equiv 1$, $g(y) = 1/(32y^2)$, $\psi(z) = -(1+v)z/2$ and $b_0 = (1+v)/2$. Choose and fix $n_0 \in \{3, 4, \cdots\}$ with

$$\frac{1}{\sqrt{n_0}} < \frac{(1-v)(8+\lambda^2)^{1/2}}{\lambda(1+v)} \quad \text{and} \quad \frac{1}{n_0} \le \frac{\lambda^2}{8+\lambda^2}. \tag{9.79}$$

Let

$$\rho_n = \frac{1}{2n(8+\lambda^2)^{1/2}} \quad \text{and} \quad k_0 = 1.$$

Clearly, (C32)–(C34), (C40), (C42) and (C45) hold. Next notice that for $n \in \mathbb{N}_0 = \{n_0, n_0+1, \cdots\}$, $1/n \le t \le 1$ and $0 < y \le \rho_n$ we have

$$q(t)f(t, y) \ge \frac{1}{32n^2y^2} - \frac{\lambda^2}{8} \ge \frac{1}{32n^2\rho_n^2} - \frac{\lambda^2}{8} = \frac{(8+\lambda^2)}{8} - \frac{\lambda^2}{8} = 1,$$

so (C44) is satisfied. Now let

$$\beta_n(t) = \frac{t}{2\lambda} + \sqrt{n_0}\,\rho_n. \tag{9.80}$$

Also notice that for $n \in \mathbb{N}_0$ that $\beta_n(t) \ge \rho_n$ for $t \in [0, 1]$ and

$$\beta_n'(1) + \psi(\beta_n(1)) = \frac{1}{2\lambda} - \frac{(1+v)}{2}\left(\frac{1}{2\lambda} + \sqrt{n_0}\rho_n\right)$$

$$= \frac{(1-v)}{4\lambda} - \frac{(1+v)}{2}\sqrt{n_0}\,\rho_n \ge 0$$

since (9.79) implies

$$\sqrt{n_0}\,\rho_n = \frac{\sqrt{n_0}}{2n(8+\lambda^2)^{1/2}} \le \frac{\sqrt{n_0}}{2n_0(8+\lambda^2)^{1/2}} = \frac{1}{2\sqrt{n_0}(8+\lambda^2)^{1/2}} \le \frac{1-v}{2\lambda(1+v)}.$$

Now if $n \in N_0$ and $t \in [1/n, 1)$ we have

$$\beta_n''(t) + q(t)f(t, \beta_n(t)) = \frac{t^2}{32\left(\frac{t}{2\lambda} + \sqrt{n_0}\rho_n\right)^2} - \frac{\lambda^2}{8} \le \frac{4t^2\lambda^2}{32t^2} - \frac{\lambda^2}{8} = 0,$$

whereas if $n \in \mathbb{N}_0$ and $t \in (0, 1/n)$ we have from (9.79) that

$$\beta_n''(t) + q(t)f(1/n, \beta_n(t)) = \frac{1}{32n^2\left(\frac{t}{2\lambda} + \sqrt{n_0\rho_n}\right)^2} - \frac{\lambda^2}{8}$$

$$\leq \frac{1}{32n^2 n_0 \rho_n^2} - \frac{\lambda^2}{8} = \frac{8 + \lambda^2}{8n_0} - \frac{\lambda^2}{8} \leq 0.$$

Thus (C41) holds. It remains to check (C43). Let $\tau = 1/(1 - b_0)$. Now if $t \in [\tau, 1)$ and $0 < y \leq \rho_{n_0}[1 - b_0(1 - \tau)]^{-1}$ then we have

$$f(t, y) \geq \frac{\tau^2}{32y^2} - \frac{\lambda^2}{8} \geq \frac{\tau^2}{32} \frac{[1 - b_0(1 - \tau)]^2}{\rho_{n_0}^2} - \frac{\lambda^2}{8}$$

$$\geq \frac{\tau^2(1 - b_0)^2}{32\rho_{n_0}^2} - \frac{\lambda^2}{8} = \frac{\tau^2(1 - b_0)^2 n_0^2(8 + \lambda^2)}{8} - \frac{\lambda^2}{8}$$

$$\geq \tau^2(1 - b_0)^2\left(1 + \frac{\lambda^2}{8}\right) - \frac{\lambda^2}{8} = \left(1 + \frac{\lambda^2}{8}\right) - \frac{\lambda^2}{8} = 1.$$

Thus (C43) holds. Existence of a solution to (9.72), (9.73) is now guaranteed from Corollary 9.1.

Now we shall apply Theorem 9.8 to show that (9.73), (9.74) has a solution. For this, let $\alpha(t)$ be as in (9.78) with b and a are defined by (9.77), and β_n is as in (9.80). Clearly (C32)–(C35), (C37)–(C40) hold. It remains to show (C36). Now as in the illustration of Theorem 9.7 we have $\alpha(0) = 0$, $\alpha'(1) + \psi(\alpha(1)) = 0$, and $a/(2\lambda) > b$ and $a^2 \leq \lambda^2/(\lambda^2 + 16b)$ implies $q(t)f(t, \alpha(t)) + \alpha''(t) \geq 0$ for $n \in \mathbb{N}_0$ and $t \in [1/n, 1)$. Further, if $n \in \mathbb{N}_0$ and $t \in (0, 1/n)$ then since $nt \leq 1$ we have

$$q(t)f(1/n, \alpha(t)) + \alpha''(t)$$

$$= \frac{1}{32n^2\left[at/(2\lambda) - bt^2\right]^2} - \frac{\lambda^2}{8} - 2b \geq \frac{1}{32n^2 t^2\left[a/(2\lambda)\right]^2} - \frac{\lambda^2}{8} - 2b$$

$$\geq \frac{1}{32[a/(2\lambda)]^2} - \frac{\lambda^2}{8} - 2b \geq \frac{\lambda^2}{8}\left(\frac{\lambda^2 + 16b}{\lambda^2} - 1\right) - 2b = 0.$$

Existence of a solution to (9.72), (9.73) now follows from Theorem 9.8.

For more details, about the above problem see [25, 89, 90].

Example 9.8 (*Deformation Shape of a Membrane Cap*) Consider the deformation shape of a membrane cap which is subjected to a uniform vertical pressure P and either a radial displacement or a radial stress on the boundary. Assuming the cap is shallow, (i.e., nearly flat), the strains are small, the pressure P is small, and the undeformed shape of the membrane is radially symmetric and described in cylindrical coordinates by $z = C(1 - r^\gamma)$ (where $0 \leq r \leq 1$ and $\gamma > 1$) where the undeformed radius is $r = 1$ and $C > 0$ is the height at the center of the cap. Then, for any radially symmetric deformed state the scaled radial stress S_r satisfies the differential equation

$$r^2 S_r'' + 3r S_r' = \frac{\lambda^2 r^{2\gamma-2}}{2} + \frac{\beta \nu r^2}{S_r} - \frac{r^2}{8 S_r^2}, \tag{9.81}$$

the regularity condition

$$S_r(r) \quad \text{bounded as} \quad r \to 0^+ \tag{9.82}$$

and the boundary condition

$$b_0 S_r(1) + b_1 S_r'(1) = A, \tag{9.83}$$

where λ and β are positive constants depending on the pressure P, the thickness of the membrane, and Young's modulus, $b_0 > 0$, $b_1 \geq 0$ and $A > 0$. For the stress problem $b_0 = 1$, $b_1 = 0$ whereas for the displacement problem $b_0 = 1 - \nu$, $b_1 = 1$ where ν $(0 \leq \nu < 0.5)$ is the Poisson ratio, see [39, 56, 75].

To move the singularity from $r = 0$ to ∞ in (9.81)–(9.83), we use the substitution $t = r^{-2}$, $u(t) = S_r(r)$ to obtain the infinite interval problem

$$u'' = \frac{1}{t^3} \left[\frac{\lambda^2}{8 t^{\gamma-2}} - \frac{1}{32 u^2} + \frac{\mu}{4u} \right], \quad 1 < t < \infty \tag{9.84}$$

$$a_0 u(1) - a_1 u'(1) = A, \quad u(t) \quad \text{bounded as} \quad t \to \infty,$$

where $\mu = \beta \nu$, $a_0 = b_0$, and $a_1 = 2 b_1$.

In what follows, we assume that $a_0 > 0$, $a_1 \geq 0$, $A > 0$, $\lambda > 0$ and $\gamma > 1$. To show (9.84) has a solution, we need the following general existence criterion for the boundary value problem:

$$y'' = q f(t, y), \quad a < t < \infty, \quad a \geq 0 \quad \text{is fixed}$$

$$-a_0 y(a) + b_0 y'(a) = c_0, \quad a_0 > 0, \quad b_0 \geq 0 \tag{9.85}$$

$$y(t) \quad \text{bounded on} \quad [a, \infty).$$

Recall that by an upper solution β to (9.85) we mean a function $\beta \in BC[a, \infty) \cap C^2(a, \infty)$, $\beta' \in C[a, \infty)$ with $\beta'' \leq q f(t, \beta)$, $a < t < \infty$, $-a_0 \beta(a) + b_0 \beta'(a) \leq c_0$, $\beta(t)$ bounded on $[a, \infty)$; here $BC[a, \infty)$ denotes the space of continuous, bounded functions from $[a, \infty)$ to $\mathbb{R}$. Similarly, by a lower solution α to (9.85) we mean a function $\alpha \in BC[a, \infty) \cap C^2(a, \infty)$, $\alpha' \in C[a, \infty)$ with $\alpha'' \geq q f(t, \alpha)$, $a < t < \infty$, $-a_0 \alpha(a) + b_0 \alpha'(a) \geq c_0$, $\alpha(t)$ bounded on $[a, \infty)$.

Theorem 9.9 ([19, 21]) *Assume that*
(C46) $f : [a, \infty) \times \mathbb{R} \to \mathbb{R}$ is continuous,
(C47) $q \in C(a, \infty)$ with $q > 0$ on (a, ∞),
(C48) $\int_a^\mu q(s) ds < \infty$ for any $\mu > a$,
(C49) there exist α, β, respectively, lower and upper solutions of (9.85) with $\alpha(t) \leq \beta(t)$ for $t \in [a, \infty)$, and

(C50) there exists a constant $M > 0$ with $|f(t, u)| \le M$ for $t \in [a, \infty)$ and $u \in [\alpha(t), \beta(t)]$.
Then, (9.85) has a solution $y \in BC[a, \infty) \cap C^2(a, \infty)$, $y' \in C[a, \infty)$ with $\alpha(t) \le y(t) \le \beta(t)$ for $t \in [a, \infty)$.

With respect to the problem (9.84), we consider the following four cases:
Case 1. $\nu > 0$, $\gamma \ge 2$. In this case, we let

$$\beta(t) = \max \left\{ \frac{A}{a_0}, \frac{1}{8\mu} \right\} \equiv \beta_0$$

and

$$\alpha(t) = \min \left\{ \frac{A}{a_0}, -\frac{\mu}{\lambda^2} + \sqrt{\frac{\mu^2}{\lambda^4} + \frac{1}{4\lambda^2}} \right\} \equiv \alpha_0.$$

Notice $\beta_0 > 0$, $\alpha_0 > 0$ and $\alpha(t) \le \beta(t)$ for $t \in [1, \infty)$. Let $q \equiv 1$ and

$$f(t, y) = \begin{cases} \dfrac{1}{t^3} \left[\dfrac{\lambda^2}{8t^{\gamma-2}} - \dfrac{1}{32\alpha_0^2} + \dfrac{\mu}{4\alpha_0} \right], & y \le \alpha_0 \\[4mm] \dfrac{1}{t^3} \left[\dfrac{\lambda^2}{8t^{\gamma-2}} - \dfrac{1}{32y^2} + \dfrac{\mu}{4y} \right], & y \ge \alpha_0. \end{cases}$$

Clearly (C46)–(C48) and (C50) hold. It remains to check (C49). First we show β is an upper solution for (9.85). If $1 < t < \infty$ then

$$\beta''(t) - q(t)f(t, \beta(t)) = \frac{1}{t^3} \left[\frac{1}{32\beta_0^2} - \frac{\lambda^2}{8t^{\gamma-2}} - \frac{\mu}{4\beta_0} \right]$$

$$= \frac{1}{4\beta_0 t^3} \left[\frac{1}{8\beta_0} - \mu \right] - \frac{\lambda^2}{8t^{\gamma+1}} \le 0.$$

In addition, $a_0\beta(1) - a_1\beta'(1) = a_0\beta_0 \ge a_0 A/a_0 = A$, so β is an upper solution for (9.85). To show α is a lower solution for (9.85) first notice that $a_0\alpha(1) - a_1\alpha'(1) = a_0\alpha_0 \le a_0 A/a_0 = A$. Also for $1 < t < \infty$ in view of $\gamma \ge 2$ we have

$$\alpha''(t) - q(t)f(t, \alpha(t)) = \frac{1}{t^3} \left[\frac{1}{32\alpha_0^2} - \frac{\lambda^2}{8t^{\gamma-2}} - \frac{\mu}{4\alpha_0} \right]$$

$$= \frac{1}{32\alpha_0^2 t^3} \left[1 - 8\mu\alpha_0 - \frac{4\alpha_0^2 \lambda^2}{t^{\gamma-2}} \right]$$

$$\ge \frac{1}{32\alpha_0^2 t^3} \left[1 - 8\mu\alpha_0 - 4\alpha_0^2 \lambda^2 \right]$$

$$= -\frac{\lambda^2}{8\alpha_0^2 t^3} \left[\alpha_0^2 + 2\frac{\mu}{\lambda^2}\alpha_0 - \frac{1}{4\lambda^2} \right]$$

$$= -\frac{\lambda^2}{8\alpha_0^2 t^3}(\alpha_0 - x_0)(\alpha_0 + x_1) \geq 0$$

since $\alpha_0 \leq x_0$; here

$$x_0 = -\frac{\mu}{\lambda^2} + \sqrt{\frac{\mu^2}{\lambda^4} + \frac{1}{4\lambda^2}} \quad \text{and} \quad x_1 = \frac{\mu}{\lambda^2} + \sqrt{\frac{\mu^2}{\lambda^4} + \frac{1}{4\lambda^2}}.$$

Thus α is a lower solution of (9.85), so (C49) holds. Theorem 9.9 guarantees that (9.85) has a solution $y \in BC[1, \infty) \cap C^2(1, \infty)$, $y' \in C[1, \infty)$ with $\alpha_0 \leq y(t) \leq \beta_0$ for $t \in [1, \infty)$. This y is a solution of (9.84).

Case 2. $\nu > 0$, $1 < \gamma < 2$. In this case, we let

$$\beta(t) = \max\left\{\frac{A}{a_0}, \frac{1}{8\mu}\right\} \equiv \beta_0$$

and

$$\alpha(t) = \frac{\alpha_1}{t^{(2-\gamma)/2}} \quad \text{where} \quad \alpha_1 = \min\left\{\frac{A}{a_0 + \frac{a_1}{2}}, -\frac{\mu}{\lambda^2} + \sqrt{\frac{\mu^2}{\lambda^4} + \frac{1}{4\lambda^2}}\right\}.$$

Notice

$$\alpha(t) \leq \alpha_1 \leq \frac{A}{a_0 + \frac{a_1}{2}} \leq \frac{A}{a_0} \leq \beta(t) \quad \text{for } t \in [1, \infty),$$

and α, β are bounded on $[1, \infty)$ (note $1 < \gamma < 2$). Let $q \equiv 1$ and

$$f(t, y) = \begin{cases} \dfrac{\lambda^2}{8t^{\gamma+1}} - \dfrac{1}{32\alpha_1^2 t^{\gamma+1}} + \dfrac{\mu}{4\alpha_1 t^{(\gamma+4)/2}}, & y \leq \alpha_1/t^{(2-\gamma)/2} \\[3mm] \dfrac{1}{t^3}\left[\dfrac{\lambda^2}{8t^{\gamma-2}} - \dfrac{1}{32y^2} + \dfrac{\mu}{4y}\right], & y \geq \alpha_1/t^{(2-\gamma)/2}. \end{cases}$$

Clearly (C46)–(C48) and (C50) hold. The same argument as in Case 1 guarantees that $\beta = \beta_0$ is an upper solution for (9.85). To show α is a lower solution for (9.85) first notice that since $1 < \gamma < 2$ that

$$a_0\alpha(1) - a_1\alpha'(1) = a_0\alpha_1 - a_1\alpha_1\left(\frac{\gamma-2}{2}\right) = \alpha_1\left[a_0 - a_1\left(\frac{\gamma-2}{2}\right)\right]$$

$$\leq \alpha_1\left(a_0 + \frac{a_1}{2}\right) \leq A.$$

Also for $1 < t < \infty$ we have since $1 < \gamma < 2$ that

$$\alpha''(t) - q(t)f(t, \alpha(t))$$

$$= \alpha_1 \left(\frac{\gamma - 2}{2}\right)\left(\frac{\gamma - 4}{2}\right)\frac{1}{t^{(6-\gamma)/2}} + \frac{1}{t^3}\left[\frac{t^{2-\gamma}}{32\alpha_1^2} - \frac{\lambda^2}{8t^{\gamma-2}} - \frac{\mu\, t^{(2-\gamma)/2}}{4\alpha_1}\right]$$

$$= \frac{\alpha_1(\gamma - 2)(\gamma - 4)}{4t^{(6-\gamma)/2}} + \frac{1}{32\alpha_1^2 t^{\gamma+1}}\left[1 - 4\lambda^2\alpha_1^2 - \frac{8\mu\alpha_1}{t^{(2-\gamma)/2}}\right]$$

$$\geq \frac{\alpha_1(\gamma - 2)(\gamma - 4)}{4t^{(6-\gamma)/2}} + \frac{1}{32\alpha_1^2 t^{\gamma+1}}\left[1 - 4\lambda^2\alpha_1^2 - 8\mu\alpha_1\right]$$

$$= \frac{\alpha_1(\gamma - 2)(\gamma - 4)}{4t^{(6-\gamma)/2}} - \frac{\lambda^2}{8\alpha_1^2 t^{\gamma+1}}\left[\alpha_1^2 + 2\frac{\mu}{\lambda^2}\alpha_1 - \frac{1}{4\lambda^2}\right]$$

$$= \frac{\alpha_1(\gamma - 2)(\gamma - 4)}{4t^{(6-\gamma)/2}} - \frac{\lambda^2}{8\alpha_1^2 t^{\gamma+1}}(\alpha_1 - x_0)(\alpha_1 + x_1) \geq 0$$

since $\alpha_1 \leq x_0$; here

$$x_0 = -\frac{\mu}{\lambda^2} + \sqrt{\frac{\mu^2}{\lambda^4} + \frac{1}{4\lambda^2}} \quad \text{and} \quad x_1 = \frac{\mu}{\lambda^2} + \sqrt{\frac{\mu^2}{\lambda^4} + \frac{1}{4\lambda^2}}.$$

Thus α is a lower solution of (9.85), so (C49) holds. Theorem 9.9 guarantees that (9.85) has a solution $y \subset BC[1, \infty) \cap C^2(1, \infty)$, $y' \in C[1, \infty)$ with

$$\frac{\alpha_1}{t^{(2-\gamma)/2}} \leq y(t) \leq \beta_0 \quad \text{for } t \in [1, \infty).$$

As a result, y is a solution of (9.84).

Case 3. $\nu = 0$, $\gamma \geq 2$. In this case we let

$$\alpha(t) = \min\left\{\frac{A}{a_0}, \frac{2}{\lambda}\right\} \equiv \alpha_2 \quad \text{and} \quad \beta(t) = \frac{1}{24d_0^2}\left(1 - \frac{1}{\sqrt{t}}\right) + d_0;$$

here $d_0 > 0$ is chosen so that

$$d_0 \geq \frac{A}{a_0} \quad \text{and} \quad a_0 d_0 - \frac{a_1}{48d_0^2} \geq A.$$

Notice $\alpha(t) \leq A/a_0 \leq d_0 \leq \beta(t)$ for $t \in [1, \infty)$, and α, β are bounded on $[1, \infty)$. Let $q \equiv 1$ and

$$f(t, y) = \begin{cases} \dfrac{1}{t^3}\left[\dfrac{\lambda^2}{8t^{\gamma-2}} - \dfrac{1}{32\alpha_2^2}\right], & y \leq \alpha_2 \\[2ex] \dfrac{1}{t^3}\left[\dfrac{\lambda^2}{8t^{\gamma-2}} - \dfrac{1}{32y^2}\right], & y \geq \alpha_2. \end{cases}$$

Clearly (C46)–(C48) and (C50) hold. It remains to check (C49). First we show α is a lower solution for (9.85). If $1 < t < \infty$ then $\gamma \geq 2$ gives

$$
\alpha''(t) - q(t) f(t, \alpha(t)) = \frac{1}{t^3}\left(\frac{1}{32\alpha_2^2} - \frac{\lambda^2}{8t^{\gamma-2}}\right)
$$

$$
\geq \frac{1}{t^3}\left(\frac{\lambda^2}{8} - \frac{\lambda^2}{8t^{\gamma-2}}\right) = \frac{1}{t^3}\frac{\lambda^2}{8}\left(1 - \frac{1}{t^{\gamma-2}}\right) \geq 0.
$$

In addition, $a_0\alpha(1) - a_1\alpha'(1) = a_0\alpha_2 \leq A$, so α is a lower solution of (9.85). To show β is an upper solution of (9.85) first notice that

$$
a_0\beta(1) - a_1\beta'(1) = a_0 d_0 - a_1 \frac{1}{48d_0^2} \geq A.
$$

Also for $1 < t < \infty$ we have

$$
\beta''(t) - q(t)\, f(t, \beta(t)) \leq -\frac{1}{32d_0^2}\frac{1}{t^{5/2}} + \frac{1}{t^3}\left(\frac{1}{32d_0^2} - \frac{\lambda^2}{8t^{\gamma-2}}\right)
$$

$$
= \frac{1}{t^3}\left(-\frac{\sqrt{t}}{32d_0^2} + \frac{1}{32d_0^2} - \frac{\lambda^2}{8t^{\gamma-2}}\right)
$$

$$
\leq \frac{1}{t^3}\left(-\frac{1}{32d_0^2} + \frac{1}{32d_0^2} - \frac{\lambda^2}{8t^{\gamma-2}}\right)
$$

$$
= \frac{1}{t^3}\left[-\frac{\lambda^2}{8t^{\gamma-2}}\right] \leq 0.
$$

Thus β is an upper solution of (9.85), so (C49) holds. Theorem 9.9 guarantees that (9.85) has a solution $y \in BC[1, \infty) \cap C^2(1, \infty)$, $y' \in C[1, \infty)$ with

$$
\alpha_2 \leq y(t) \leq \frac{1}{24d_0^2}\left(1 - \frac{1}{\sqrt{t}}\right) + d_0 \quad \text{for } t \in [1, \infty).
$$

This y is a solution of (9.84).

Case 4. $\nu = 0$, $1 < \gamma < 2$. In this case, we let

$$
\alpha(t) = \frac{\alpha_3}{t^{(2-\gamma)/2}} \quad \text{where } \alpha_3 = \min\left\{\frac{A}{a_0 + \frac{a_1}{2}},\ \frac{1}{2\lambda}\right\}
$$

and

$$
\beta(t) = \max\left\{\frac{A}{a_0},\ \frac{1}{2\lambda}\right\} \equiv \beta_1.
$$

Note that

$$
\alpha(t) \leq \frac{A}{a_0 + \frac{a_1}{2}} \leq \frac{A}{a_0} \leq \beta(t) \quad \text{for } t \in [1, \infty),
$$

and α, β are bounded on $[1, \infty)$. Let $q \equiv 1$ and

$$f(t, y) = \begin{cases} \dfrac{\lambda^2}{8t^{\gamma+1}} - \dfrac{1}{32\alpha_3^2 t^{\gamma+1}}, & y \le \alpha_3/t^{(2-\gamma)/2} \\[2mm] \dfrac{1}{t^3}\left[\dfrac{\lambda^2}{8t^{\gamma-2}} - \dfrac{1}{32y^2} \right], & y \ge \alpha_3/t^{(2-\gamma)/2}. \end{cases}$$

Clearly (C46)–(C48) and (C50) hold. In addition, notice that $a_0\beta(1) - a_1\beta'(1) = a_0\beta_1 \ge A$, and for $1 < t < \infty$ we have since $1 < \gamma < 2$ that

$$\beta''(t) - q(t)f(t, \beta(t)) = \frac{1}{t^3}\left[\frac{1}{32\beta_1^2} - \frac{\lambda^2}{8t^{\gamma-2}} \right]$$

$$\le \frac{1}{t^3}\left[\frac{\lambda^2}{8} - \frac{\lambda^2}{8t^{\gamma-2}} \right] = \frac{\lambda^2}{8t^3}\left[1 - t^{2-\gamma} \right] \le 0.$$

Thus, β is an upper solution for (9.85). To show α is a lower solution for (9.85) notice, since $1 < \gamma < 2$ that

$$a_0\alpha(1) - a_1\alpha'(1) = a_0\alpha_3 - a_1\alpha_3\left(\frac{\gamma-2}{2} \right) = \alpha_3\left[a_0 - a_1\left(\frac{\gamma-2}{2} \right) \right]$$

$$\le \alpha_3\left(a_0 + \frac{a_1}{2} \right) \le A,$$

and for $1 < t < \infty$ we have

$$\alpha''(t) - q(t)\, f(t, \alpha(t))$$

$$= \alpha_3\left(\frac{\gamma-2}{2} \right)\left(\frac{\gamma-4}{2} \right)\frac{1}{t^{(6-\gamma)/2}} + \frac{1}{t^3}\left[\frac{t^{2-\gamma}}{32\alpha_3^2} - \frac{\lambda^2}{8t^{\gamma-2}} \right]$$

$$\ge \frac{1}{t^{\gamma+1}}\left[\frac{1}{32\alpha_3^2} - \frac{\lambda^2}{8} \right] \ge 0,$$

since $\alpha_3 \le 1/(2\lambda)$. Thus, α is a lower solution of (9.85), so (C49) holds. Theorem 9.9 guarantees that (9.85) has a solution $y \in BC[1, \infty) \cap C^2(1, \infty)$, $y' \in C[1, \infty)$ with

$$\frac{\alpha_3}{t^{(2-\gamma)/2}} \le y(t) \le \beta_1 \quad \text{for} \quad t \in [1, \infty).$$

This y is a solution of (9.84).

Example 9.9 (*Plasma Physics*) In 1982, in the study of plasma physics, Gregus [66] formulated the boundary value problem

$$\frac{1}{t^2}(t^2 y')' = \sinh(y) - \frac{1}{\alpha \beta^3} e^{-t/\beta}, \quad 0 < t < \infty$$

$$y'(0) = 0, \quad \lim_{t \to \infty} y(t) = 0, \quad \lim_{t \to \infty} y'(t) = 0$$

(9.86)

with α and β positive parameters.

Motivated by the problem (9.86), we state the following existence principle for the general boundary value problem

$$y'' + \frac{\gamma}{t} y' = qf(t, y), \quad 0 < t < \infty, \quad \gamma > 1$$

$$y'(0) = 0, \quad \lim_{t \to \infty} y(t) = 0.$$

(9.87)

Theorem 9.10 ([19]) *Assume that*
(C51) $f : [0, \infty) \times \mathbb{R} \to \mathbb{R}$ is continuous,
(C52) $q \in C(0, \infty)$ with $q > 0$ on $(0, \infty)$,
(C53) $\int_0^\mu s^\gamma q(s)ds < \infty$, $\int_0^\mu t^{-\gamma} \int_0^t s^\gamma q(s)ds dt < \infty$ and $\lim_{t \to \infty} \int_\mu^t s^{-\gamma} \times \int_\mu^s \tau^\gamma q(\tau)d\tau ds = \infty$ for any $\mu > 0$,
(C54) $f(t, 0) \le 0$ for $t \in (0, \infty)$,
(C55) there exists $r_0 \ge 0$ with $f(t, r_0) \ge 0$ for $t \in (0, \infty)$,
(C56) there exists $M > 0$ with $|f(t, u)| \le M$ for $t \in [0, \infty)$ and $u \in [0, r_0]$,
(C57) $\int_0^\infty s^\gamma q(s)|f(s, 0)|ds < \infty$ and $\int_0^\infty sq(s)|f(s, 0)|ds < \infty$,
(C58) $\lim_{t \to 0^+} t^2 q(t) f(t, 0)$ exists,
(C59) $f(t, u) - f(t, 0) \ge 0$ for $t \in [0, \infty)$ and $u \in [0, r_0]$,
(C60) for any constant $A \in (0, r_0]$ there exists a constant $K > 0$ (which may depend on A) and a constant $c_2 > 0$ (which may depend on A) with $f(t, u) \ge K$ for $A \le u \le r_0$ and $t \ge c_2$,
(C61) there exists $h : (0, \infty) \to (0, \infty)$ with $|f(t, u)| \le h(t)$ for $t \in (0, \infty)$ and $u \in [0, w(t)]$; here

$$w(t) = -\frac{t^{1-\gamma}}{\gamma - 1} \int_0^t s^\gamma q(s) f(s, 0)ds - \frac{1}{\gamma - 1} \int_t^\infty sq(s) f(s, 0)ds,$$

(C62) $\lim_{t \to 0^+} th(t)q(t) = 0$.
Then, (9.87) has a solution $y \in C^1[0, \infty) \cap C^2(0, \infty)$ with $0 \le y(t) \le w(t)$ for $t \in [0, \infty)$. In addition, assume
(1) $\lim_{t \to \infty} q(t)h(t) = 0$ if $\int_0^\infty s^\gamma q(s)h(s)ds = \infty$, then $\lim_{t \to \infty} y'(t)/t = 0$,
(2) $\lim_{t \to \infty} tq(t)h(t) = 0$ if $\int_0^\infty s^\gamma q(s)h(s)ds = \infty$, then $\lim_{t \to \infty} y'(t) = 0$,
(3) $\lim_{t \to \infty} w'(t) = 0$ and $\sup_{t \in [0,\infty)} q(t) < \infty$, then $\lim_{t \to \infty} y'(t) = 0$.

We shall apply Theorem 9.10 to show that (9.86) has a solution $y \in C^1[0, \infty) \cap C^2(0, \infty)$ with

$$0 \le y(t) \le \frac{2\beta^3 \sigma \left(1 - e^{-t/\beta}\right)}{t} - \sigma \beta^2 e^{-t/\beta} \quad \text{for} \quad t \in [0, \infty);$$

(9.88)

here $\sigma = 1/(\alpha\beta^3)$. For this, we let

$$\gamma = 2, \quad q \equiv 1, \quad r_0 = \sigma \quad \text{and} \quad f(t, y) = \sinh(y) - \frac{1}{\alpha\beta^3} e^{-t/\beta}.$$

Clearly, conditions (C51)–(C54) are satisfied. Now since

$$f(t, r_0) = \sinh(r_0) - r_0\, e^{-t/\beta} = \sum_{k=0}^{\infty} \frac{1}{(2k+1)!} r_0^{2k+1} - r_0\, e^{-t/\beta}$$

$$= \sum_{k=1}^{\infty} \frac{1}{(2k+1)!} r_0^{2k+1} + r_0\left(1 - e^{-t/\beta}\right) \geq 0,$$

so (C55) holds. Also,

$$|f(t, u)| \leq \sinh(r_0) + r_0 \quad \text{for} \quad t \in [0, \infty) \quad \text{and} \quad u \in [0, r_0],$$

so (C56) is satisfied. Further, (C57) since $\int_0^\infty s^2 e^{-s/\beta} ds < \infty$ and $\int_0^\infty s \times e^{-s/\beta} ds < \infty$, (C58) since $\lim_{t\to 0^+} t^2 e^{-t/\beta} = 0$, and (C59) since $f(t, u) - f(t, 0) = \sinh u \geq 0$ for $u \geq 0$ hold. If $A \in (0, \sigma]$ notice that for $t \in [0, \infty)$ and $A \leq u \leq \sigma$,

$$f(t, u) = \sinh(u) - \sigma\, e^{-t/\beta} \geq \sinh(A) - \sigma\, e^{-t/\beta}.$$

Also since $\lim_{t\to\infty} e^{-t/\beta} = 0$ there exists $c_2 > 0$ with

$$f(t, u) \geq \frac{\sinh(A)}{2} \quad \text{for} \quad t \geq c_2 \quad \text{and} \quad A \leq u \leq \sigma,$$

so (C60) is satisfied. Conditions (C61) and (C62) hold with $h(t) = \sinh(w(t)) + \sigma\, e^{-t\beta}$. Here it is immediate to obtain

$$w(t) = \frac{2\,\beta^3\,\sigma\,\left(1 - e^{-t/\beta}\right)}{t} - \sigma\beta^2\, e^{-t/\beta}.$$

Now Theorem 9.10 guarantees that (9.86) without condition $\lim_{t\to\infty} y'(t) = 0$ has a solution $y \in C^1[0, \infty) \cap C^2(0, \infty)$ with $0 \leq y(t) \leq w(t)$ for $t \in [0, \infty)$.

Next, notice that

$$\int_0^\infty s^2 q(s)h(s)ds = \int_0^\infty s^2 \sinh(w(s))ds + \sigma \int_0^\infty s^2 e^{-s/\beta} ds = \infty$$

since $\sinh u \geq u$ for $u \geq 0$ and

$$\int_0^\infty s^2 \sinh(w(s))ds \;\geq\; \int_0^t s^2 w(s)ds$$
$$= \sigma\beta^2 \left(\beta t^2 + \beta t^2 e^{-t/\beta} + 4\beta^2 t e^{-t/\beta} + 4\beta^3 e^{-t/\beta} - 4\beta^3\right).$$

Further, we have

$$\lim_{t\to\infty} q(t)h(t) \;=\; \lim_{t\to\infty}\left[\sinh(w(t)) + \sigma e^{-t/\beta}\right] \;=\; \sinh(0) \;=\; 0.$$

Thus from the conclusion (1) it follows that $\lim_{t\to\infty} y'(t)/t = 0$.

Finally, we show that $\lim_{t\to\infty} y'(t) = 0$. For this in view of conclusion (3) it suffices to check that $\lim_{t\to\infty} w'(t) = 0$. This holds immediately since

$$w'(t) \;=\; \frac{2\beta^3\sigma}{t^2}\left[\frac{1}{\beta}te^{-t/\beta} - \left(1 - e^{-t/\beta}\right)\right] + \sigma\beta e^{-t/\beta}.$$

Thus y is a solution of (9.86) and (9.88) holds.

Example 9.10 (*Theory of Colloids*) In the theory of colloids [29], it is possible to relate particle stability with the charge on the colloidal particle. We model the particle and its attendant electrical double layer using Poisson's equation for a flat plate. If Ψ is the potential, ρ the charge density, D the dielectric constant, and y the displacement, then we have

$$\frac{d^2\Psi}{dy^2} = -\frac{4\pi\rho}{D}.$$

We assume the ions are point charged and their concentrations in the double layer satisfies the Boltzmann distribution

$$c_i \;=\; c_i^* \exp\left(\frac{-z_i e\Psi}{\kappa T}\right),$$

where c_i is the concentration of ions of type i, $c_i^* = \lim_{\Psi\to 0} c_i$, κ the Boltzmann constant, T the absolute temperature, e the electrical charge, and z the valency of the ion. In the neutral case, we have

$$\rho \;=\; c_+ z_+ e + c_- z_- e, \quad\text{or}\quad \rho \;=\; ze(c_+ - c_-),$$

where $z = z_+ - z_-$. Then, we have using

$$c_+ \;=\; c\exp\left(\frac{-ze\Psi}{\kappa T}\right) \quad\text{and}\quad c_- \;=\; c\exp\left(\frac{ze\Psi}{\kappa T}\right),$$

that

$$\frac{d^2\Psi}{dy^2} \;=\; \frac{8\pi cze}{D}\sinh\left(\frac{ze\Psi}{\kappa T}\right)$$

where the potential initially takes some positive value $\Psi(0) = \Psi_0$ and tends to zero as the distance from the plate increases, i.e., $\Psi(\infty) = 0$. Using the transformation

$$\phi(y) = \frac{ze\Psi(y)}{\kappa T} \quad \text{and} \quad x = \sqrt{\frac{4\pi cz^2 e^2}{\kappa T D}} y,$$

the problem becomes [53]

$$\frac{d^2\phi}{dx^2} = 2 \sinh(\phi), \quad 0 < x < \infty$$

$$\phi(0) = c_1, \quad \lim_{x\to\infty} \phi(x) = 0 \tag{9.89}$$

where $c_1 = ze\Psi_0/(\kappa T) > 0$. From a physical point of view, we wish the solution ϕ in (9.89) to also satisfy $\lim_{x\to\infty} \phi'(x) = 0$.

To show (9.89) has a solution, we need the following general existence principle for the boundary value: problem

$$\frac{1}{p}(py')' = qf(t, y), \quad 0 < t < \infty$$

$$-a_0 y(0) + b_0 \lim_{t\to 0^+} p(t)y'(t) = c_0, \quad a_0 > 0, \ b_0 \geq 0, \ c_0 \leq 0 \tag{9.90}$$

$$\lim_{t\to\infty} y(t) = 0.$$

Theorem 9.11 ([12]) *Assume that*
(C63) $f : [0, \infty) \times \mathbb{R} \to \mathbb{R}$ *is continuous,*
(C64) $q \in C(0, \infty)$ *with* $q > 0$ *on* $(0, \infty)$,
(C65) $p \in C[0, \infty) \cap C^1(0, \infty)$ *with* $p > 0$ *on* $(0, \infty)$ *and* $\int_0^\infty ds/p(s) = \infty$,
(C66) $\int_0^\mu ds/p(s) < \infty$ *and* $\int_0^\mu p(s)q(s)ds < \infty$ *for any* $\mu > 0$,
(C67) $f(t, 0) \leq 0$ *for* $t \in (0, \infty)$,
(C68) *there exists* $r_0 \geq -c_0/a_0$ *with* $f(t, r_0) \geq 0$ *for* $t \in (0, \infty)$,
(C69) *there exists* $M > 0$ *with* $|f(t, u)| \leq M$ *for* $t \in [0, \infty)$ *and* $u \in [0, r_0]$,
(C70) *there exists a constant* $m > 0$ *with* $q(t)p^2(t)[f(t, u) - f(t, 0)] \geq m^2 u$ *for* $t \in (0, \infty)$ *and* $u \in [0, r_0]$,
(C71) $\int_0^\infty p(x) \exp\left(-m \int_0^x ds/p(s)\right) q(x)|f(x, 0)|dx < \infty$,
(C72) $\lim_{t\to\infty} p^2(t)q(t)f(t, 0) = 0$, *and*
(C73) $\lim_{t\to\infty} \left(B_0 \int_\mu^t (1/p(s)) \int_\mu^s (1/p(x))dxds + C_0 \int_\mu^t ds/p(s)\right) = \infty$ *for any constants* $B_0 > 0$, $C_0 \in \mathbb{R}$ *and* $\mu > 0$.
Then, (9.90) has a solution $y \in C[0, \infty) \cap C^2(0, \infty)$ *with* $py' \in C[0, \infty)$ *and* $0 \leq y(t) \leq w(t)$ *for* $t \in [0, \infty)$, *where* w *is the nonnegative solution of the problem*

$$\frac{1}{p}(pw')' - \frac{m^2}{p^2}w = g(t) = q(t)f(t,0)$$

$$-a_0 w(0) + b_0 \lim_{t \to 0^+} p(t)w'(t) = c_0 \qquad (9.91)$$

$$\lim_{t \to \infty} w(t) = 0,$$

i.e.,

$$
w(t) = \frac{1}{a_0 + b_0 m} \exp\left(-m \int_0^t \frac{ds}{p(s)}\right)
$$
$$
\times \left[c_0 + b_0 \int_0^\infty p(x) \exp\left(-m \int_0^x \frac{ds}{p(s)}\right) g(x)dx\right]
$$
$$
- \int_0^t \frac{1}{p(\zeta)} \exp\left(-m \int_\zeta^t \frac{ds}{p(s)}\right) \qquad (9.92)
$$
$$
\times \left(\int_\zeta^\infty p(x) \exp\left(-m \int_\zeta^x \frac{ds}{p(s)}\right) g(x)dx\right) d\zeta.
$$

If in addition
(C74) $f(t, u) \geq 0$ for $t \in [0, \infty)$ and $u \in [0, w(t)]$, and
(C75) $\lim_{t \to \infty} p(t) \in (0, \infty]$
then $\lim_{t \to \infty} y'(t) = 0$.

In terms of our notation problem, (9.89) can be written as

$$
y'' = 2 \sinh(y), \quad 0 < t < \infty
$$
$$
y(0) = c > 0, \quad \lim_{t \to \infty} y(t) = 0. \qquad (9.93)
$$

We will apply Theorem 9.11 to show that (9.93) has a solution $y \in C[0, \infty) \cap C^2(0, \infty)$ with

$$
0 \leq y(t) \leq ce^{-t} \quad \text{for} \quad t \in [0, \infty). \qquad (9.94)
$$

To see this let $p = 1$, $q = 1$, $a_0 = 1$, $c_0 = -c$, $b_0 = 0$, and $r_0 = c$. Clearly (C63)–(C69), (C70) since $f(t, u) - f(t, 0) = \sinh u \geq u$ for $u \geq 0$, (C71)–(C73) hold. Thus Theorem 9.11 guarantees that (9.93) has a solution $y \in C[0, \infty) \cap C^2(0, \infty)$ with $0 \leq y(t) \leq w(t)$ for $t \in [0, \infty)$. It is immediate from (9.92) (since $g = 0$) that $w(t) = ce^{-t}$ for $t \in [0, \infty)$. Finally, we remark that this solution y satisfies $\lim_{t \to \infty} y'(t) = 0$. To see this we need only to check that (C74) and (C75) hold, but these are immediate.

Next, we shall deduce the existence of a solution of (9.93) by employing the following existence principle for the boundary value problem

$$\frac{1}{p}(py')' = qf(t, y), \quad a < t < \infty \quad (a \geq 0 \text{ is fixed})$$

$$-a_0 y(a) + b_0 \lim_{t \to a^+} p(t)y'(t) = c_0, \quad a_0 > 0, \ b_0 \geq 0 \tag{9.95}$$

$$\lim_{t \to \infty} y(t) = A_0.$$

Theorem 9.12 ([14]) *Assume that*
(C76) $q \in C(a, \infty)$ with $q > 0$ on (a, ∞),
(C77) $p \in C[a, \infty) \cap C^1(a, \infty)$ with $p > 0$ on (a, ∞),
(C78) $\int_a^\mu ds/p(s) < \infty$ and $\int_a^\mu p(s)q(s)ds < \infty$ for any $\mu > a$,
(C79) there exists a function $\alpha \in BC[a, \infty) \cap C^2(a, \infty)$, $p\alpha' \in C[a, \infty)$ with
$(1/p)(p\alpha')' \geq qf(t, \alpha)$, $a < t < \infty$, $-a_0\alpha(a) + b_0 \lim_{t \to a^+} p(t)\alpha'(t) \geq c_0$ and
$\alpha(t)$ is bounded on $[a, \infty)$,
(C80) there exists a function $\beta \in BC[a, \infty) \cap C^2(a, \infty)$, $p\beta' \in C[a, \infty)$ with
$(1/p)(p\beta')' \leq qf(t, \beta)$, $a < t < \infty$, $-a_0\beta(a) + b_0 \lim_{t \to a^+} p(t)\beta'(t) \leq c_0$ and
$\beta(t)$ is bounded on $[a, \infty)$, also $\alpha(t) \leq \beta(t)$ for $t \in [a, \infty)$,
(C81) $\lim_{t \to \infty} \alpha(t) = A_0 = \lim_{t \to \infty} \beta(t)$,
(C82) for each $N \in \{1, 2, \cdots\}$, $f : [a, N] \times [\alpha_N, \beta_N] \to \mathbb{R}$ is continuous; here
$\alpha_N = \min_{t \in [a,N]} \alpha(t)$ and $\beta_N = \max_{t \in [a,N]} \beta(t)$, and
(C83) there exists a constant $M_0 > 0$ with $|f(t, u)| \leq M_0$ for $t \in [a, \infty)$ and $u \in$
$[\alpha(t), \beta(t)]$.
Then, (9.95) has a solution $y \in BC[a, \infty) \cap C^2(a, \infty)$, $py' \in C[a, \infty)$ with $\alpha(t) \leq$
$y(t) \leq \beta(t)$ for $t \in [a, \infty)$.

To apply Theorem 9.12 for the problem (9.93), we let $p = q = 1$, $a_0 = 1$, $c_0 = -c$, $b_0 = 0$ and $f(t, y) = 2 \sinh y$. Also we let

$$\alpha(t) = c \exp\left(-t\sqrt{\frac{2 \sinh(c)}{c}}\right) \tag{9.96}$$

and

$$\beta(t) = c \exp\left(-t\sqrt{2}\right). \tag{9.97}$$

Clearly (C76)–(C78), (C82) and (C83) hold. To check (C79) first notice that

$$\alpha'' = \frac{2 \sinh(c)}{c} \alpha, \quad \alpha(0) = c.$$

Thus in view of $\sinh(x)/x$ nondecreasing for $x > 0$,

$$\alpha'' - qf(t, \alpha) = \alpha'' - 2 \sinh(\alpha) \geq \alpha'' - 2\frac{\sinh(c)}{c}\alpha = 0$$

for $t \in (0, \infty)$.
Next to show (C80) we note that

$$\beta'' = 2\beta, \quad \beta(0) = c.$$

Thus in view of $\sinh x \geq x$ for $x \geq 0$,

$$\beta'' - qf(t, \beta) = \beta'' - 2\sinh(\beta) \leq \beta'' - 2\beta = 0$$

for $t \in (0, \infty)$. Further $\alpha(t) \leq \beta(t)$ for $t \in [0, \infty)$ is obvious. The existence of a solution y to (9.93) follows from Theorem 9.12 since $\lim_{t \to \infty} \alpha(t) = 0 = \lim_{t \to \infty} \beta(t)$ (condition (C81)). Note as well that $\alpha(t) \leq y(t) \leq \beta(t)$ for $t \in [0, \infty)$ where α (respectively β) is given in (9.96) (respectively (9.97)).

Example 9.11 (*Flow and Heat Transfer Over a Stretching Sheet*) The boundary value problem

$$y'' = \frac{-B(1 - e^{-t})y' + (Re^{-t} - A)y - (y')^2}{1 + y}, \quad a < t < \infty \tag{9.98}$$

$$y(a) = b, \quad \lim_{t \to \infty} y(t) = 0$$

arises [105] in the flow and heat transfer over a stretching sheet; here

$$a > 0, \quad b > 0, \quad B > 0 \quad \text{are constants} \tag{9.99}$$

with

$$A < 0 \quad \text{if} \quad R > 0 \quad \text{whereas} \quad Re^{-a} - A > 0 \quad \text{if} \quad R \leq 0. \tag{9.100}$$

We shall establish the existence of a solution of (9.98) by employing the following general result for the boundary value problem:

$$\frac{1}{p}(py')' = qf(t, y, py'), \quad a < t < \infty \quad (a \geq 0 \text{ is fixed})$$

$$-a_0 y(a) + b_0 \lim_{t \to a^+} p(t)y'(t) = c_0, \quad a_0 > 0, \ b_0 \geq 0, \ c_0 \leq 0 \tag{9.101}$$

$$\lim_{t \to \infty} y(t) = 0.$$

Theorem 9.13 ([7]) *Assume that*
(C84) $f : [a, \infty) \times \mathbb{R}^2 \to \mathbb{R}$ is continuous,
(C85) $q \in C(a, \infty)$ with $q > 0$ on (a, ∞),
(C86) $p \in C[a, \infty) \cap C^1(a, \infty)$ with $p > 0$ on (a, ∞),
(C87) $\int_a^\mu ds/p(s) < \infty$ and $\int_a^\mu p(s)q(s)ds < \infty$ for any $\mu > a$,
(C88) $f(t, 0, 0) \leq 0$ for $t \in (a, \infty)$,
(C89) there exists $r_0 \geq -c_0/a_0$ with $f(t, r_0, 0) \geq 0$ for $t \in (a, \infty)$,
(C90) there exists a continuous $\psi : [0, \infty) \to (0, \infty)$ with $|f(t, u, v)| \leq \psi(|v|)$ for $(t, u) \in [a, \infty) \times [0, r_0]$,

(C91) p^2q is bounded on $[a, \infty)$ with

$$r_0 \sup_{t \in [a,\infty)} p^2(t)q(t) < \int_{d_0}^{\infty} \frac{u}{\psi(u)} du;$$

here

$$d_0 = \frac{-c_0 + a_0 r_0}{b_0} \text{ if } b_0 > 0 \text{ and } d_0 = \left[r_0 - \frac{c_0}{a_0}\right] \sup_{t \in [a, a+1]} p(t) \text{ if } b_0 = 0,$$

(C92) $f(t, u, 0) > 0$ for $t \in (a, \infty)$, $u \in (0, r_0]$,
(C93) for any $\gamma \in (0, r_0]$ there exists a constant $K > 0$ (which may depend on γ), a function $h \in BC[a, \infty)$ with $h'(t) \leq 0$ for $t > a$, a bounded differentiable function $g : [0, r_0] \to [0, \infty)$ with $g(w) > 0$ for $w \in (0, r_0]$ and a constant $c_1 > a$ (which may depend on γ) with

$$p(t)q(t)f(t, u, v) \geq \frac{-\frac{g'(u)}{p(t)}v^2 + \frac{h(t)}{p(t)}v + Kp(t)q(t)}{g(u)}$$

for $\gamma \leq u \leq r_0$, $v \in [-M, M]$ and $t \geq c_1$; here

$$M = J^{-1}\left(r_0 \sup_{t \in [a,\infty)} p^2(t)q(t)\right) \quad \text{and} \quad J(z) = \int_{d_0}^{z} \frac{u}{\psi(u)} du,$$

(C94) $\lim_{t \to \infty} \int_{c_2}^{t} p(s)q(s)ds = \infty$ for any $c_2 > a$.
Then, (9.101) has a solution $y \in BC[a, \infty) \cap C^2(a, \infty)$, $py' \in BC[a, \infty)$ with $0 \leq y(t) \leq r_0$ and $|p(t)y'(t)| \leq M$ for $t \in [a, \infty)$.

To show that (9.98) has a solution $y \in BC[a, \infty) \cap C^2(a, \infty)$ with $0 \leq y(t) \leq b$ for $t \in [a, \infty)$ in the above theorem we let $p = q = 1$, $a_0 = -1$, $b_0 = 0$, $c_0 = -b$ and

$$f(t, u, v) = \begin{cases} \dfrac{-B(1 - e^{-t})v + (Re^{-t} - A)u - v^2}{1 + u} & \text{if } u \geq 0 \\ -B(1 - e^{-t})v - v^2 & \text{if } u < 0. \end{cases}$$

It is immediate that (C84)–(C88) hold. Let $r_0 = b$, then

$$f(t, b, 0) = \frac{(Re^{-t} - A)b}{1 + b} \geq 0 \text{ for } t \geq a$$

in view of (9.100), and hence (C89) holds. Let $\psi(v) = Bv + v^2 + (Re^{-a} - A)b$ then since $\int_{d_0}^{\infty}(u/\psi(u))du = \infty$ conditions (C90) and (C91) are satisfied. Conditions (C92) and (C94) are also clear. So, it remains to check (C93). Let

$$g(u) = 1 + u \quad \text{and} \quad h(t) = -B(1 - e^{-t}).$$

Note $h'(t) = -Be^{-t} \le 0$ for $t \ge a$. Let $\gamma \in (0, r_0]$, $c_1 = 2a$ and notice that for any $\gamma \le u \le r_0$, $v \in \mathbb{R}$, and $t \ge c_1$, that (note $g'(u) = 1$),

$$g(u)f(t, u, v) = -g'(u)v^2 + h(t)v + (Re^{-t} - A)u \ge -g'(u)v^2 + h(t)v + K$$

where

$$K = \begin{cases} -A\gamma & \text{if } R > 0 \\ (Re^{-2a} - A)\gamma & \text{if } R \le 0. \end{cases}$$

Thus (C93) holds. The existence of a solution now follows from Theorem 9.13.

Example 9.12 (*Flow of a Gas Through a Semi-infinite Porous Medium*) In the study of the unsteady flow of a gas through a semi-infinite porous medium initially filled with gas at a uniform pressure $p_0 > 0$, at time $t = 0$, the pressure at the outflow face is suddenly reduced from p_0 to $p_1 \ge 0$ ($p_1 = 0$ in the case of diffusion into a vacuum) and is thereafter maintained at this lower pressure. The unsteady isothermal flow of gas is described by the nonlinear partial differential equation

$$\nabla^2(p^2) = 2A\frac{\partial p}{\partial t},$$

where the constant A is given by the properties of the medium. In the one-dimensional medium extending from $x = 0$ to $x = \infty$, this reduces to

$$\frac{\partial}{\partial x}\left(p\frac{\partial p}{\partial x}\right) = A\frac{\partial p}{\partial t}$$

with the boundary conditions

$$p(x, 0) = p_0, \quad 0 < x < \infty$$
$$p(0, t) = p_1 \, (< p_0), \quad 0 \le t < \infty.$$

To obtain a similarity solution, we introduce the new independent variable

$$z = \frac{x}{2}\sqrt{\frac{A}{p_0 t}}$$

and the dimension-free dependent variable w, defined by

$$w(z) = \alpha^{-1}\left(1 - \frac{p^2(x)}{p_0^2}\right) \quad \text{where} \quad \alpha = 1 - \frac{p_1^2}{p_0^2}.$$

In terms of the new variable, Kidder's problem [33, 77, 85] takes the form

$$w''(z) + \frac{2z}{\sqrt{1 - \alpha w(z)}} w'(z) = 0, \quad 0 < z < \infty$$

$$w(0) = 1, \quad w(\infty) = 0.$$

(9.102)

Making the change of variable $u(z) = 1 - \alpha w(z)$, we have the boundary value problem

$$u''(z) + \frac{2z}{\sqrt{u(z)}} u'(z) = 0, \quad 0 < z < \infty$$

$$u(0) = 1 - \alpha, \quad 0 < \alpha \le 1, \quad u(\infty) = 1.$$

(9.103)

Motivated by the above problem, we shall present three existence principles. Out first result is for the nonsingular boundary value problem

$$y'' + qf(t, y, y') = 0, \quad 0 < t < \infty$$

$$y(0) = c_0, \quad 0 < c_0 < 1, \quad \lim_{t \to \infty} y(t) = 1.$$

(9.104)

Theorem 9.14 ([10, 14]) *Assume that*
(C95) $f : [0, \infty) \times [c_0, \infty) \times \mathbb{R} \to \mathbb{R}$ *is continuous (nonlinearity is nonsingular),*
(C96) $q \in C[0, \infty)$ *with* $q > 0$ *on* $(0, \infty)$,
(C97) $zf(t, y, z) \ge 0$ *on* $[0, \infty) \times [c_0, \infty) \times \mathbb{R}$,
(C98) there exists a continuous function $\tau : [0, \infty) \to (0, \infty)$ *with* $f(t, y, z) \le \tau(z)$
on $[0, \infty) \times [c_0, 1] \times [0, \infty)$,
(C99) $\int_0^\infty du/\tau(u) > \int_0^{1-c_0} du/\tau(u) + \int_0^1 q(s)ds$,
(C100) q *is nondecreasing on* $(0, \infty)$, *and*
(C101) for constants $H > c_0$, $K > 0$ *there exists a function* $\psi_{H,K}$ *continuous on*
$[0, \infty)$ *and positive and nondecreasing on* $(0, \infty)$, *and a constant* $1 \le r < 2$ *with*
$f(t, y, z) \ge \psi_{H,K} z^r$ *on* $[0, \infty) \times [c_0, H] \times [0, K]$.
Then, (9.104) has a solution $y \in C^2[0, \infty)$ *with* $1 \ge y(t) \ge c_0$ *for* $t \in [0, \infty)$.

In terms of our notation problem (9.103) with $\alpha < 1$ is the same as

$$y'' + \frac{2t}{\sqrt{y}} y' = 0, \quad 0 < t < \infty$$

$$y(0) = 1 - \alpha, \quad 0 < \alpha < 1, \quad \lim_{t \to \infty} y(t) = 1.$$

(9.105)

We shall apply Theorem 9.14 to conclude that (9.105) has a solution $y \in C^2[0, \infty)$ with $y(t) \ge 1 - \alpha$ for $t \in [0, \infty)$. For this, we let

$$c_0 = 1 - \alpha, \quad q(t) = 2t, \quad f(t, y, z) = z/y^{1/2} \quad \text{and} \quad \tau(z) = (z + 1)/c_0^{1/2}.$$

Clearly, (C95)–(C98), (C99) since $\int_0^\infty du/\tau(u) = \infty$ and (C100) are satisfied. Also (C101) holds with $\psi_{H,K} = H^{-1/2}$ and $r = 1$. Now Theorem 9.14 guarantees that (9.105) has the desired solution.

Our next result is for the singular boundary value problem

$$y'' + qf(t, y, y') = 0, \quad 0 < t < \infty$$

$$y(0) = 0, \quad \lim_{t \to \infty} y(t) = 1. \tag{9.106}$$

Theorem 9.15 ([10, 14]) *Assume that in addition to (C96) and (C100) the following hold*

(C102) $f : [0, \infty) \times (0, \infty) \times \mathbb{R} \to \mathbb{R}$ is continuous (nonlinearity is singular at $y = 0$),

(C103) $zf(t, y, z) \geq 0$ on $[0, \infty) \times (0, \infty) \times \mathbb{R}$,

(C104) there exists a continuous function $\tau : [0, \infty) \to (0, \infty)$, and a continuous nonincreasing function $g : (0, \infty) \to (0, \infty)$ with $f(t, y, z) \leq g(y)\tau(z)$ on $[0, \infty) \times (0, 1] \times [0, \infty)$,

(C105) $\int_0^1 g(u)du < \infty$,

(C106) $\int_0^\infty du/\tau(u) > \int_0^1 du/\tau(u) + q(1) \int_0^1 g(v)dv$, and

(C107) for constants $H > 0$, $K > 0$ there exist a function $\psi_{H,K}$ continuous on $[0, \infty)$ and positive and nondecreasing on $(0, \infty)$, and a constant $1 \leq r < 2$ with $f(t, y, z) \geq \psi_{H,K}z^r$ on $[0, \infty) \times (0, H] \times [0, K]$.

Then, (9.106) has a solution $y \in C^1[0, \infty) \cap C^2(0, \infty)$ with $1 \geq y(t) > 0$ for $t \in (0, \infty)$.

In terms of our notation, problem (9.103) with $\alpha = 1$ is the same as

$$y'' + \frac{2t}{\sqrt{y}}y' = 0, \quad 0 < t < \infty$$

$$y(0) = 0, \quad \lim_{t \to \infty} y(t) = 1. \tag{9.107}$$

We shall apply Theorem 9.15 to show that (9.107) has a solution $y \in C^1[0, \infty) \cap C^2(0, \infty)$ with $y(t) > 0$ for $t \in (0, \infty)$. For this, we let

$$q(t) = 2t, \quad f(t, y, z) = z/y^{1/2}, \quad g(y) = 1/y^{1/2} \quad \text{and} \quad \tau(z) = z + 1.$$

Clearly, (C96), (C100), (C102)–(C106) are satisfied. Also (C107) holds with $\psi_{H,K} = H^{-1/2}$ and $r = 1$. The existence of a required solution of (9.107) now follows from Theorem 9.15.

Our final result is for the boundary value problem

$$\frac{1}{p}(py')' = qf(t, y, py'), \quad a < t < \infty \quad (a \geq 0 \text{ is fixed})$$

$$-a_0 y(a) + b_0 \lim_{t \to a^+} p(t)y'(t) = c_0, \quad a_0 > 0, \, b_0 \geq 0 \tag{9.108}$$

$$\lim_{t \to \infty} y(t) = A_0.$$

Theorem 9.16 ([10, 14]) *Assume that*

(C108) $q \in C(a, \infty)$ with $q > 0$ on (a, ∞),

(C109) $p \in C[a, \infty) \cap C^1(a, \infty)$ with $p > 0$ on (a, ∞),

(C110) $\int_a^\mu ds/p(s) < \infty$ and $\int_a^\mu p(s)q(s)ds < \infty$ for any $\mu > a$,

(C111) there exists a function $\alpha \in BC[a, \infty) \cap C^2(a, \infty)$, $p\alpha' \in BC[a, \infty)$ with $(1/p)(p\alpha')' \geq qf(t, \alpha, p\alpha')$, $a < t < \infty$, $-a_0\alpha(a) + b_0 \lim_{t \to a^+} p(t)\alpha'(t) \geq c_0$ and $\alpha(t)$ is bounded on $[a, \infty)$,

(C112) there exists a function $\beta \in BC[a, \infty) \cap C^2(a, \infty)$, $p\beta' \in BC[a, \infty)$ with $(1/p)(p\beta')' \leq qf(t, \beta, p\beta')$, $a < t < \infty$, $-a_0\beta(a) + b_0 \lim_{t \to a^+} p(t)\beta'(t) \leq c_0$ and $\beta(t)$ is bounded on $[a, \infty)$, also $\alpha(t) \leq \beta(t)$ for $t \in [a, \infty)$,

(C113) $\lim_{t \to \infty} \alpha(t) = A_0 = \lim_{t \to \infty} \beta(t)$,

(C114) for each $N \in \{1, 2, \cdots\}$, $f : [a, N] \times [\alpha_N, \beta_N] \times \mathbb{R} \to \mathbb{R}$ is continuous; here $\alpha_N = \min_{t \in [a, N]} \alpha(t)$ and $\beta_N = \max_{t \in [a, N]} \beta(t)$, and

(C115) for each $N \in \{1, 2, \cdots\}$, $zf(t, y, z) \leq 0$ for $(t, y, z) \in [a, N] \times [\alpha(t), \beta(t)] \times \mathbb{R}$.

(I) $b_0 > 0$. *Then, (9.108) has a solution* $y \in BC[a, \infty) \cap C^2(a, \infty)$, $py' \in BC[a, \infty)$ with $\alpha(t) \leq y(t) \leq \beta(t)$ for $t \in [a, \infty)$.

(II) $b_0 = 0$. *Assume in addition that*

(C116) there exists a continuous $\psi : [0, \infty) \to (0, \infty)$ with $|f(t, y, z)| \leq \psi(|z|)$ for $(t, y) \in [a, \infty) \times [\alpha(t), \beta(t)]$, and

(C117) $$\int_0^\infty \frac{u}{\psi(u)} du > \int_0^{d_0} \frac{u}{\psi(u)} du + \int_a^{a+1} p(s)q(s)ds; \text{ here}$$

$$d_0 = \left[\max\{|\alpha(a+1)|, |\beta(a+1)|\} + \frac{|c_0|}{a_0} \right] \sup_{t \in [a, a+1]} p(t).$$

Then, (9.108) has a solution $y \in BC[a, \infty) \cap C^2(a, \infty)$, $py' \in BC[a, \infty)$ with $\alpha(t) \leq y(t) \leq \beta(t)$ for $t \in [a, \infty)$.

In terms of our notation, problem (9.102) with $\alpha < 1$ is the same as

$$y'' + \frac{2t}{\sqrt{1 - \lambda y}} y' = 0, \quad 0 < t < \infty$$
$$y(0) = 1, \quad \lim_{t \to \infty} y(t) = 0 \tag{9.109}$$

with $0 < \lambda < 1$. We will use Theorem 9.16(II) to show (9.109) has a solution.

Let $p = 1$, $q(t) = 2t$, $a_0 = 1$, $c_0 = -1$,

$$f(t, y, z) = -\frac{z}{\sqrt{1 - \lambda y}} \quad \text{and} \quad \psi(z) = \frac{z + 1}{\sqrt{1 - \lambda}}.$$

Also we let

$$\alpha(t) = 1 - \frac{2(1 - \lambda)^{1/4}}{\sqrt{\pi}} \int_0^t \exp\left(-(1 - \lambda)^{-1/2} x^2\right) dx \tag{9.110}$$

and

$$\beta(t) = 1 - \frac{2}{\sqrt{\pi}} \int_0^t \exp\left(-x^2\right) dx. \tag{9.111}$$

Notice α, β, α', $\beta' \in BC[0, \infty)$ and clearly (C108)–(C110) hold. To check (C111) notice first that

$$\alpha'' + \frac{2t}{\sqrt{1-\lambda}}\alpha' = 0, \quad \alpha(0) = 1.$$

Thus in view of $1 - \lambda \le 1 - \lambda\alpha(t) \le 1$, $t \in [0, \infty)$, and $\alpha' \le 0$ on $(0, \infty)$,

$$\alpha'' - qf(t, \alpha, \alpha') = \alpha'' + \frac{2t}{\sqrt{1-\lambda\alpha}}\alpha' \ge \alpha'' + \frac{2t}{\sqrt{1-\lambda}}\alpha' = 0 \text{ for } t \in (0, \infty).$$

Next to show (C112) we note that

$$\beta'' + 2t\beta' = 0, \quad \beta(0) = 1.$$

Hence, in view of $1 - \lambda \le 1 - \lambda\beta(t) \le 1$, $t \in [0, \infty)$ and $\beta' \le 0$ on $(0, \infty)$,

$$\beta'' - qf(t, \beta, \beta') = \beta'' + \frac{2t}{\sqrt{1-\lambda\beta}}\beta' \le \beta'' + 2t\beta' = 0 \text{ for } t \in (0, \infty).$$

Further, $\alpha(t) \le \beta(t)$ for $t \in [0, \infty)$ is obvious. In addition (C114), (C115), and (C116) hold; notice that for each $N \in \{1, 2, \cdots\}$, $\alpha_N = \min_{t \in [0, N]} \alpha(t) = \alpha(N) \in (0, 1)$, $\beta_N = \max_{t \in [0, N]} \beta(t) = \beta(0) = 1$ and $0 < \lambda < 1$. Finally (C117) is true since $\int_0^\infty [u/\psi(u)]du = \infty$. The existence of a solution y to (9.109) follows from Theorem 9.16(II) since $\lim_{t \to \infty} \alpha(t) = 0 = \lim_{t \to \infty} \beta(t)$ (condition (C113)). Note as well that $\alpha(t) \le y(t) \le \beta(t)$ for $t \in [0, \infty)$ where α (respectively β) is given in (9.110) (respectively (9.111)).

Example 9.13 (*Draining Flow*) Differential equations arise naturally in the study of flow of a thin film of viscous fluid over a solid surface. When such a film drains down a vertical wall and the effects of surface tension, gravity, and viscosity are taken into account one encounters the equation

$$\frac{d^3u}{dx^3} = f(u)$$

for the flow profile $u(x)$ in a coordinate frame moving with the fluid, see [41, 74, 106, 111, 112]. When the surface is dry this leads to the function

$$f(u) = \frac{1}{u^2} \tag{9.112}$$

(note that (9.112) occurs also in the spreading of oil drops on horizontal surfaces). The problem in its entirety involves looking for a smooth function $u(x)$ defined on

$(-\infty, \infty)$ with

$$u''' = \frac{1}{u^2}$$
$$u(0) = 1, \quad u'(0) = 0 \tag{9.113}$$
$$\lim_{x \to -\infty} u''(x) = 0.$$

If u is a solution of (9.113) then $u'''(x) > 0$ for all x, together with $\lim_{x \to -\infty} u''(x) = 0$ gives $u''(x) > 0$ for all $x \in (-\infty, \infty)$. Also $u'(x) > 0$ if $x > 0$ and $u'(x) < 0$ if $x < 0$, and so we have $u(x) \geq 1$ for all $x \in (-\infty, \infty)$ with $\lim_{x \to -\infty} u(x) = \infty$. We may introduce u as an independent variable and introduce the function $y(u) = [u'(x)]^2$ as a dependent variable. Carrying out the transformation leads to the boundary value problem

$$y'' + \frac{2}{u^2} \frac{1}{\sqrt{y}} = 0, \quad 1 < u < \infty$$
$$y(1) = 0, \quad \lim_{u \to \infty} y'(u) = 0 \tag{9.114}$$
$$y > 0 \quad \text{on} \quad (1, \infty).$$

We shall establish the existence of solutions of (9.114) by employing the following general existence theorem for the singular boundary value problem:

$$y'' + qf(t, y) = 0, \quad a < t < \infty$$
$$y(a) = 0, \quad \lim_{t \to \infty} y'(t) = 0. \tag{9.115}$$

Theorem 9.17 ([5]) *Assume that*
(C118) $f : [a, \infty) \times (0, \infty) \to [0, \infty)$ *is continuous,*
(C119) $q \in C(a, \infty)$ *with* $q > 0$ *on* (a, ∞),
(C120) $\int_b^\infty q(s)ds < \infty$ *for any* $b > a$,
(C121) $0 \leq f(t, u) \leq g(u)$ *on* $(a, \infty) \times (0, \infty)$ *with* $g > 0$ *continuous and nonincreasing on* $(0, \infty)$,
(C122) $\int_a^\infty q(s)g(k_0(s - a))ds < \infty$ *for any* $k_0 > 0$, *and*
(C123) for any fixed $k > a$ *if* $y \in C[a, k]$ *satisfies*

$$y(t) \leq G^{-1}\left\{ G(1) + \int_a^t \int_s^\infty q(x)dxds \right\}$$

for $t \in [a, k]$ *then there exists a continuous function* $\psi : [a, \infty) \to (0, \infty)$ *(independent of* k*) with* $f(t, y(t)) \geq \psi(t)$ *for* $t \in (a, k)$; *here* $G(z) = \int_0^z du/g(u)$.
Then, (9.115) has a solution $y \in C^1[a, \infty) \cap C^2(a, \infty)$ *with* $y' \in BC[a, \infty)$ *and* $y(t) > 0$ *for* $t \in (a, \infty)$.

We shall apply Theorem 9.17 to show that the problem

$$y'' + \frac{2}{t^2}\frac{1}{\sqrt{y}} = 0, \quad 1 < t < \infty$$

$$y(1) = 0, \quad \lim_{t \to \infty} y'(t) = 0 \tag{9.116}$$

has a solution $y \in C^1[1, \infty) \cap C^2(1, \infty)$ with $y' \in BC[1, \infty)$ and $y(t) > 0$ for $t \in (1, \infty)$. For this let $a = 1$, $q(t) = 2/t^2$ and $f(t, y) = g(y) = 1/\sqrt{y}$. Clearly (C118)–(C122) hold. To check (C123) we fix $k > a$ and $y \in C[a, k]$ with

$$y(t) \leq G^{-1}\left\{ G(1) + \int_a^t \int_s^\infty q(x)dxds \right\} \quad \text{for} \quad t \in [a, k].$$

This means that

$$y(t) \leq [1 + 3\ln(t)]^{2/3} \quad \text{for} \quad t \in [1, k].$$

Thus (C123) is immediate since

$$f(t, y(t)) = \frac{1}{\sqrt{y(t)}} \geq \frac{1}{[1 + 3\ln(t)]^{1/3}} \equiv \psi(t) \quad \text{for} \quad t \in [1, k].$$

Theorem 9.17 now guarantees that (9.116) has the desired solution.

Example 9.14 (*Homann Flow*) In the axisymmetric stagnation flow, i.e., Homann flow (Fritz Homann was a fishing trawler that was built in 1930) the Navier–Stokes equation can be reduced to the third order Falkner–Skan, after V. M. Falkner and Sylvia W. Skan in 1930, equation (for $f(\eta)$)

$$f''' + ff'' + \frac{1}{2}\left[1 - (f')^2\right] = 0, \quad 0 < \eta < \infty \tag{9.117}$$

with boundary conditions

$$f(0) = 0, \quad f'(0) = 0 \quad \text{and} \quad f'(\infty) = 1, \tag{9.118}$$

see [103, 116]. Assume $f(\eta)$ is a solution of (9.117), (9.118) and $f''(\eta) > 0$ for all $\eta \geq 0$. Then, $\eta = g(t)$, the inverse function to $t = f'(\eta)$, exists and is strictly increasing on $(0, 1)$ with $g(0) = 0$ and

$$t = f'(g(t)) \quad \text{for all} \quad t \in (0, 1).$$

Differentiating with respect to t, we obtain

$$w(t) \equiv f''(g(t)) = \frac{1}{g'(t)}, \quad 0 < t < 1.$$

For simplicity, a prime will denote differentiation with respect to t or η. Substituting $\eta = g(t)$ into (9.117), and using

$$w'(t) = f'''(g(t))g'(t) = \frac{f'''(g(t))}{w(t)},$$

we find

$$w'(t)w(t) + f(g(t))w(t) + \frac{1}{2}(1 - t^2) = 0, \quad 0 < t < 1. \tag{9.119}$$

Dividing by w and differentiating with respect to t, we get

$$w''(t) = \frac{tw(t) + \frac{1}{2}(1 - t^2)w'(t)}{w^2(t)} - \frac{t}{w(t)} = \frac{(1 - t^2)w'(t)}{2w^2(t)}, \quad 0 < t < 1.$$

Notice also that (9.118) and (9.119) imply

$$w'(0)w(0) = -\frac{1}{2} \quad \text{and} \quad w(1) = 0.$$

As a result, w satisfies

$$w''(t) - \frac{1}{2}\frac{(1 - t^2)w'(t)}{w^2(t)} = 0, \quad 0 < t < 1$$

$$w'(0)w(0) = -\frac{1}{2} \quad \text{and} \quad w(1) = 0. \tag{9.120}$$

From a physical point of view, it is of interest to look for solutions to (9.120) with $w(t) > 0$ for $t \in [0, 1)$. It is clear that if w satisfies the integral equation

$$w(t) = \int_t^1 \frac{(1 - s)\left(\frac{1}{2} + \frac{3}{2}s\right)}{w(s)} ds + (1 - t)\int_0^t \frac{s}{w(s)} ds, \quad 0 < t < 1 \tag{9.121}$$

with $w(t) > 0$, $t \in [0, 1)$ and $w(1) = 0$, then w is a solution of (9.120).

Clearly (9.121) motivates the study of the integral equation

$$y(t) = \int_0^1 k(t, s)\left[g(y(s)) + h(y(s))\right] ds, \quad t \in [0, 1]. \tag{9.122}$$

Here g may be singular at $y = 0$. We shall establish the existence of solutions of (9.120) by employing the following general existence theorem for the singular integral equation (9.122).

Theorem 9.18 ([8, 20]) *Assume that*

(C124) $g > 0$ is continuous and nonincreasing on $(0, \infty)$,

(C125) $h \geq 0$ is continuous on $[0, \infty)$ and h/g is nondecreasing on $[0, \infty)$,
(C126) $k_t(s) = k(t, s) \in L^1[0, 1]$ for each $t \in [0, 1]$,
(C127) the map $t \mapsto k_t$ is continuous from $[0, 1]$ to $L^1[0, 1]$,
(C128) for each $t \in [0, 1]$, $k(t, s) \geq 0$ for a.e. $s \in [0, 1]$,
(C129) for t_1, $t_2 \in (0, 1)$ with $t_1 < t_2$ we have $k(t_1, s) \geq k(t_2, s)$ for a.e. $s \in [0, 1]$,
(C130) $\int_0^1 k(0, s)g(\alpha(s))ds < \infty$ where $\alpha(s) = G^{-1}\left(\int_s^1 k(s, x)dx\right)$ for $s \in [0, 1]$
and $G(z) = z/g(z)$ for $z > 0$,
(C131) there exists $r \in C[0, 1]$ with $\int_0^1 [k(t, s) - k(x, s)]g(\alpha(s))ds \leq |r(x) - r(t)|$
for t, $x \in [0, 1]$ with $t < x$, and
(C132) if $z > 0$ satisfies $z \leq a + b[1 + h(z)/g(z)]$ for constants $a \geq 0$, $b \geq 0$ then
there exists a constant M (which may depend on a and b) with $z \leq M$.
Then, (9.122) has a solution $y \in C[0, 1]$ with $y(t) \geq \alpha(t)$ for $t \in [0, 1]$ (here α is
as in (C130)).

As an application of Theorem 9.18 we shall show that the singular integral equation
(9.121) on $[0, 1]$ has a solution. Notice that (9.121) is of the form (9.122) with
$g(y) = 1/y$, $h = 0$ and

$$
k(t, s) = \begin{cases} (1 - s)\left(\dfrac{1}{2} + \dfrac{3}{2}s\right), & s > t \\ (1 - t)s, & s < t. \end{cases}
$$

Clearly (C124)–(C128), and (C132) hold. In addition, (C129) is immediate since if
t_1, $t_2 \in (0, 1)$ with $t_1 < t_2$ then

$$
k(t_1, s) - k(t_2, s) \begin{cases} = s(t_2 - t_1) \geq 0, & 0 < s < t_1 \\ = (1 - s)\left(\dfrac{1}{2} + \dfrac{3}{2}s\right) - (1 - t_2)s \\ \geq (1 - s)\left(\dfrac{1}{2} + \dfrac{3}{2}s\right) - (1 - s)s \\ = \dfrac{1}{2}(1 - s^2) \geq 0, & t_1 < s < t_2 \\ = 0, & t_2 < s < 1. \end{cases}
$$

In our problem $G(z) = z^2$ so for $t \in [0, 1]$, we have

$$
\alpha(t) = \sqrt{\int_t^1 (1 - s)\left(\frac{1}{2} + \frac{3}{2}s\right) ds} ,
$$

so

$$
\alpha(t) = \sqrt{\frac{(1 - t)^2}{4} + \frac{3t(1 - t)^2}{4} + \frac{(1 - t)^3}{4}}, \quad t \in [0, 1]. \tag{9.123}
$$

Note that $\alpha(t) \geq (1 - t)/2$ for $t \in [0, 1]$ and

$$\int_0^1 k(0, s)g(\alpha(s))ds \le \int_0^1 (1-s)\left(\frac{1}{2}+\frac{3}{2}s\right)\frac{2}{1-s}ds = \int_0^1 (1+3s)ds < \infty,$$

so (C130) holds. It remains to check (C131). Let t, $x \in [0, 1]$ with $t < x$. Then, since $\alpha(t) \ge (1-t)/2$, $t \in [0, 1]$, we have

$$\int_0^1 [k(t, s) - k(x, s)]g(\alpha(s))ds$$

$$= (x-t)\int_0^t sg(\alpha(s))ds + \int_t^x \left[(1-s)\left(\frac{1}{2}+\frac{3}{2}s\right) - (1-x)s\right]g(\alpha(s))ds$$

$$= \int_t^x \int_0^z sg(\alpha(s))dsdz - \int_t^x s(x-s)g(\alpha(s))ds$$

$$\quad + \int_t^x \left[(1-s)\left(\frac{1}{2}+\frac{3}{2}s\right) - (1-x)s\right]g(\alpha(s))ds$$

$$= \int_t^x \int_0^z sg(\alpha(s))dsdz + \frac{1}{2}\int_t^x (1-s)(1+s)g(\alpha(s))ds$$

$$\le 2\int_t^x \int_0^z \frac{s}{1-s}dsdz + \int_t^x (1+s)ds$$

$$= 2\int_t^x [-z - \ln(1-z)]dz + \int_t^x (1+s)ds$$

$$= r(x) - r(t),$$

where

$$r(z) = -\frac{z^2}{2} + 2(1-z)\ln(1-z) + 3z.$$

Notice that $r \in C[0, 1]$ and by l'Hopital's rule

$$\lim_{z \to 1}(1-z)\ln(1-z) = \lim_{z \to 1}\frac{\ln(1-z)}{(1-z)^{-1}} = \lim_{z \to 1}(1-z) = 0.$$

Thus (C131) holds. The existence of a solution y to (9.121) now follows from Theorem 9.18. Note as well that $y(t) \ge \alpha(t)$ for $t \in [0, 1]$ where α is given in (9.123).

We also note that y is a solution of (9.120). This is immediate once we check $y(1) = 0$. Notice that for $t \in (0, 1)$ that

$$(1-t)\int_0^t \frac{s}{y(s)}ds \le 2(1-t)\int_0^t \frac{s}{1-s}ds = 2(1-t)[-t - \ln(1-t)],$$

and as a result

$$\lim_{t \to 1}(1-t)\int_0^t \frac{s}{y(s)}ds = 0.$$

Thus $y(1) = 0$.

Example 9.15 (*Pseudoplastic Fluids*) The boundary layer equations for steady flow over a semi-infinite plate are

$$U \frac{\partial U}{\partial X} + V \frac{\partial U}{\partial Y} = \frac{1}{\rho} \frac{\partial \tau_{XY}}{\partial Y}$$

$$\frac{\partial U}{\partial X} + \frac{\partial V}{\partial Y} = 0,$$

where the X and Y axes are taken along and perpendicular to the plate, ρ is the density, U and V are the velocity components parallel and normal to the plate and the shear stress $\tau_{XY} = K (\partial U / \partial Y)^n$. The case $n = 1$ corresponds to a Newtonian fluid and for $0 < n < 1$ the power law relation between shear stress and rate of strain describes pseudoplastic non-Newtonian fluids. The fluid has zero velocity on the plate and the flow approaches stream conditions far from the plate, i.e.,

$$U(X, 0) = V(X, 0) = 0, \quad U(X, \infty) = U_\infty,$$

where U_∞ is the uniform potential flow. The above results (if we use stream function-similarity variables) in a third order infinite interval problem

$$F''' + F(F'')^{2-n} = 0, \quad F(0) = F'(0) = 0, \quad F'(\infty) = 1,$$

see [2, 46, 86, 104, 122]. Now using the Luigi Crocco (1909–1986) type transformation $u = F'$ and $G = F''$, we find

$$G^n G'' + (n - 1) G^{n-1} (G')^2 + u = 0, \quad G'(0) = 0, \quad G(1) = 0.$$

Setting $y = G^n$, we obtain

$$y^{1/n} y'' + nu = 0, \quad 0 < u < 1$$
$$y'(0) = y(1) = 0.$$
(9.124)

We shall prove the existence of solutions of (9.124) by employing the following general existence result for the mixed boundary value problem

$$\frac{1}{p} (py')' + qf(t, y) = 0, \quad 0 < t < 1$$
$$\lim_{t \to 0^+} p(t) y'(t) = y(1) = 0.$$
(9.125)

Theorem 9.19 ([16]) *Assume that*
(C133) $p \in C[0, 1] \cap C^1(0, 1)$ with $p > 0$ on $(0, 1)$,
(C134) $q \in C(0, 1)$ with $q > 0$ on $(0, 1)$,
(C135) $\int_0^1 p(s)q(s)ds < \infty$ and $\int_0^1 (1/p(t)) \int_0^t p(s)q(s)ds dt < \infty$,
(C136) $f : [0, 1] \times (0, \infty) \to \mathbb{R}$ is continuous,

(C137) there exists $n_0 \in \{1, 2, \cdots\}$ and associated with each $m \in \mathbb{N}_0 = \{n_0, n_0 + 1, \cdots\}$, there exists $\alpha_m \in C[0, 1] \cap C^2(0, 1)$, $p\alpha'_m \in AC[0, 1]$, with $p(t)q(t)f(t, \alpha_m(t)) + (p(t)\alpha'_m(t))' \geq 0$ for $t \in (0, 1)$, $\lim_{t \to 0^+} p(t)\alpha'_m(t) \geq 0$ and $0 < \alpha_m(1) \leq 1/m$,
(C138) there exists $\alpha \in C[0, 1]$, $\alpha > 0$ on $[0, 1)$ and $\alpha(t) \leq \alpha_m(t)$, $t \in [0, 1]$ for each $m \in \mathbb{N}_0$,
(C139) there exists $\beta \in C[0, 1] \cap C^2(0, 1)$, $p\beta' \in AC[0, 1]$ with $p(t)q(t) \times f(t, \beta(t)) + (p(t)\beta'(t))' \leq 0$ for $t \in (0, 1)$, $\lim_{t \to 0^+} p(t)\beta'(t) \leq 0$ and $\beta(1) \geq \beta_0 > 0$,
(C140) $\alpha_m(t) \leq \beta(t)$, $t \in [0, 1]$ for each $m \in \mathbb{N}_0$, and
(C141) $0 \leq f(t, y) \leq g(y)$ on $[0, 1] \times (0, a_0]$ with $g > 0$ continuous and nonincreasing on $(0, \infty)$, where $a_0 = \sup_{t \in [0,1]} \beta(t)$, or
(C142) $f(t, x) - f(t, y) > 0$ for $0 < x < y$ and each fixed $t \in (0, 1)$.
Then, (9.125) has a solution $y \in C[0, 1] \cap C^2(0, 1)$ with $y(t) \geq \alpha(t)$ for $t \in [0, 1]$.

Notice that the problem (9.124) can be written as

$$y'' + \frac{\nu t}{y^{1/\nu}} = 0, \quad 0 < t < 1$$
$$y'(0) = y(1) = 0,$$
(9.126)

where $0 < \nu \leq 1$. We will show using Theorem 9.19 that (9.126) has a solution.
First we choose $n_0 \in \{1, 2, \cdots\}$ so that

$$\frac{\nu}{6} + \frac{1}{n_0} \leq 1 \quad \text{and} \quad \left(\frac{\nu}{6} - 1\right)\frac{1}{\nu + 1} + \frac{1}{n_0} \leq 0.$$
(9.127)

Let $p = 1$, $q(t) = 2t$ and clearly (C133)–(C136) hold. Also let

$$\alpha_m(t) = \frac{\nu}{6}(1 - t^3) + \frac{1}{m}$$
(9.128)

$$\alpha(t) = \frac{\nu}{6}(1 - t^3)$$
(9.129)

and

$$\beta(t) = 1 - \frac{\nu}{\nu + 1}t^3.$$
(9.130)

To check (C137), for $m \in \mathbb{N}_0 = \{n_0, n_0 + 1, \cdots\}$, notice that $\alpha_m(1) = 1/m$, $\alpha'_m(0) = 0$ and

$$\alpha''_m + qf(t, \alpha_m) = -\nu t + \frac{\nu t}{[\alpha_m(t)]^{1/\nu}} \geq -\nu t + \nu t = 0 \quad \text{for } t \in (0, 1),$$

since $\alpha_m(t) \leq \nu/6 + 1/n_0 \leq 1$, $t \in [0, 1]$ from (9.127). Thus (C137) holds and (C138) is immediate. To check (C139) notice that $\beta(1) = 1 - [\nu/(\nu + 1)] \equiv \beta_0$,

$\beta'(0) = 0$ and

$$\beta'' + qf(t, \beta) = \frac{-6\nu t}{\nu + 1} + \frac{\nu t}{[\beta(t)]^{1/\nu}} \le \frac{-6\nu t}{\nu + 1} + \nu t (\nu + 1)^{1/\nu}$$

$$= \nu t \left\{ \frac{-6}{\nu + 1} + (\nu + 1)^{1/\nu} \right\} \le 0 \quad \text{for} \quad t \in (0, 1),$$

since $\beta(t) \ge 1/(\nu + 1)$ for $t \in [0, 1]$, and $(\nu + 1)^{(\nu+1)/\nu} \le 4 \le 6$ for $0 < \nu \le 1$ (note with $f(x) = (x + 1)^{(x+1)/x}$ we have $f(0^+) = e$, $f(1) = 4$ and $f'(x) \ge 0$ on $(0, 1)$). Thus (C139) holds. In addition (C140) is true since (9.127) implies for $m \in \mathbb{N}_0$ that

$$\alpha_m(t) = \frac{\nu}{6}(1 - t^3) + \frac{1}{m} \le \frac{\nu}{6}\left(1 - \frac{\nu}{\nu + 1}t^3\right) + \frac{1}{n_0}$$

$$= \frac{\nu}{6}\beta(t) + \frac{1}{n_0} = \beta(t) + \left\{ \frac{1}{n_0} + \left(\frac{\nu}{6} - 1\right)\beta(t) \right\}$$

$$\le \beta(t) + \left\{ \frac{1}{n_0} + \left(\frac{\nu}{6} - 1\right)\frac{1}{\nu + 1} \right\}$$

$$\le \beta(t) \quad \text{for} \quad t \in (0, 1)$$

since $\nu/(\nu + 1) \le 1$ and

$$\left(\frac{\nu}{6} - 1\right)\frac{1}{\nu + 1} + \frac{1}{n_0} \le 0.$$

Finally (C141) with $g(y) = y^{-1/\nu}$ (or (C142) since if $0 < x < y$ then $x^{1/\nu} < y^{1/\nu}$) holds. The existence of a solution y to (9.126) now follows from Theorem 9.19. Note as well that $y(t) \ge \alpha(t)$ for $t \in [0, 1]$, where α is given in (9.129).

Example 9.16 (*Percolation of Water*) Suppose we are given a cylindrical reservoir filled with water surrounded by dry soil. The bottom of the reservoir touches a horizontal impermeable layer having a plane upper boundary. In such a situation, the flow percolates from the reservoir into the surrounding unsaturated region. If we suppose that the reservoir is refilled continuously with water and that the surrounding area is sufficiently big, then the process described above may be observed over a long time. Describing the time dependence of this phenomena leads to a nonstationary percolation problem. Throughout, we assume the soil is homogeneous and isotropic so that the coefficients of hydraulic conductivity K and of permeability of the soil m are constant. Neglecting the effects of capillarity we suppose the water table divides the soil into two regions, a saturated one and a dry one. Using the hydraulic model of filtration the nonstationary percolation is described by the Joseph Valentin Boussinesq (1842–1929) equation

$$\frac{1}{2}\Delta_{x,y}h^2 = \frac{m}{K}h_t, \tag{9.131}$$

where the upper boundary of the impermeable layer is taken as the (x, y) plane and h denotes the height of the saturated region (so $z = h(x, y, t)$ is the water table at time t). Also $\Delta_{x,y}$ denotes the Laplace operator (in the plane). We assume the z-axis is at the same time the axis of the reservoir whose radius is one, and hence $h|_{r=1} = \alpha$ (α a constant), so for simplicity we can assume

$$h|_{r=1} = 1. \tag{9.132}$$

One expects the percolation process to have radial symmetry so we are interested only in solutions of (9.131) of the form

$$h = h(r, t) \quad \text{where} \quad r = \sqrt{x^2 + y^2}.$$

Using polar coordinates we obtain

$$\frac{1}{2}(h^2)_{rr} + \frac{1}{2r}(h^2)_r = h_t, \tag{9.133}$$

where for simplicity we assume $m/K = 1$. Also, we assume the reach of the saturated region is finite at each $t > 0$ and

$$h|_{r=r_0(t)} = 0, \tag{9.134}$$

where $r_0(t)$ is the reach of the infiltrating water at t. We are interested in solutions of (9.133) for $1 < r < r_0(t)$, $t > 0$ which satisfy (9.132) and (9.134). Here we are interested in solutions of the form

$$h(r, t) = p(s) \quad \text{where} \quad s = \frac{r}{r_0(t)}.$$

In this situation, we also assume

$$r_0(t)r_0'(t) = c,$$

where c is a positive constant. Now since at the beginning the soil is dry we have $r_0(0) = 1$, and so we have

$$r_0(t) = \sqrt{1 + 2ct}.$$

Fix $t > 0$. With the above substitution, it is easy to see that our problem reduces to

$$(p')^2 + pp'' + \frac{1}{s}pp' + csp' = 0, \quad ' = \frac{d}{ds}, \quad b < s < 1 \quad \text{with} \quad b = \frac{1}{r_0(t)}$$
$$p(b) = 1, \quad p(1) = 0.$$

$$\tag{9.135}$$

If we make the change of variables

$$s = b^x \quad \text{and} \quad q(x) = p(s), \quad 0 < x < 1$$

then (9.135) reduces to

$$(q^2)'' = 2BA^{-x}q', \quad ' = \frac{d}{dx}, \quad 0 < x < 1 \tag{9.136}$$
$$q(0) = 0, \quad q(1) = 1,$$

where $A = r_0^2(t) = 1 + 2ct$ and $B = -c\ln(b) = -c\ln(\sqrt{A}) = c\ln(A)/2$ and $t > 0$ is fixed. Finally letting $u(x) = B^{-1}q(x)$ our problem reduces to

$$(u^2)'' = 2A^{-x}u', \quad 0 < x < 1 \tag{9.137}$$
$$u(0) = 0, \quad u(1) = B^{-1}$$

where $A = 1 + 2ct$ and $B = c\ln(\sqrt{A})$ for each $t > 0$ fixed. For simplicity, we assume $c = 1$ so to solve our percolation problem we need to solve

$$(u^2)'' = 2A^{-x}u', \quad 0 < x < 1 \tag{9.138}$$
$$u(0) = 0, \quad u(1) = B^{-1},$$

where $A = 1 + 2t$ and $B = \ln(\sqrt{A})$ for each $t > 0$ fixed. For more details about this problem see [62].

For the existence of a solution of (9.138), we shall establish from the following general results for the boundary value problems:

$$(G'(y) + py^m)' + qf(t, y) = p'y^m, \quad ' = \frac{d}{dt}, \quad 0 < t < 1 \tag{9.139}$$
$$y(0) = 0, \quad y(1) = b_0 > 0$$

and

$$(y^m y')' + p(y^m)' + qf(t, y) = 0, \quad ' = \frac{d}{dt}, \quad 0 < t < 1 \tag{9.140}$$
$$y(0) = 0, \quad y(1) = b_0 > 0$$

where $G(z) = \int_0^z g(x)dx$ and

$$g(x) = \begin{cases} x^m, & x \geq 0 \\ -x^m, & x < 0 \end{cases}$$

with $m > 0$ odd. It follows that

$$G(z) = \begin{cases} \dfrac{z^{m+1}}{m+1}, & z \geq 0 \\ \dfrac{-z^{m+1}}{m+1} = -\dfrac{|z|^{m+1}}{m+1}, & z < 0. \end{cases}$$

By a solution to (9.139), we mean a function $y \in C[0, 1]$ with $G(y) \in C^1[0, 1]$, $G'(y) + p\, y^m \in AC[0, 1] \cap C^1(0, 1)$ which satisfies $y(0) = 0$, $y(1) = b_0$ and the differential equation (9.139) on $(0, 1)$. Similarly, by a solution to (9.140), we mean a function $y \in C[0, 1] \cap C^1(0, 1]$ with $y^m y' \in C^1(0, 1)$ which satisfies $y(0) = 0$, $y(1) = b_0$ and the differential equation (9.140) on $(0, 1)$.

Theorem 9.20 ([26]) *Suppose that*
(C143) $f : [0, 1] \times \mathbb{R} \to \mathbb{R}$ is continuous,
(C144) $q \in C(0, 1) \cap L^1[0, 1]$ with $q > 0$ on $(0, 1)$,
(C145) $p \in C^1[0, 1]$ with $p \leq 0$ on $[0, 1]$,
(C146) $f(t, 0) \geq 0$ for $t \in (0, 1)$, and
(C147) $f(t, b_0) \leq 0$ for $t \in (0, 1)$.
Then, (9.139) has a solution y with $0 \leq y(t) \leq b_0$ for $t \in [0, 1]$.

Theorem 9.21 ([26]) *Suppose that in addition to (C143)–(C147) the following hold:*
(C148) there exists $\alpha \in C[0, 1]$ with $G(\alpha) \in C^1[0, 1]$, $G'(\alpha) + p\alpha^m \in AC[0, 1] \cap C^1(0, 1)$ with $b_0 \geq \alpha > 0$ on $(0, 1]$, $\alpha(0) = 0$, $\alpha(1) \leq b_0$, and $(G'(\alpha) + p\alpha^m)' + q(t) f(t, \alpha) \geq p'(t)\alpha^m(t)$ on $(0, 1)$,
(C149) for each $t \in (0, 1)$ we have $q(t)[f(t, y) - f(t, \alpha(t))] \geq 0$ for $0 \leq y \leq \alpha(t)$, and
(C150) $p' > 0$ on $(0, 1)$.
Then, (9.139) has a solution y with $\alpha(t) \leq y(t) \leq b_0$ for $t \in [0, 1]$. In addition $y \in C^1(0, 1]$ with $G'(y) = y^m y'$ on $(0, 1)$ and y is a solution of (9.140).

Theorem 9.22 ([26]) *Suppose that in Theorem 9.21 the condition (C148) is replaced by*
(C151) there exists $\alpha \in C[0, 1] \cap C^1(0, 1)$ with $G(\alpha) \in C^1[0, 1]$, $\alpha^m \alpha' \in C^1(0, 1)$, $b_0 \geq \alpha > 0$ on $(0, 1]$, $\alpha(0) = 0$, $\alpha(1) \leq b_0$ and $(\alpha^m \alpha')' + p(\alpha^m)' + q(t) f(t, \alpha) \geq 0$ on $(0, 1)$.
Then, (9.140) has a solution y with $\alpha(t) \leq y(t) \leq b_0$ for $t \in [0, 1]$.

We note that for $t = 1$, $u = y$ the problem (9.138) is the same as

$$(y^2)'' = 2 \times 3^{-x} y', \quad 0 < x < 1$$
$$y(0) = 0, \quad y(1) = B^{-1}, \tag{9.141}$$

where $B = \ln(\sqrt{3})$.

We will use Theorem 9.22 to show that (9.141) has a solution. To see this, consider the problem

$$(yy')' - 3^{-x} y' = 0, \quad 0 < x < 1$$
$$y(0) = 0, \quad y(1) = B^{-1}. \tag{9.142}$$

We let $m = 1$, $p = -3^{-x}$, $q \equiv 0$, $f(x, y) \equiv 0$, $b_0 = B^{-1} = 2/\ln(3)$ and

$$g(z) = \begin{cases} z, & z \geq 0 \\ -z, & z < 0. \end{cases}$$

Clearly (C143)–(C147), (C149), and (C150) hold. Let

$$\alpha(x) = \frac{3^x - 1}{\ln(3)} \quad \text{for} \quad 0 \leq x \leq 1.$$

Now $\alpha(0) = 0$ and $\alpha(1) = 2/\ln(3) = B^{-1}$. Also $\alpha' = 3^x$, $\alpha\alpha' = (3^{2x} - 3^x)/\ln(3)$, so for $x \in (0, 1)$ we have

$$(\alpha\alpha')' - 3^{-x}\alpha' = (2 \times 3^{2x} - 3^x) - 1 = 3^x(2 \times 3^x - 1) - 1 \geq 0.$$

Thus (C151) holds and so Theorem 9.22 guarantees that there exists a solution y to (9.142) with

$$\frac{3^x - 1}{\ln(3)} \leq y(x) \leq \frac{2}{\ln(3)} \quad \text{for} \quad x \in [0, 1]. \tag{9.143}$$

Also since $y' \in C(0, 1)$ we have $(y^2)' = 2yy'$ on $(0, 1)$ and so y is a solution of (9.141).

Notice that, in terms of our initial problem (9.135), inequality (9.143) becomes

$$\frac{1}{2}\left(\frac{3}{r^2} - 1\right) \leq h(r, 1) \leq 1 \quad \text{for} \quad r \in [1, \sqrt{3}]. \tag{9.144}$$

Example 9.17 (Slender Dry Patch in a Liquid Film) Consider a thin film of viscous liquid with constant density ρ and viscosity μ flowing down a planer substrate inclined at an angle α $(0 < \alpha \leq \pi/2)$ to the horizontal. We adopt Cartesian coordinates (x, y, z) with the x-axis down the greatest slope and the z-axis normal to the plane. With the usual lubrication approximation the height of the free surface $z = h(x, y, z)$ satisfies

$$3\mu h_t = \nabla \cdot [h^3 \nabla(\rho g h \cos(\alpha) - \sigma \nabla^2 h)] - \rho g \sin(\alpha)[h^3]_x, \tag{9.145}$$

where t denotes time, g the magnitude of acceleration due to gravity, and σ the coefficient of surface tension. We are interested in solutions symmetric about $y = 0$, and seek a steady-state solution for a slender dry patch for which the length scale down the plane (i.e., in the x-direction) is much greater than in the transverse direction (i.e., in the y-direction), so the Eq. (9.145) is approximated by

$$[h^3(\rho g h \cos(\alpha) - \sigma h_{yy})_y]_y - \rho g \sin(\alpha)[h^3]_x = 0. \tag{9.146}$$

The velocity component down the plane is $u(x, y, z) = \rho g \sin(\alpha)(2hz - z^2)/(2\mu)$ and so for a slender dry patch of semi-width $y_e = y_e(x)$ the average volume flux around the dry patch per unit width in the transverse direction down the plane (denoted by $Q(x)$) is approximately

$$Q = \frac{\rho g \sin(\alpha)}{3\mu} \lim_{y\to\infty} \frac{1}{y} \int_{y_e(x)}^{y} [h(x, w)]^3 \, dw. \tag{9.147}$$

We seek a similarity solution to Eq. (9.146) of the form $h = f(x)H(\eta)$ where $\eta = y/y_e(x)$. Note $H(1) = 0$ and (9.146) takes the form

$$\rho g \cos(\alpha) f^2 y_e^2 (H^3 H')' - \sigma f^2 (H^3 H''')'$$
$$-3\rho g \sin(\alpha) y_e^3 H^2 (f'Hy_e - fH'y_e'\eta) = 0 \tag{9.148}$$

with the corresponding expression for Q being

$$Q = \frac{\rho g \sin(\alpha)}{3\mu} f^3 \lim_{\eta\to\infty} \frac{1}{\eta} \int_1^{\eta} [H(w)]^3 \, dw.$$

For weak surface-tension effects, the second term in (9.148) can be neglected and so the only relevant similarity solution is given (after a suitable choice of origin in x) by

$$f(x) = b(cx)^m \quad \text{and} \quad y_e(x) = (cx)^k,$$

where the coefficients b and c and the exponents m and k are constants with $m = 2k - 1$. In this case, $\alpha \neq \pi/2$ and so we may choose without loss of generality $b = ck \tan \alpha$ and so (9.148) becomes

$$((H' + \eta)'H^3)' - \left(7 - \frac{3}{k}\right)H^3 = 0. \tag{9.149}$$

The unknown exponent k is determined by the requirement that the average volume flux per unit width around the dry patch, Q, is independent of x. This is possible only if $m = 0$ and $H \sim H_0 > 0$ (a constant) as $\eta \to \infty$. Thus

$$Q = \frac{\rho g \sin(\alpha)}{3\mu} (bH_0)^3 \quad \text{and so} \quad m = 0, \quad k = \frac{1}{2}.$$

Setting $k = 1/2$ in (9.149) yields

$$(H^3 H')' + \eta(H^3)' = 0. \tag{9.150}$$

Also the solutions to (9.150) must satisfy the boundary condition $H(1) = 0$ and the far-field condition $\lim_{\eta\to\infty} H(\eta) = H_0$. As a result, one is interested in the boundary value problem, see [118]

$$(H^3 H')' + \eta(H^3)' = 0, \quad 1 < \eta < \infty$$
$$H(1) = 0, \quad \lim_{\eta \to \infty} H(\eta) = H_0 > 0. \tag{9.151}$$

For the existence of a solution of (9.151) we shall establish from the following general results for the boundary value problems

$$(G'(y) + py^m)' + qf(t, y) = p'y^m, \quad a < t < \infty$$
$$y(a) = 0, \quad y \text{ bounded on } [a, \infty) \tag{9.152}$$

and

$$(g(y)y')' + p(y^m)' + qf(t, y) = 0, \quad a < t < \infty$$
$$y(a) = 0, \quad y \text{ bounded on } [a, \infty) \tag{9.153}$$

where $G(z) = \int_0^z g(x)dx$ and

$$g(x) = \begin{cases} x^m, & x \ge 0 \\ -x^m, & x < 0 \end{cases}$$

with $m > 0$ odd. It follows that

$$G(z) = \begin{cases} \dfrac{z^{m+1}}{m+1}, & z \ge 0 \\ \dfrac{-z^{m+1}}{m+1} = -\dfrac{|z|^{m+1}}{m+1}, & z < 0. \end{cases}$$

By a solution to (9.152), we mean a function $y \in BC[a, \infty)$ (bounded continuous functions on $[0, \infty)$) with $G(y) \in C^1[a, \infty)$, $G'(y) + p\, y^m \in AC_{loc}[a, \infty) \cap C^1(a, \infty)$ which satisfies $y(a) = 0$ and the differential equation in (9.152) on (a, ∞). Similarly, by a solution to (9.153) we mean a function $y \in BC[a, \infty) \cap C^1(a, \infty)$ with $y^m y' \in C^1(a, \infty)$ which satisfies $y(a) = 0$ and the differential equation in (9.153) on (a, ∞).

Theorem 9.23 ([18]) *Suppose the following conditions are satisfied:*
(C152) $f : [a, \infty) \times \mathbb{R} \to \mathbb{R}$ is continuous,
(C153) $q \in C(a, \infty) \cap L_{loc}^1[a, \infty)$ with $q > 0$ on (a, ∞),
(C154) $p \in C^1[a, \infty)$ with $p \ge 0$ on $[a, \infty)$,
(C155) $f(t, 0) \ge 0$ for $t \in (a, \infty)$,
(C156) there exists $b_0 > 0$ with $f(t, b_0) \le 0$ for $t \in (a, \infty)$, and
(C157) there exists $\mu \in L_{loc}^1[a, \infty)$ with $|f(t, u)| \le \mu(t)$ for a.e. $t \in [a, \infty)$ and $u \in [0, b_0]$.
Then, (9.152) has a solution y with $0 \le y(t) \le b_0$ for $t \in [a, \infty)$.

Theorem 9.24 ([18]) *Suppose that in addition to (C152)–(C157) the following hold:*

(C158) there exists $\alpha \in BC[a, \infty)$ with $G(\alpha) \in C^1[a, \infty)$, $G'(\alpha) + p\,\alpha^m \in AC_{loc}[a, \infty) \cap C^1(a, \infty)$ with $b_0 \geq \alpha > 0$ on (a, ∞), $\alpha(a) = 0$ and $(G'(\alpha) + p\alpha^m)'(t) + q(t)f(t, \alpha) \geq p'(t)\alpha^m(t)$ on (a, ∞),
(C159) for each $t \in [a, \infty)$ we have $q(t)[f(t, y) - f(t, \alpha(t))] \geq 0$ for $0 \leq y \leq \alpha(t)$, and
(C160) $p' > 0$ on (a, ∞).
Then, (9.152) has a solution y with $0 \leq y(t) \leq b_0$ for $t \in [a, \infty)$. In addition $y \in C^1(a, \infty)$ with $G'(y) = y^m y'$ on (a, ∞) and y is a solution of (9.153).

Remark 9.1 If $\lim_{t \to \infty} \alpha(t) = b_0$ (here b_0 is as in (C156)), then the solution y to (9.152) (guaranteed from Theorem 9.24) is a solution of the boundary value problem

$$(g(y)y')' + p(y^m)' + qf(t, y) = 0, \quad a < t < \infty$$
$$y(a) = 0, \quad \lim_{t \to \infty} y(t) = b_0. \tag{9.154}$$

Theorem 9.25 ([18]) Suppose that in Theorem 9.24 the condition (C158) is replaced by
(C161) there exists $\alpha \in BC[a, \infty) \cap C^1(a, \infty)$ with $G(\alpha) \in C^1[a, \infty)$, $\alpha^m \alpha' \in C^1(a, \infty)$, $b_0 \geq \alpha > 0$ on (a, ∞), $\alpha(a) = 0$ and $(\alpha^m \alpha')' + p(\alpha^m)' + q(t)f(t, \alpha) \geq 0$ on (a, ∞).
Then, (9.153) has a solution y with $\alpha(t) \leq y(t) \leq b_0$ for $t \in [a, \infty)$.

Remark 9.2 If $\lim_{t \to \infty} \alpha(t) = b_0$ (here b_0 is as in (C156)) then the solution y to (9.153) (guaranteed from Theorem 9.25) is a solution of (9.154).

In terms of our notation, problem (9.151) is the same as

$$(y^3 y')' + t(y^3)' = 0, \quad 1 < t < \infty$$
$$y(1) = 0, \quad \lim_{t \to \infty} y(t) = H_0 > 0. \tag{9.155}$$

We will now use Theorem 9.25 (with Remark 9.2) to show that (9.155) has a solution. To see this we consider

$$(y^3 y' + ty^3)' = y^3, \quad 1 < t < \infty$$
$$y(1) = 0, \quad y \text{ bounded on } [1, \infty). \tag{9.156}$$

Clearly, $y \equiv 0$ is a solution of (9.156). Let $m = 3$, $a = 1$, $p = t$, $q \equiv 0$, $f(t, y) \equiv 0$, $b_0 = H_0$, and

$$g(z) = \begin{cases} z^3, & z \geq 0 \\ -z^3 = |z|^3, & z < 0. \end{cases}$$

Clearly (C152)–(C157), (C159), and (C160) hold. Let

$$\alpha(t) = A \int_1^t \sigma(s)ds,$$

where

$$\sigma(t) = \exp\left(-\frac{3t^2}{2H_0}\right) \text{ and } A = \frac{H_0}{\int_1^\infty \sigma(s)ds}.$$

Note that $\alpha(1) = 0$ and

$$\alpha' = A\sigma.$$

Also for $t \in (1, \infty)$ we have

$$(\alpha^3\alpha')' + t(\alpha^3)' = A^4 \left(\int_1^t \sigma(s)ds\right)^2 \left[3\,[\sigma(t)]^2 - \frac{3t}{H_0}\sigma(t)\int_1^t \sigma(s)ds\right]$$

$$+3tA^3\left(\int_1^t \sigma(s)ds\right)^2 \sigma(t)$$

$$= 3tA^3\left(\int_1^t \sigma(s)ds\right)^2 \sigma(t)\left[1 - \frac{A}{H_0}\int_1^t \sigma(s)ds\right]$$

$$+3A^4\left(\int_1^t \sigma(s)ds\right)^2 [\sigma(t)]^2 \geq 0,$$

since

$$\frac{A}{H_0}\int_1^t \sigma(s)ds = \frac{\int_1^t \sigma(s)ds}{\int_1^\infty \sigma(s)ds} \leq 1.$$

Thus, (C161) holds and hence Theorem 9.25 guarantees that (9.156) has a solution y with $\alpha(t) \leq y(t) \leq H_0$ for $t \in [1, \infty)$. Also since $\lim_{t\to\infty} \alpha(t) = H_0$ this y is a solution of (9.155).

Example 9.18 (Semiconductor Theory) Suppose a, b, ℓ, α, β, γ, δ, A, B, C, D, F are positive numbers. A fourth-order boundary value problem which arose in the theory of semiconductors [96, 97] can be written as

$$y^{(4)} - a[y'' + b(y'y''' + y''^2)] = 0, \quad 0 \leq t \leq \ell \tag{9.157}$$

$$\begin{aligned} y(0) &= \alpha\{Fy'''(0) - By'(0)[C + Dy''(0)]\} \\ y'(0) &= \beta\{Fy'''(0) - By'(0)[C + Dy''(0)]\} \\ y(\ell) &= A - \gamma\{Fy'''(\ell) - By'(\ell)[C + Dy''(\ell)]\} \\ y'(\ell) &= \delta\{Fy'''(\ell) - By'(\ell)[C + Dy''(\ell)]\}. \end{aligned} \tag{9.158}$$

A solution of (9.157), (9.158) represents potential and the function z defined in terms of y by

$$z = Fy''' - By'(C + Dy'') \tag{9.159}$$

means current in a semiconductor.

Integrating (9.157) twice and assuming that

$$Fab - BD = 0, \tag{9.160}$$

the problem (9.157), (9.158) can be reduced to four different types of boundary value problems:

1. $\beta(Fa - BC) - 1 = \delta(Fa - BC) - 1 = 0$, then (9.157), (9.158) reduces to

$$y'' = a\left(y + \frac{b}{2}y'^2\right) + c_2 \tag{9.161}$$

$$\frac{\alpha}{\beta}y'(0) - y(0) = 0$$
$$\frac{\gamma}{\delta}y'(\ell) + y(\ell) - A = 0 \tag{9.162}$$

and the associated function $z = y'/\beta$.

2. $\beta(Fa - BC) - 1 = 0 \neq \delta(Fa - BC) - 1$, then (9.157), (9.158) reduces to (9.161),

$$\frac{\alpha}{\beta}y'(0) - y(0) = 0$$
$$y(\ell) = A, \quad y'(\ell) = 0 \tag{9.163}$$

and the associated function $z = y'/\beta$.

3. $\beta(Fa - BC) - 1 \neq 0 = \delta(Fa - BC) - 1$, then (9.157), (9.158) reduces to (9.161),

$$y(0) = y'(0) = 0$$
$$\frac{\gamma}{\delta}y'(\ell) + y(\ell) - A = 0 \tag{9.164}$$

and the associated function $z = y'/\delta$.

4. $\beta(Fa - BC) - 1 \neq 0 \neq \delta(Fa - BC) - 1$, then (9.157), (9.158) reduces to

$$y'' = a\left(y + \frac{b}{2}y'^2\right) + c_1 t + c_2 \tag{9.165}$$

$$y(0) = \frac{\alpha F c_1}{1 - \beta(Fa - BC)}, \qquad y'(0) = \frac{\beta F c_1}{1 - \beta(Fa - BC)}$$
$$y(\ell) = A - \frac{\gamma F c_1}{1 - \delta(Fa - BC)}, \qquad y'(\ell) = \frac{\delta F c_1}{1 - \delta(Fa - BC)} \tag{9.166}$$

and the associated function $z = (Fa - BC)y' + Fc_1$.

In all the above problems c_1, $c_2 \in \mathbb{R}$ are arbitrary. It is clear that the boundary value problem (9.165), (9.162) contains other problems as special cases.

We shall prove the existence and uniqueness of solutions of (9.165), (9.162) by using the general theory for the boundary value problem

$$y'' = f(t, y, y') \tag{9.167}$$

$$g[y(0), y'(0)] = 0 = h[y(\ell), y'(\ell)]. \tag{9.168}$$

In what follows, we shall assume that the function $f = f(t, y, y')$ is defined and continuous on the set $\mathcal{S} = \{(t, y, y') : 0 \leq t \leq \ell, |y| + |y'| < \infty\}$. By a lower solution of (9.167) we mean a function $\phi \in C^2[0, \ell]$ such that $\phi'' \geq f(t, \phi, \phi')$, $t \in [0, \ell]$, and similarly, $\psi \in C^2[0, \ell]$ is an upper solution of (9.167) if $\psi'' \leq f(t, \psi, \psi')$, $t \in [0, \ell]$.

We also recall that the function $f = f(t, y, y')$ is said to satisfy a Mitio Nagumo (1905–1995) condition with respect to the pair ϕ, $\psi \in C[0, \ell]$ where $\phi \leq \psi$ on $[0, \ell]$ if there is a positive function $h \in C[0, \infty)$ such that $|f(t, y, y')| \leq h(|y'|)$ for all $t \in [0, \ell]$, $\phi(t) \leq y \leq \psi(t)$, $|y'| < \infty$ and

$$\int_\lambda^\infty \frac{s \, ds}{h(s)} > \max_{t \in [0, \ell]} \psi(t) - \min_{t \in [0, \ell]} \phi(t)$$

with $\lambda \ell = \max\{|\phi(\ell) - \psi(0)|, |\phi(0) - \psi(\ell)|\}$.

If ϕ is a lower and ψ an upper solution of (9.167) and $\phi(0) < \psi(0)$ ($\phi(\ell) < \psi(\ell)$), then H_1 (H_2) denotes the class of all continuous functions $g = g(y, y')$ ($h = h(y, y')$) defined on $[\phi(0), \psi(0)] \times \mathbb{R}$ ($[\phi(\ell), \psi(\ell)] \times \mathbb{R}$) which are nondecreasing in y' and satisfy $g[\phi(0), \phi'(0)] \geq 0$, $g[\psi(0), \psi'(0)] \leq 0$ ($h[\phi(\ell), \phi'(\ell)] \leq 0$, $h[\psi(\ell), \psi'(\ell)] \geq 0$). Further, H_3 (H_4) represents the class of all continuous functions $g = g(y, y')$ ($h = h(y, y')$) defined on $\mathbb{R}^2$ which are nonincreasing in y (nondecreasing in y) and nondecreasing in y'.

Theorem 9.26 ([57]) *Suppose that*
(C162) f is defined and continuous on $\mathcal{S}$,
(C163) there exist ϕ, ψ, respectively, lower and upper solutions of (9.167) with $\phi(t) \leq \psi(t)$ for $t \in [0, \ell]$, $\phi(0) < \psi(0)$, $\phi(\ell) < \psi(\ell)$,
(C164) f satisfies a Nagumo condition with respect to the pair ϕ, ψ, and
(C165) $g \in H_1$, $h \in H_2$.
Then, (9.167), (9.168) has a solution $y \in C^2[0, \ell]$ with $\phi(t) \leq y(t) \leq \psi(t)$ for $t \in [0, \ell]$.

Theorem 9.27 ([96, 97]) *Suppose that*
(C166) $f_1 = f_1(t, y, y')$, $f_2 = f_2(t, y, y')$ are defined and continuous, and $f_1 \leq f_2$ ($f_1 \geq f_2$) on $\mathcal{S}$,
(C167) $\lim_{y \to -\infty} f_2(t, y, y') = -\infty$ ($\lim_{y \to \infty} f_2(t, y, y') = \infty$) uniformly in t, y' on any compact subset of $[0, \ell] \times \mathbb{R}$,
(C168) f_2 satisfies a Nagumo condition with respect to any pair ϕ, $\psi \in C[0, \ell]$, $\phi \leq \psi$, and
(C169) $g \in H_3$, $h \in H_4$.
Then, if there is a solution y_1 of the problem (9.168),

$$y'' = f_1(t, y, y'), \tag{9.169}$$

there also exists a solution y_2 of the problem (9.168),

$$y'' = f_2(t, y, y')$$ (9.170)

and a number $c > 0$ such that $y_1 - c \le y_2 \le y_1$ ($y_1 \le y_2 \le y_1 + c$) on $[0, \ell]$.

Theorem 9.28 ([96, 97]) *Suppose that in addition to (C162),*
(C170) f is increasing in y on S, and
(C171) g (h) is decreasing in y (increasing in y) and nondecreasing in y' on $\mathbb{R}^2$.
Then, (9.167), (9.168) has at most one solution.

Clearly, the problem (9.165), (9.162) is the same as (9.167), (9.168) with

$$f = a\left(y + \frac{b}{2}y'^2\right) + c_1 t + c_2$$
$$g(y, y') = \frac{\alpha}{\beta}y' - y, \quad h(y, y') = \frac{\gamma}{\delta}y' + y - A.$$ (9.171)

For these functions f, g, h the conditions (C162), (C170), and (C171) are obviously satisfied, and hence the problem (9.165), (9.162) has at most one solution.

Next, we note that this function f satisfies Nagumo condition with respect to any pair of functions ϕ, $\psi \in C[0, \ell]$, $\phi \le \psi$. This follows from the estimate $|f| \le M + (ab/2)y'^2$ which holds with a suitable constant $M > 0$ for all $t \in [0, \ell]$, $\phi(t) \le y \le \psi(t)$, arbitrary y', and the fact that

$$\int^\infty \frac{s\,ds}{M + \frac{ab}{2}s^2} = \infty.$$

Thus the condition (C164) is satisfied.

Now we shall show that the functions

$$\phi(t) = -\frac{c_1}{a}t + q_1, \quad \psi(t) = -\frac{c_1}{a}t + q_2,$$ (9.172)

where

$$q_1 = \min\left\{\frac{\alpha}{\beta}\left(-\frac{c_1}{a}\right), \ A + \frac{c_1}{a}\left(\frac{\gamma}{\delta} + \ell\right), \ -\frac{c_2}{a} - \frac{bc_1^2}{2a^2}\right\}$$

and

$$q_2 = \max\left\{\frac{\alpha}{\beta}\left(-\frac{c_1}{a}\right), \ A + \frac{c_1}{a}\left(\frac{\gamma}{\delta} + \ell\right), \ -\frac{c_2}{a} - \frac{bc_1^2}{2a^2}\right\}$$

satisfy the conditions (C163) and (C165). For this, we note that

$$f(t, \phi(t), \phi'(t)) = a \left(-\frac{c_1}{a} t + q_1 + \frac{b}{2} \left(-\frac{c_1}{a} \right)^2 \right) + c_1 t + c_2$$

$$= a q_1 + \frac{b c_1^2}{2a} + c_2 \leq 0 = \phi''(t),$$

and similarly,

$$f(t, \psi(t), \psi'(t)) = a q_2 + \frac{b c_1^2}{2a} + c_2 \geq 0 = \psi''(t).$$

The inequalities $\phi(t) \leq \psi(t)$, $t \in [0, \ell]$, $\phi(0) < \psi(0)$, $\phi(\ell) < \psi(\ell)$ are obvious. Next, we note that

$$g[\phi(0), \phi'(0)] = \frac{\alpha}{\beta} \left(-\frac{c_1}{a} \right) - q_1 \geq 0,$$

$$g[\psi(0), \psi'(0)] = \frac{\alpha}{\beta} \left(-\frac{c_1}{a} \right) - q_2 \leq 0,$$

$$h[\phi(\ell), \phi'(\ell)] = \frac{\gamma}{\delta} \left(-\frac{c_1}{a} \right) - \frac{c_1}{a} \ell + q_1 - A \leq 0,$$

$$h[\psi(\ell), \psi'(\ell)] = \frac{\gamma}{\delta} \left(-\frac{c_1}{a} \right) - \frac{c_1}{a} \ell + q_2 - A \geq 0.$$

Thus, in view of Theorem 9.26, problems (9.165), (9.162) have a solution $y \in C^2[0, \ell]$ with $-(c_1/a)t + q_1 \leq y(t) \leq -(c_1/a)t + q_2$ for $t \in [0, \ell]$. In what follows, we shall denote this solution by $y(c_1, c_2, t)$.

Finally, as an application of Theorem 9.27, we shall show that if $c_1' t + c_2' > c_1 t + c_2$ on $[0, \ell]$ and

$$c = \max_{t \in [0, \ell]} \frac{1}{a} [c_1' t + c_2' - (c_1 t + c_2)], \qquad (9.173)$$

then

$$y(c_1, c_2, t) - c \leq y(c_1', c_2', t) < y(c_1, c_2, t) \quad \text{on} \quad [0, \ell]. \qquad (9.174)$$

For this, we note that the functions $f_1 = a \left(y + \frac{b}{2} y'^2 \right) + c_1 t + c_2$, $f_2 = a \left(y + \frac{b}{2} y'^2 \right) + c_1' t + c_2'$, and the functions involved in the boundary conditions (9.162) satisfy all assumptions of Theorem 9.27. Thus by the uniqueness of $y(c_1', c_2', t)$ there is a $c > 0$ such that $y(c_1, c_2, t) - c \leq y(c_1', c_2', t) \leq y(c_1, c_2, t)$ for $t \in [0, \ell]$. A direct calculation shows that c given in (9.173) can be chosen. To show the strict inequality in (9.174), consider the case $y(c_1', c_2', t_0) = y(c_1, c_2, t_0)$ at a point $t_0 \in [0, \ell]$. Then, it is necessary that $y'(c_1', c_2', t_0) = y'(c_1, c_2, t_0)$ and hence $y''(c_1', c_2', t_0)$

$> y''(c_1, c_2, t_0)$. But, this contradicts the inequality $y(c_1', c_2', t) \leq y(c_1, c_2, t)$. Hence, (9.174) holds.

Several other results for the boundary value problems (9.157), (9.158) and (9.165), (9.162), and the related function z defined in (9.159) are given in Problems 9.27–9.33.

Problems

9.1 ([3, 33]) Consider the boundary value problem

$$y'' + f(t, y, y') = 0 \tag{9.175}$$

$$y(a) = A, \quad y(b) = B \tag{9.176}$$

where f is continuous on $[a, b] \times \mathbb{R}^2$ and is bounded for all values of its arguments, i.e., $|f(t, y, y')| \leq M$. Show that (9.175), (9.176) has a solution $y \in C^{(2)}[a, b]$. In particular, show that the simple pendulum equation

$$y'' + \frac{mg}{\ell} \sin(y(t)) = \phi(t), \tag{9.177}$$

together with the boundary conditions (9.176) has a solution provided the driving function $\phi(t)$ is continuous on $[a, b]$.

9.2 ([33]) Suppose $f(t, y, y')$ is continuous and satisfies the generalized Lipschitz condition

$$G_1(y - x, y' - x') \leq f(t, y, y') - f(t, x, x') \leq G_2(y - x, y' - x'),$$

where

$$G_1(y, y') = \begin{cases} K_1 y + L_1 y', & y \geq 0, \quad y' \geq 0 \\ K_1 y + L_2 y', & y \geq 0, \quad y' \leq 0 \\ K_2 y + L_2 y', & y \leq 0, \quad y' \leq 0 \\ K_2 y + L_1 y', & y \leq 0, \quad y' \geq 0 \end{cases}$$

and

$$G_2(y, y') = \begin{cases} K_2 y + L_2 y', & y \geq 0, \quad y' \geq 0 \\ K_2 y + L_1 y', & y \geq 0, \quad y' \leq 0 \\ K_1 y + L_1 y', & y \leq 0, \quad y' \leq 0 \\ K_1 y + L_2 y', & y \leq 0, \quad y' \geq 0 \end{cases}$$

on $[a, b] \times \mathbb{R}^2$. We define

$$\alpha(L, K) = \begin{cases} \dfrac{2}{\sqrt{4K - L^2}} \cos^{-1}\left(\dfrac{L}{2\sqrt{K}}\right) & \text{if } 4K - L^2 > 0 \\[2ex] \dfrac{2}{\sqrt{L^2 - 4K}} \cosh^{-1}\left(\dfrac{L}{2\sqrt{K}}\right) & \text{if } 4K - L^2 < 0, \\ & \quad L > 0, \quad K > 0 \\[2ex] \dfrac{2}{L} & \text{if } 4K - L^2 = 0, \quad L > 0 \\[2ex] \infty & \text{otherwise} \end{cases}$$

$$\beta(L, K) = \begin{cases} \dfrac{2}{\sqrt{4K - L^2}} \cos^{-1}\left(\dfrac{-L}{2\sqrt{K}}\right) & \text{if } 4K - L^2 > 0 \\[2ex] \dfrac{2}{\sqrt{L^2 - 4K}} \cosh^{-1}\left(\dfrac{-L}{2\sqrt{K}}\right) & \text{if } 4K - L^2 < 0, \\ & \quad L < 0, \quad K > 0 \\[2ex] -\dfrac{2}{L} & \text{if } 4K - L^2 = 0, \quad L < 0 \\[2ex] \infty & \text{otherwise.} \end{cases}$$

Show that
(1) the differential equation (9.175) together with the boundary conditions

$$y(a) = A, \quad y'(b) = m \tag{9.178}$$

has a unique solution provided $0 < b - a < \alpha(L_2, K_2)$,
(2) the differential equation (9.175) together with the boundary conditions

$$y'(a) = m, \quad y(b) = B \tag{9.179}$$

has a unique solution provided $0 < b - a < \beta(L_1, K_2)$,
(3) the boundary value problem (9.175), (9.176) has a unique solution provided $0 < b - a < \alpha(L_2, K_2) + \beta(L_1, K_2)$.
These results are the best possible in the sense that in case of equality either existence or uniqueness fails.

In particular, show that the problem (9.177), (9.176) has a unique solution provided $\phi(t)$ is continuous on $[a, b]$, and

$$0 < b - a < \alpha(0, K_2) + \beta(0, K_2) = \pi\sqrt{\frac{\ell}{mg}}. \tag{9.180}$$

Further, show that a violation of inequality (9.180) may lead to nonuniqueness of solutions.

9.3 ([33]) Assume that $f(t, y, y')$ is continuous on $[a, b] \times \mathbb{R}^2$, and that the differential equation (9.175) has the following properties:

(i) all initial value problems have unique solutions which exist throughout the interval $[a, b]$, and

(ii) two different solutions of (9.175) cannot agree in value at more than one point of $[a, b]$, i.e., solutions of (9.175) with $u(t_1) = \alpha$, $u(t_2) = \beta$, where $a \le t_1 < t_2 \le b$, if exist, are unique.

Further, let $z(t)$ be a twice continuously differentiable function on $[a, b]$ satisfying

$$z''(t) + f(t, z(t), z'(t)) \ge 0. \tag{9.181}$$

Show that

(1) if $y(t)$ is a solution of (9.175) which agrees with $z(t)$ in both value and slope at some point $t_0 \in [a, b]$, then

$$z(t) \ge y(t), \quad t \ne t_0, \tag{9.182}$$

(2) if $y(t)$ is a solution of (9.175) which agrees with $z(t)$ in value at a and at b, then

$$z(t) \le y(t), \quad t \ne a, b. \tag{9.183}$$

The inequality sign may be reversed throughout.

9.4 ([33]) Assume that in addition to assumptions in Problem 9.3, no two solutions of (9.175) can agree in value at one point as well as in slope at another point to the right of the first. Show that

(1) if $y(t)$ is a solution of (9.175) which agrees with $z(t)$ in both value and slope at a, then

$$z'(t) \ge y'(t), \quad t \ne a, \tag{9.184}$$

(2) if $y(t)$ is a solution of (9.175) which agrees with $z(t)$ in value at a and slope at b, then

$$z(t) \le y(t), \quad t \ne a. \tag{9.185}$$

If $f(t, y, y')$ is increasing in y then also $z'(t) \le y'(t)$, $t \ne b$. The inequality sign may be reversed throughout.

9.5 ([33]) Assume that in addition to assumptions in Problem 9.3, no two solutions of (9.175) can agree in value at one point as well as in slope at another point to the left of the first. Show that

(1) if $y(t)$ is a solution of (9.175) which agrees with $z(t)$ in both value and slope at b, then

$$z'(t) \le y'(t), \quad t \ne b, \tag{9.186}$$

(2) if $y(t)$ is a solution of (9.175) which agrees with $z(t)$ in value at b and slope at a, then

$$z(t) \le y(t), \quad t \ne b. \tag{9.187}$$

If $f(t, y, y')$ is increasing in y then also $z'(t) \geq y'(t)$, $t \neq a$. The inequality sign may be reversed throughout.

9.6 ([33]) Consider a long, thin cantilever beam of length L and flexural rigidity B subjected to a concentrated vertical load P at the free end. The resultant displacement can be described in terms of the arc length s and slope angle $\phi(s)$ where ϕ is determined by the boundary value problem

$$\frac{d^2\phi}{ds^2} + \frac{P}{B}\cos(\phi(s)) = 0 \tag{9.188}$$

$$\phi(0) = 0, \quad \phi'(L) = 0. \tag{9.189}$$

The solution of (9.188), (9.189) in terms of elliptic integrals is known, see Bisshopp and Drucker [43]. From physical grounds, it is clear that $\pi/2 \geq \phi(s) \geq 0$. One of the quantities of interest is $(L - \Delta)/L$, where Δ is the horizontal component of the displacement of the loaded end of the beam and is given in terms of ϕ by

$$\frac{L - \Delta}{L} = \frac{B}{PL}\phi'(0).$$

Consider the modified differential equation

$$\phi''(s) + f(\phi(s)) = 0, \tag{9.190}$$

where

$$f(\phi) = \begin{cases} P/B & \text{if } \phi < 0 \\ (P/B)\cos(\phi) & \text{if } 0 \leq \phi \leq \pi/2 \\ 0 & \text{if } \phi > \pi/2. \end{cases}$$

Show that
(1) in (9.190) the function $f(\phi)$ satisfies generalized Lipschitz condition with $L_1 = L_2 = 0$, $K_1 = -P/B$, $K_2 = 0$,
(2) for (9.190) conditions (i) and (ii) of Problem 9.3 are satisfied throughout the interval $[0, L]$,
(3) the unique solution $\phi(s)$ of (9.190), (9.189) satisfies

$$u_1(s) = L\sqrt{\frac{P}{B}}\frac{\sinh\left(s\sqrt{P/B}\right)}{\cosh\left(L\sqrt{P/B}\right)} - \frac{P}{2B}s^2 \leq \phi(s) \leq \frac{PL}{B}s - \frac{Ps^2}{2B} = u_2(s),$$

(4) $1 = \dfrac{B}{PL}u_2'(0) \geq \dfrac{B}{PL}\phi'(0) = \dfrac{L - \Delta}{L} \geq \dfrac{B}{PL}u_1'(0) = \dfrac{1}{\cosh\left(L\sqrt{P/B}\right)}$,
(5) the unique solution $\phi(s)$ of (9.190), (9.189) also satisfies $0 \leq \phi(s) \leq \pi/2$ for $0 \leq s \leq L$ and hence is a solution of (9.188), (9.189).

9.7 ([33]) The problem for the stationary temperature distribution in a bar whose ends $t = \pm 1$ are kept at the temperature $y = 1$ and which transfers heat to the environment at the temperature $y = 0$ proportional to $(y + y^2/4)$ is

$$y'' - \left(y + \frac{1}{4}y^2\right) = 0 \tag{9.191}$$

$$y(-1) = 1, \quad y(1) = 1. \tag{9.192}$$

On physical grounds, it is expected that a solution $y(t)$ of (9.191), (9.192) exists such that $0 \leq y(t) \leq 1$ for all $-1 \leq t \leq 1$. Consider the modified differential equation

$$y'' + f(y) = 0, \tag{9.193}$$

where

$$f(y) = \begin{cases} \dfrac{1}{4} - \dfrac{3}{2}y & \text{if } 1 \leq y \\[2mm] -y\left(1 + \dfrac{1}{4}y\right) & \text{if } 0 \leq y \leq 1 \\[2mm] -y & \text{if } y \leq 0. \end{cases}$$

Show that
(1) in (9.193) the function $f(y)$ satisfies generalized Lipschitz condition with $L_1 = L_2 = 0$, $K_1 = -3/2$, $K_2 = -1$,
(2) problems (9.193), (9.192) have exactly one solution $y(t)$ such that

$$0 < \text{sech}(\sqrt{3/2})\cosh(\sqrt{3/2})t \leq y(t) \leq \text{sech}(1)\cosh(t) \leq 1, \quad -1 \leq t \leq 1$$

(3) problems (9.191), (9.192) have two solutions $y_1(t) = y(t)$ and $y_2(t)$, both parabolic, where $y_2(t)$ drops below up to -17.

9.8 Show that problems (9.30), (9.31) for each $n \geq 0$ has a solution $\theta(x)$ such that

$$0 \leq \theta(x) \leq 2\left[\ln(\cosh(c)) - \ln(\cosh(cx))\right]$$

provided $\sqrt{A/[2(1+n)]}\cosh(c) \leq c$.

9.9 ([72, 76]) Show that
(1) if $0 \leq n \leq 1$ there exists $A^* > 0$ such that problems (9.30), (9.31) have exactly one solution for $A = A^*$ and exactly two solutions for each $A \in (0, A^*)$,
(2) if $1 < n < 9$, problems (9.30), (9.31) have an unbounded continuum of solutions which oscillates around the line $A = 2(n-1)$ with the amplitude of oscillation tending to zero as $\theta(0) \to \infty$,
(3) if $n \geq 9$, problems (9.30), (9.31) have a unique solution for each $A \in (0, 2(n-1))$ and no solution for $A \geq 2(n-1)$. Further, $\|\theta\| \to \infty$ as $A \to 2(n-1)$.

9.10 The boundary value problem

$$\frac{d^2\theta}{dr^2} + \frac{2}{r}\frac{d\theta}{dr} + \frac{q}{k} = 0, \quad 0 < r < R \tag{9.194}$$

$$\theta(0) \text{ finite}, \quad -k\frac{d\theta}{dr} = \beta(\theta - \theta_a) \text{ at } r = R \tag{9.195}$$

arises in the study of the distribution of heat sources in the human head, see Thron [110]. Here q is the heat production rate per unit volume, θ is the absolute temperature, r is the radial distance from the center, k is the thermal conductivity (average) inside the head, β is a heat exchange coefficient, and θ_a is the ambient temperature. Show that

(1) If $q = q_1 r^2$ (q_1 is a constant) considered by Flesch [59], then the general solution of Eq. (9.194) is

$$\theta(r) = c_1 + c_2\frac{1}{r} - \frac{1}{20}\frac{q_1}{k}r^4,$$

and hence find c_1 and c_2 satisfying the boundary conditions (9.195).
(2) If $q = q_2 r^3$ (q_2 is a constant) considered by Flesch [59], then the general solution of the Eq. (9.194) is

$$\theta(r) = c_1 + c_2\frac{1}{r} - \frac{1}{30}\frac{q_2}{k}r^5,$$

and hence find c_1 and c_2 satisfying the boundary conditions (9.195).
(3) If $q = \alpha e^{-N\theta/\alpha}$, $\alpha > 0$, $N > 0$ considered by Anderson and Arthurs [31], then with the substitution $r = Rx$, $\phi = N(\theta - \theta_a)/\alpha$, problem (9.194), (9.195) can be written as

$$-\frac{1}{x^2}\frac{d}{dx}\left(x^2\frac{d\phi}{dx}\right) = f(\phi), \quad 0 < x < 1 \tag{9.196}$$
$$\phi(0) \text{ finite}, \quad \phi'(1) + b\phi(1) = 0,$$

where

$$f(\phi) = ae^{-\phi}, \quad a = \frac{NR^2}{k}e^{-N\theta_a/\alpha}, \quad b = \frac{\beta R}{k}.$$

Discuss existence results for the problem (9.196); to help the reader consider, for example (note one could also consider a more general situation),

$$-\frac{1}{x^2}\frac{d}{dx}\left(x^2\frac{d\phi}{dx}\right) = f(\phi), \quad 0 < x < 1 \tag{9.197}$$
$$\lim_{x\to 0^+} x^2\phi'(x) = 0, \quad \phi'(1) + b\phi(1) = 0,$$

(4) If $f(\phi) = a(1 - \phi)$, similar to the case considered by Gray [65], the solution of the problem (9.196) can be written as

$$\phi(\theta) \;=\; 1 + \frac{B}{x}\sinh(\sqrt{a}x),$$

where

$$B \;=\; -b/\big[(b-1)\sinh(\sqrt{a}) + \sqrt{a}\cosh(\sqrt{a})\big],$$

9.11 The boundary value problem

$$\frac{d^2u}{dx^2} + \frac{2}{x}\frac{du}{dx} = \frac{\alpha u}{u + K_m}, \quad 0 < x < 1$$

$$u'(0) = 0, \quad u'(1) = H(1 - u(1)) \tag{9.198}$$

arises in the study of steady-state oxygen diffusion in a cell with Michaelis–Menten (after Leonor Michaelis 1875–1949 and Maud Leonora Menten 1879–1960) uptake kinetics, see [30, 32, 71, 80, 81]. Here α, H and K_m are positive constants involving the reaction rate, permeability, and Michaelis constant. Show that problem (9.198) has a unique solution $u(x)$ satisfying $0 \le u(x) \le 1$. Further, show that

$$u(x) \;\le\; \frac{H}{k\cosh(k) - (1 - H)\sinh(k)}\,\frac{\sinh(kx)}{x}, \quad 0 < x < 1,$$

where $k^2 = \alpha/(1 + K_m)$.

9.12 ([85]) In an analysis of the heat transfer in the radial flow between parallel circular disks, the following infinite interval problem occurs:

$$\frac{d^2 f}{d\eta^2} + \eta^2 \frac{df}{d\eta} - 3\alpha\eta f = 0, \quad f(0) = 1, \quad f(\infty) = 0,$$

where α is a physical constant. Discuss existence results for this problem.

9.13 ([85]) In an analysis of the problem of phase change of solids with temperature-dependent thermal conductivity, the temperature distribution θ is found by solving the following infinite interval problem:

$$\frac{d}{d\eta}\left\{(1 + \beta\theta)\frac{d\theta}{d\eta}\right\} + 2\eta\frac{d\theta}{d\eta} = 0, \quad \theta(0) = 0, \quad \theta(\infty) = 1,$$

where β is a physical constant. Discuss existence results for this problem.

9.14 ([91, 117]) Related to Example 9.13 the following boundary value, problem arises in modeling the flows draining down a dry vertical wall

$$y''' = f(y), \quad x > 0$$

$$y(0) = 0, \quad y(\infty) = 1, \quad y'(\infty) = 0, \quad y''(\infty) = 0. \tag{9.199}$$

In (9.199), the case $f(y) = y^{-3}(1 - y)$ is the simplest and most important. With respect to (9.199) assume that there exists a positive number λ and a function $g(y)$ such that $f(y) = (1 - y)^\lambda g(y)$, where $g(y)$ is defined, continuous, and non-increasing on $(0, 1]$ and $g(1) \geq 1$. Show that problem (9.199) has a solution $y(x) \in C[0, \infty) \cap C^{(3)}(0, \infty)$. Moreover, if $\lambda > 1$, then

$$y(x) \geq 1 - \frac{1}{(1 + (\rho - 1)Ax)^{1/(\rho-1)}} \quad \text{for all} \quad x \geq 0;$$

if $\lambda = 1$, then

$$y(x) \geq 1 - e^{-x} \quad \text{for all} \quad x \geq 0,$$

and, if $\lambda \in (0, 1)$ then there exists an $x_1 \in (0, \bar{x}_1)$, where $\bar{x}_1 = 1/(1 - \rho)A$ such that $y(x) = 1$ for all $x \geq x_1$. Here,

$$\rho = \frac{\lambda + 2}{3} > \frac{2}{3} \quad \text{and} \quad A = \frac{1}{\sqrt[3]{\rho(2\rho - 1)}}.$$

If $g(y) \equiv 1$, then

$$y(x) = \begin{cases} 1 - \dfrac{1}{[1 + (\rho - 1)Ax]^{1/(\rho-1)}}, & \lambda > 1 \\ 1 - e^{-x}, & \lambda = 1 \\ 1 - [1 - (1 - \rho)Ax]_+^{1/(1-\rho)}, & 0 < \lambda < 1, \end{cases}$$

is a solution of (9.199). Here $[a]_+ = \max\{a, 0\}$.

9.15 ([68, 116, 120, 121]) The boundary value problem (for $f(\eta)$)

$$f''' + ff'' + \lambda\left[1 - (f')^2\right] = 0, \quad 0 < \eta < \infty \tag{9.200}$$

$$f(0) = 0, \quad f'(0) = 0 \quad \text{and} \quad f'(\infty) = 1 \tag{9.201}$$

is of great importance in the boundary layer theory in fluid mechanics. The cases $\lambda = 0$ and $\lambda = 1/2$ (see Problem 6.16 and Example 9.14) of (9.200) are often called the Blasius and Homann differential equations. Using the same transformations as in Example 9.14, problems (9.200), (9.201) become

$$w''(t) + \lambda\left(\frac{1 - t^2}{w(t)}\right)' + \frac{t}{w(t)} = 0, \quad 0 < t < 1 \tag{9.202}$$

$$w'(0)w(0) = -\lambda \quad \text{and} \quad w(1) = 0.$$

Show that
(1) for each fixed $\lambda \geq 0$, problem (9.202) has a unique positive solution $w(t, \lambda)$, i.e., $w \in C[0, 1] \cap C^2[0, 1)$, $w(t, \lambda) > 0$ for all $t \in [0, 1)$, with

$$A(1-t) \leq w(t, \lambda) \leq A^{-1}(1-t)[2\lambda + 1 - \ln(1-t)], \quad t \in [0, 1]$$

where $A = [(4\lambda + 1)/6]^{1/2}$, further

$$w(t, \lambda_1) \geq w(t, \lambda_2), \quad t \in [0, 1] \ \text{ if } \ \lambda_1 \geq \lambda_2 \geq 0$$

and

$$w(0, \lambda_1) > w(0, \lambda_2), \quad \text{if } \ \lambda_1 > \lambda_2 > 0,$$

(2) for each fixed $\lambda \geq 0$, the problem (9.200), (9.201) has a unique positive solution $f(t, \lambda)$, which satisfies

$$0 < f'(\eta, \lambda) < 1, \quad f''(\eta, \lambda) > 0, \quad \eta > 0, \tag{9.203}$$

$$f'(\eta, \lambda_1) > f'(\eta, \lambda_2), \quad \eta > 0 \ \text{ if } \ \lambda_1 > \lambda_2 \geq 0$$

and

$$\exp\left(-(2\lambda + 1)(e^{\eta/A} - 1)\right) \leq 1 - f'(\eta, \lambda) \leq e^{-A\eta}, \quad \eta \geq 0,$$

(3) for each fixed $\lambda \leq -1/2$, the problem (9.202) has no positive solution,

(4) for each fixed $\lambda \leq -1/2$, the problem (9.200), (9.201) has no solution satisfying (9.203),

(5) for each fixed $\lambda \in (\lambda_0, 0)$, the problem (9.202) has a positive solution, where $\lambda_0 = \max\{\lambda : H(\lambda) = 0, \ -1/2 < \lambda < 0\}$, and

$$H(\lambda) = \frac{2(-4\lambda + 3)\lambda}{-\lambda + 1/2}\sqrt{\frac{-\lambda}{-\lambda + 1/2}} + \frac{(2\lambda + 1)^3}{(\lambda + 1)^2} + \frac{3(2\lambda + 1)}{\lambda + 1}\left(\frac{-\lambda}{\lambda + 1}\right)^2,$$

$$H(0) = 1, \quad H\left(-\frac{1}{2}\right) = -\frac{5\sqrt{2}}{2},$$

(6) for each fixed $\lambda \in (\lambda_0, 0)$, problems (9.200), (9.201) has a solution which satisfies (9.203),

(7) for each fixed $\lambda \in (-1/2, 0)$, problems (9.200), (9.201) have a solution $f(\eta, \lambda) \in C^1[0, \infty)$ which satisfies the differential equation (9.200) only a.e. in $(0, \infty)$, and $0 < f'(\eta, \lambda) < 1, \ \eta > 0$.

9.16 ([85]) The analysis of the performance of solid-propellant rockets leads to the following problem:

$$\frac{1 + \epsilon}{1 + \beta} \frac{N}{R} \frac{d^2 N}{d\eta^2} + f\frac{dN}{d\eta} = 0$$

$$N(0) = 0, \quad N(\infty) = 1,$$

where the function f which relates to the flow around the probe (stagnation point flow in the viscous flow theory) is given by the solution of (9.200), (9.201) with $\lambda = 1/2$. Discuss existence results for this problem.

9.17 ([94]) The flow of an electrically conducting fluid up a hot vertical plate in the presence of strong magnetic field normal to the plate is governed by the boundary value problem

$$f''' + \frac{1}{2} f f'' = 0, \quad f(0) = 0, \quad f'(0) = 1, \quad f'(\infty) = 0.$$

Discuss existence results for this problem.

9.18 ([115]) The following boundary value problem describes an axisymmetric flow due to stretching of flat surface:

$$f''' + 2 f f'' - (f')^2 = 0, \quad f(0) = 0, \quad f'(0) = 1, \quad f'(\infty) = 0.$$

Discuss existence results for this problem.

9.19 ([1]) The natural convection (the movement caused within a fluid by the tendency of hotter and therefore less dense material to rise, and colder, denser material to sink under the influence of gravity, which consequently results in transfer of heat) at high Prandtl's number (the Ludwig Prandtl (1875–1953) number is a dimensionless number approximating the ratio of momentum diffusivity to thermal diffusivity) on a horizontal plate is governed by the boundary value problem

$$f''' + \frac{3}{5} f f'' - \frac{1}{5} (f')^2 = 0, \quad f(0) = 0, \quad f'(0) = 1, \quad f'(\infty) = 0.$$

Discuss existence results for this problem.

9.20 ([34]) The two-dimensional flow due to stretching wall is described by the boundary value problem

$$f''' + f f'' - \beta (f')^2 = 0, \quad f(0) = 0, \quad f'(0) = 1, \quad f'(\infty) = 0,$$

where $\beta = 2m/(1+m)$ is a parameter. Discuss existence results for this problem.

9.21 ([1]) The study of condensation and natural convection at high Prandtl's number on a vertical plate leads to the boundary value problem

$$f''' + 3 f f'' - 2 (f')^2 = 0, \quad f(0) = c, \quad f'(0) = 1, \quad f'(\infty) = 0,$$

where c is suction parameter. Discuss existence results for this problem.

9.22 ([40]) The following boundary value problem is encountered in the study of free convection along a vertical flat plate embedded in a porous medium:

$$f''' + \frac{\alpha + 1}{2} ff'' - \alpha(f')^2 = 0, \quad f(0) = 0, \quad f'(0) = 1, \quad f'(\infty) = 0,$$

where α pertains to the temperature distribution prescribed on the wall. Discuss existence results for this problem.

9.23 ([44]) The description of convection in a porous medium leads to the following boundary value problem

$$f''' + Aff'' + B(f')^2 = 0, \quad f(0) = 0, \quad f''(0) = -1, \quad f'(\infty) = 0,$$

where A, B are constants. Discuss existence results for this problem.

9.24 ([37]) The boundary value problem encountered in the study of thermal capillary flows in viscous layers (the flow in the pre surface layer) is

$$5f''' + 6ff'' + 3(f')^2 = 0, \quad f(0) = 0, \quad f''(0) = -1, \quad f'(\infty) = 0.$$

Discuss existence results for this problem.

9.25 ([37]) The boundary value problem encountered in the study of thermal capillary flows in viscous layers (the axially symmetric case) is

$$f''' + 2ff'' - (f')^2 = 0, \quad f(0) = 0, \quad f''(0) = -1, \quad f'(\infty) = 0.$$

Discuss existence results for this problem.

9.26 ([42, 95]) The following boundary value problem occurs in the plane hydrodynamic jet theory:

$$3\epsilon \frac{d^3 y}{dx^3} + y \frac{d^2 y}{dx^2} + \left(\frac{dy}{dx}\right)^2 = 0$$

$$y(0) = y''(0) = 0, \quad y'(\infty) = 0.$$

Show that $y = a \tanh(ax/(6\epsilon))$, where a is an arbitrary constant, is a solution of this problem.

9.27 Let for fixed c_1, c_2, $y(c_1, c_2, t)$ be the solution of the problem (9.165), (9.162). Show that for any two numbers c_1, d there exist uniquely determined numbers c_2', c_2'' such that $y(c_1, c_2', 0) = d = y(c_1, c_2'', \ell)$.

9.28 Let for fixed c_1, c_2, $y(c_1, c_2, t)$ be the solution of the problem (9.165), (9.162). Show that
(1) if $y''(c_1, c_2, t) \geq 0$ in $[0, \ell]$, then

$$y(c_1, c_2, 0) \leq \frac{\alpha}{\beta} y(c_1, c_2, \ell) \Big/ \left(\ell + \frac{\alpha}{\beta}\right),$$

$$y(c_1, c_2, \ell) \leq \frac{\gamma}{\delta} y(c_1, c_2, 0) \Big/ \Big(\ell + \frac{\gamma}{\delta} \Big)$$

$$+ A \left[\Big(\ell + \frac{\alpha}{\beta} \Big) \Big(\ell + \frac{\gamma}{\delta} \Big) - \frac{\alpha \gamma}{\beta \delta} \right] \Big/ \left[\Big(\frac{\alpha}{\beta} + \ell + \frac{\gamma}{\delta} \Big) \Big(\ell + \frac{\gamma}{\delta} \Big) \right],$$

(2) if $y''(c_1, c_2, t) \leq 0$ in $[0, \ell]$, then

$$y(c_1, c_2, 0) \geq \frac{\alpha}{\beta} y(c_1, c_2, \ell) \Big/ \Big(\ell + \frac{\alpha}{\beta} \Big),$$

$$y(c_1, c_2, \ell) \geq \frac{\gamma}{\delta} y(c_1, c_2, 0) \Big/ \Big(\ell + \frac{\gamma}{\delta} \Big).$$

9.29 Let $Fab - BD = 0$, $\beta = \delta$, $Fa - BC = 1/\beta$. Show that
(1) for each number y_0 there exists a unique solution $y_{y_0}(t)$ of the problem (9.157), (9.158) such that $y_{y_0}(0) = y_0$,
(2) if $y_0 < y_1$, then $y_{y_0}(t) < y_{y_1}(t)$ on $[0, \ell]$ and $y_{y_0}(t)$ continuously depends on y_0,
(3) there is a unique c_0, $0 < c_0 < A$ such that for $y_0 > c_0$, or for $y_0 < 0$ there exists a number $d_0 = d_0(y_0)$, $0 < d_0 < \ell$ for which the following is true:
When $y_0 > c_0$, then $y'_{y_0}(t) > 0$ for $0 \leq t < d_0$ and $y'_{y_0}(t) < 0$ for $d_0 < t \leq \ell$. If $y_0 < 0$, then $y'_{y_0}(t) < 0$ for $0 \leq t < d_0$ and $y'_{y_0}(t) > 0$ for $d_0 < t \leq \ell$.
When $0 \leq y_0 \leq c_0$, then $y'_{y_0}(t) \geq 0$ on $[0, \ell]$ and there is at most one zero point of $y'_{y_0}(t)$.

9.30 Let $Fab - BD = 0$, $\beta \neq \delta$, $Fa - BC = 1/\beta$ ($Fa - BC = 1/\delta$). Show that there exists a unique solution $y(t)$ of the problem (9.157), (9.158). Further, show that this solution satisfies the inequalities $y(t) \geq 0$, $y'(t) \geq 0$, $y''(t) \leq 0$ ($y(t) \geq 0$, $y'(t) \geq 0$, $y''(t) \leq 0$) for $t \in [0, \ell]$, and the only zero of $y'(t)$ is at ℓ (at 0).

9.31 Let $Fab - BD = 0$ and let either (1) $1 - \beta(Fa - BC) > 0$, $1 - \delta(Fa - BC) > 0$, $\ell \leq a\alpha F/[1 - \beta(Fa - BC)]$, or (2) $1 - \beta(Fa - BC) < 0$, $1 - \delta (Fa - BC) < 0$. Show that there exists a unique solution $y(t)$ of the problem (9.157), (9.158). Further, in the case (1) the corresponding associated function $z(t)$ defined in (9.159) is positive, while in the case (2) $z(t)$ is either positive or its only local extremum is a local minimum. When ℓ is sufficiently small, this minimum is positive.

9.32 Let $Fab - BD = 0$, $1 - \beta(Fa - BC) < 0$, $1 - \delta(Fa - BC) > 0$. Further, let

$$\frac{\gamma}{1 - \delta(Fa - BC)} > - \Big(1 + \frac{\ell \beta}{\alpha} \Big) \frac{1}{1 - \beta(Fa - BC)},$$

or

$$\frac{\gamma}{1 - \delta(Fa - BC)} < - \frac{\gamma}{\delta} \frac{\alpha}{1 - \beta(Fa - BC)} \frac{\alpha}{\ell + (\gamma/\delta)}.$$

Show that there exists at least one solution $y(t)$ of the boundary value problem (9.157), (9.158). In the former case (the latter case), the associated function $z(t)$

defined in (9.159) satisfies the inequalities $z(0) < 0$, $z(\ell) > 0$, $z'(t) \geq 0$ for all $t \in [0, \ell]$ ($z(0) > 0$, $z(\ell) < 0$, $z'(t) \leq 0$ for $t \in [0, \ell]$).

9.33 Let $Fab - BD = 0$, $1 - \beta(Fa - BC) > 0$, $1 - \delta(Fa - BC) < 0$. Further, let

$$\frac{\gamma}{1 - \delta(Fa - BC)} < -\left(1 + \frac{\ell \beta}{\alpha}\right) \frac{\alpha}{1 - \beta(Fa - BC)},$$

or

$$\frac{\gamma}{1 - \delta(Fa - BC)} > \max\left\{-\frac{\gamma}{\delta\left(\ell + \frac{\gamma}{\delta}\right)[1 - \beta(Fa - BC)]}, -\frac{\gamma}{\delta}\frac{\beta}{1 - \beta(Fa - BC)}\right\}$$

Show that there exists at least one solution $y(t)$ of the boundary value problem (9.157), (9.158). In the former case (the latter case), the associated function $z(t)$ defined in (9.159) satisfies the inequalities $z(0) < 0$, $z(\ell) > 0$, and either $z'(t) \geq 0$ for $t \in [0, \ell]$, or there is a subinterval $[0, \ell_1]$ such that $z'(t) \geq 0$, $t \in [0, \ell_1]$ and $z'(t) \leq 0$, $t \in [\ell_1, \ell]$ ($z(0) > 0$, $z(\ell) < 0$, and either $z'(t) \leq 0$ for $t \in [0, \ell]$, or there is a subinterval $[0, \ell_1]$ such that $z'(t) \leq 0$, $t \in [0, \ell_1]$ and $z'(t) \geq 0$, $t \in [\ell_1, \ell]$).

9.34 ([28, 51, 109]) Certain chemical reactions in tubular reactors can be mathematically described by the boundary value problem

$$y'' = (y' - g(y))/c, \quad 0 \leq x \leq 1 \tag{9.204}$$

$$y'(0) - \ell\, y(0) = 0 = y'(1), \tag{9.205}$$

where y is the temperature in the reactor,

$$g(s) = d(q - s)e^{-k/(1+s)}, \quad 0 \leq s \leq q \tag{9.206}$$

is the rate of chemical production of the species in the reactor and c, ℓ, d and q are known positive constants.

In (9.204) assume that (i) g is twice continuously differentiable, (ii) $g(s) > 0$ for $s < q$, $g(s) < 0$ for $s > q$, and (iii) $\int^q ds/g(s) = \infty$. Note that g given by (9.206) satisfies (i)–(iii).

For the problem (9.204), (9.205) show that
(1) any solution y satisfies $0 < y < q$ and $y' \geq 0$ on $[0, 1]$,
(2) there is a minimal solution y_m and a maximal solution y_M,
(3) $y_m = y_M$ if

$$\frac{d}{ds}\left(\frac{g(s)}{s}\right) < 0 \quad \text{for} \ \ 0 < s < q$$

and hence the solution is unique,
(4) the set

$$S = \{\ell \in \mathbb{R} : \exists\, a,\, b,\ 0 < b < a < e \text{ such that } g(b)/b < \ell < g(a)/a\} \neq \emptyset$$

if and only if

$$\frac{d}{ds}\left(\frac{g(s)}{s}\right) > 0 \text{ for some } s, \quad 0 < s < q$$

(5) if $\ell \in S$ and $c > 0$ is sufficiently small, there are at least three solutions of (9.204), (9.205).

9.35 ([50, 67, 69, 70, 79, 92]) The generalized Painlevé equation

$$y'' = y^{2n+1} - (t - c)^{2k+1} y^{2(n-k)-1}, \quad t \in (0, \infty) \tag{9.207}$$

models a superheating field attached to a semi-infinite superconductor. Show that for any $0 < \beta < 1$, $M > 0$ and n, $k \in \mathbb{N}$ with $k < n$ there exists a strictly positive solution $y(t)$ of (9.207) satisfying

$$y'(0) = 0, \quad \lim_{t \to \infty} \frac{y(t)}{t^\beta} = M \text{ and } \lim_{t \to \infty} t^\beta y'(t) = \beta M.$$

9.36 The boundary value problem

$$\frac{d^2\theta}{dX^2} - \epsilon\theta^4 = 0, \quad \theta'(0) = 0, \quad \theta(1) = 1$$

occurs in heat transfer. Show that

$$\theta(X) = 1 + \epsilon\frac{1}{2}(X^2 - 1) + \epsilon^2\frac{1}{6}(X^4 - 6X^2 + 5) + O(\epsilon^3).$$

9.37 The boundary value problem

$$(1 + \epsilon\theta)\frac{d^2\theta}{dX^2} + \epsilon\left(\frac{d\theta}{dX}\right)^2 - N^2\theta = 0, \quad \theta'(0) = 0, \quad \theta(1) = 1$$

occurs in heat transfer. Show that

$$\theta(X) = \operatorname{sech}(N)\cosh(NX) + \epsilon\frac{1}{3}\operatorname{sech}^2(N)(\cosh(2N)\operatorname{sech} N \cosh(NX)$$

$$- \cosh(2NX)) + \epsilon^2\frac{1}{6}\operatorname{sech}^3(N)\left[\left(\frac{4}{3}\operatorname{sech}^2(N)\cosh^2(2N) - \frac{1}{2}N\tanh(N)\right.\right.$$

$$\left.- \frac{9}{8}\operatorname{sech}(N)\cosh(3N)\right)\cosh(NX) - \frac{4}{3}\operatorname{sech}(N)\cosh(2N)\cosh(2NX)$$

$$\left.+ \frac{9}{8}\cosh(3NX) + \frac{1}{2}NX\sinh(NX)\right] + O(\epsilon^3).$$

9.38 ([36, 49, 61, 107]) The equation

$$\frac{\partial u}{\partial t} = \frac{\partial^6 u}{\partial x^6} + A\frac{\partial^4 u}{\partial x^4} + B\frac{\partial^2 u}{\partial x^2} + u - u^3 \tag{9.208}$$

is a model for describing the behavior of phase fronts in materials that are undergoing a transition between the liquid and solid state. In this equation A and B are positive constants and satisfy the inequality $A^2 < 4B$. Clearly, the stationary solutions of (9.208) are given by the sixth-order differential equation

$$\frac{d^6 u}{dx^6} + A\frac{d^4 u}{dx^4} + B\frac{d^2 u}{dx^2} + u - u^3 = 0. \tag{9.209}$$

For (9.209), the natural boundary conditions (known as Lidstone conditions [23]) are

$$u^{(2i)}(0) = u^{(2i)}(L) = 0, \quad i = 0, 1, 2. \tag{9.210}$$

Show that

(1) the problem (9.209), (9.210) has only the trivial solution if

$$0 < L \le L_1 = \pi \max\left\{ \left(B - \frac{A^2}{4} \right)^{1/2}, \left(1 - \frac{A^2}{4B} \right)^{1/6} \right\},$$

(2) the problem (9.209), (9.210) has $m \in \mathbb{N}$ distinct pairs of nontrivial solution s if $L > mL_2$, $(L_2 > L_1)$ where

$$L_2^2 = M^{1/3} + \frac{1}{9}B^2\pi^4 M^{-1/3} + \frac{1}{3}B\pi^2$$

and

$$M = \left(\frac{1}{2} + \frac{1}{27}B^3 + \frac{1}{18}\sqrt{81 + 12B^3} \right)\pi^6.$$

References

1. J.A.D. Ackroyd, ZAMP **29**, 729–741 (1978)
2. A. Acrivos, M.J. Shah, E.E. Peterson, A.I.Ch.E.J. **6**, 312–317 (1960)
3. R.P. Agarwal, *Boundary Value Problems for Higher Order Differential Equations* (World Scientific, Singapore, 1986)
4. R.P. Agarwal, D. O'Regan, IMA J. Appl. Math. **65**, 173–198 (2000)
5. R.P. Agarwal, D. O'Regan, IMA J. Appl. Math. **66**, 621–635 (2001)
6. R.P. Agarwal, D. O'Regan, *Infinite Interval Problems for Differential, Difference and Integral Equations* (Kluwer Academic Publishers, Dordrecht, 2001)
7. R.P. Agarwal, D. O'Regan, Mathematika **48**, 273–292 (2001)

8. R.P. Agarwal, D. O'Regan, Dyn. Contin. Discret. Impuls. Syst. Ser. B: Appl. Algorithms **9**, 481–488 (2002)
9. R.P. Agarwal, D. O'Regan, Math. Probl. Eng. **8**, 135–142 (2002)
10. R.P. Agarwal, D. O'Regan, Stud. Appl. Math. **108**, 245–257 (2002)
11. R.P. Agarwal, D. O'Regan, NoDEA - Nonlinear Differ. Equ. Appl. **9**, 419–440 (2002)
12. R.P. Agarwal, D. O'Regan, Math. Probl. Eng. **8**, 143–150 (2002)
13. R.P. Agarwal, D. O'Regan, *Singular Differential and Integral Equations with Applications* (Kluwer Academic Publishers, Dordrecht, 2002)
14. R.P. Agarwal, D. O'Regan, Mathematika **49**, 129–140 (2002)
15. R.P. Agarwal, D. O'Regan, Int. J. Nonlinear Mech. **38**, 1369–1376 (2003)
16. R.P. Agarwal, D. O'Regan, J. Aust. Math. Soc. Ser. B - Appl. Math. **45**, 167–179 (2003)
17. R.P. Agarwal, D. O'Regan, Dyn. Contin. Discret. Impuls. Syst. **10**, 965–972 (2003)
18. R.P. Agarwal, D. O'Regan, Methods Appl. Anal. **10**, 363–376 (2003)
19. R.P. Agarwal, D. O'Regan, Stud. Appl. Math. **111**, 339–358 (2003)
20. R.P. Agarwal, D. O'Regan, ZAMM **83**, 344–350 (2003)
21. R.P. Agarwal, D. O'Regan, Int. J. Nonlinear Mech. **39**, 779–784 (2004)
22. R.P. Agarwal, D. O'Regan, *Handbook of Differential Equations, Ordinary Differential Equations*, vol. 1, ed. by A. Canada, P. Drabek, A. Fonda (Elsevier B.V., Amsterdam, 2004), pp. 1–68
23. R.P. Agarwal, P.J.Y. Wong, in *Approximation Theory, in Memory of A.K. Varma*, ed. by N.K. Govil et al. (Marcal Dekker, New York, 1998), pp. 1–41
24. R.P. Agarwal, D. O'Regan, P.J.Y. Wong, *Positive Solutions of Differential, Difference and Integral Equations* (Kluwer Academic Publishers, Dordrecht, 1999)
25. R.P. Agarwal, D. O'Regan, V. Lakshmikantham, Nonlinear Anal. Real-World Appl. **4**, 223–244 (2003)
26. R.P. Agarwal, D. O'Regan, R. Precup, Nonlinear Anal.: Real World Appl. **6**, 123–131 (2005)
27. R.P. Agarwal, D. O'Regan, P.K. Palamides, Math. Methods Appl. Sci. **29**, 49–66 (2006)
28. R.P. Agarwal, H.B. Thompson, C.C. Tisdell, Dyn. Syst. Appl. **16**, 595–609 (2007)
29. A.E. Alexander, P. Johnson, *Colloid Science*, vol. 1 (Oxford at the Clarendon Press, London, 1949), pp. 100–120
30. N. Anderson, A.M. Arthurs, Bull. Math. Biol. **42**, 131–135 (1980)
31. N. Anderson, A.M. Arthurs, Bull. Math. Biol. **43**, 341–346 (1981)
32. N. Anderson, A.M. Arthurs, Bull. Math. Biol. **47**, 145–153 (1985)
33. P. Bailey, L. Shampine, P. Waltman, *Nonlinear Two Point Boundary Value Problems* (Academic, New York, 1968)
34. W.H.H. Banks, Journal de Mecanique Theorique et Appliquee **2**, 375–392 (1983)
35. H. Bateman, *Partial Differential Equations of Mathematical Physics* (Cambridge University Press, London, 1959)
36. P. Bates, P. Fife, R. Gardner, C.K.R.J. Jones, SIAM J. Math. Anal. **28**, 60–93 (1997)
37. V.A. Batyshchev, J. Appl. Math. Mech. **55**, 315–321 (1991)
38. J. Baxley, SIAM J. Appl. Math. **48**, 497–505 (1988)
39. J. Baxley, S.B. Robinson, J. Comput. Appl. Math. **88**, 203–224 (1998)
40. Z. Belhachmi, B. Brighi, K. Taous, Eur. J. Appl. Math. **12**, 513–528 (2001)
41. F. Bernis, L.A. Peletier, SIAM J. Math. Anal. **27**, 515–527 (1996)
42. W.G. Bickley, Philos. Mag. **23**, 727 (1937)
43. K.E. Bisshopp, D.C. Drucker, Q. Appl. Math. **3**, 272–275 (1945)
44. P.A. Blythe, P.G. Daniels, P.G. Simpkins, Proc. R. Soc. Lond. A **380**, 119–136 (1982)
45. L. Bougoffa, Appl. Math. Lett. **21**, 275–278 (2008)
46. A. Callegari, M.B. Friedman, J. Math. Anal. Appl. **21**, 510–529 (1968)
47. P.L. Chambré, J. Chem. Phys. **20**, 1795 (1952)
48. C.Y. Chan, Y.C. Hon, Q. Appl. Math. **46**, 711–726 (1988)
49. J.V. Chaparova, L.A. Peletier, S.A. Tersian, Appl. Math. Lett. **17**, 1207–1212 (2004)
50. S.J. Charman, SIAM J. Appl. Math. **55**, 1233–1258 (1995)
51. D.S. Cohen, SIAM J. Appl. Math. **20**, 1–13 (1971)

52. A. Constantin, Annali di Mathematica pura ed applicata **176**, 379–394 (1999)
53. M. Countryman, R. Kannan, Comput. Math. Appl. **28**, 121–130 (1994)
54. H.T. Davis, *Introduction to Nonlinear Differential and Integral Equations* (Dover, New York, 1962)
55. R.W. Dickey, Arch. Ration. Mech. Anal. **26**, 219–236 (1967)
56. R.W. Dickey, Q. Appl. Math. **47**, 571–581 (1989)
57. L.H. Erbe, J. Differ. Equ. **7**, 459–472 (1970)
58. E. Fermi, Rend Accad. Naz. del Lincei Cl. Sci. Fis. Mat. e Nat. **6**, 602–607 (1927)
59. U. Flesch, J. Theor. Biol. **54**, 285–287 (1975)
60. R.L. Fosdick, K.R. Rajagopal, Proc. R. Soc. Lond. Ser. A **369**, 351–377 (1980)
61. R.A. Gardner, C.K.R.T. Jones, Indiana Univ. Math. J. **38**, 1197–1222 (1989)
62. J. Goncerzewicz, H. Marcinkowska, W. Okrasinski, K. Tabisz, Zastosow Mat. **16**, 246–261 (1978)
63. V. Goyal, Appl. Math. Lett. **19**, 1406–1408 (2006)
64. A. Granas, R.B. Guenther, J.W. Lee, ZAMM **61**, 204–205 (1981)
65. B.F. Gray, J. Theor. Biol. **82**, 473–476 (1980)
66. M. Gregus, Acta Math. Univ. Comen. **40**, 161–168 (1982)
67. M. Guedda, Electron. J. Differ. Equ. **49**, 1–4 (2001)
68. P. Hartman, *Ordinary Differential Equations* (Wiley, New York, 1964)
69. S.P. Hasting, J.B. McLeon, Arch. Ration. Mech. Anal. **73**, 31–51 (1980)
70. B. Helffer, F.B. Weissler, Eur. J. Appl. Math. **9**, 223–243 (1998)
71. P. Hiltmann, P. Lory, Bull. Math. Biol. **45**, 661–664 (1983)
72. J. Jacobsen, K. Schmitt, in *Handbook of Differential Equations, Ordinary Differential Equations*, vol. 1, ed. by A. Canada, P. Drabek, A. Fonda (Elsevier B.V, Amsterdam, 2004), pp. 359–435
73. J. Janus, J. Myjak, Nonlinear Anal. **23**, 953–970 (1994)
74. D. Jiang, R.P. Agarwal, Appl. Math. Lett. **15**, 445–451 (2002)
75. K.N. Johnson, Q. Appl. Math. **55**, 537–550 (1997)
76. D.D. Joseph, T.S. Lundgren, Arch. Ration. Mech. Anal. **49**, 241–269 (1973)
77. R.E. Kidder, J. Appl. Mech. **27**, 329–332 (1957)
78. H. Lempke, J. Reine Angew. Math. **142**, 118 (1913)
79. D. Levi, P. Winternitz, NATO ASI Ser. Ser. B: Phys. **278** (1990)
80. S.H. Lin, J. Theor. Biol. **60**, 449–457 (1976)
81. D.L.S. McElwain, J. Theor. Biol. **71**, 255–263 (1978)
82. A. Misir, A. Tiryaki, Appl. Math. Lett. **21**, 1204–1208 (2008)
83. P. Moon, D.E. Spencer, *Field Theory Handbook* (Springer, Berlin, 1961)
84. J.W. Mooney, Q. Appl. Math. **36**, 305–314 (1978)
85. T.Y. Na, *Computational Methods in Engineering Boundary Value Problems* (Academic, New York, 1979)
86. A. Nachman, A. Callegari, SIAM J. Appl. Math. **38**, 275–282 (1980)
87. S.K. Ntouyas, P.K. Palamides, Nonlinear Oscil. **4**, 326–344 (2001)
88. D. O'Regan, *Theory of Singular Boundary Value Problems* (World Scientific, Singapore, 1994)
89. D. O'Regan, Proc. Edinb. Math. Soc. **39**, 505–523 (1996)
90. D. O'Regan, Nonlinear Anal. **47**, 1163–1174 (2001)
91. D. O'Regan, Appl. Math. Comput. **205**, 438–441 (2008)
92. P.K. Palamides, Math. Comput. Model. **38**, 177–189 (2003)
93. O.W. Richardson, *The Emission of Electricity from Hot Bodies* (Longmans, Green, New York, 1921)
94. N. Riley, J. Fluid Mech. **18**, 577–586 (1964)
95. H. Schlichting, *Boundary Layer Theory*, 4th edn. (McGraw-Hill, New York, 1960)
96. V. Seda, in *Proceedings of the Equadiff III* (1973), pp. 145–153
97. V. Seda, Acta F.R.N. Univ. Comen. Math. **30**, 95–119 (1975)
98. V. Seda, Acta Math. Univ. Comen. **XXXIX**, 97–113 (1980)

99. S.K. Sen, H. Agarwal, Comput. Math. Appl. **49**, 1499–1514 (2005)
100. S.K. Sen, H. Agarwal, Int. J. Comput. Math. **83**, 663–674 (2006)
101. S.K. Sen, H. Agarwal, T. Samanta, Comput. Math. Appl. **51**, 1021–1046 (2006)
102. J.Y. Shin, J. Korean Math. Soc. **32**, 761–773 (1995)
103. J.Y. Shin, J. Math. Anal. Appl. **212**, 443–451 (1997)
104. E. Soewono, K. Vajravelu, R. Mohapatra, J. Math. Anal. Appl. **159**, 251–270 (1991)
105. E. Soewona, K. Vajravelu, R.N. Mohapatra, Nonlinear Anal. **18**, 93–98 (1992)
106. L.H. Tanner, J. Phys. D: Appl. Phys. **12**, 1473–1484 (1979)
107. S. Tersian, J. Chaparova, J. Math. Anal. Appl. **272**, 223–239 (2002)
108. L.H. Thomas, Proc. Camb. Philos. Soc. **23**, 542–548 (1927)
109. H.B. Thompson, Nonlinear Anal. **53**, 97–110 (2003)
110. H.L. Thron, Pfüg. Arch. Ges. Physiol. **263**, 109
111. W.C. Troy, SIAM J. Math. Anal. **24**, 155–171 (1993)
112. E.O. Tuck, L.W. Schwartz, SIAM Rev. **32**, 453–469 (1990)
113. K. Vajravelu, J.R. Cannon, D. Rollins, J. Math. Anal. Appl. **250**, 204–221 (2000)
114. G.W. Walker, Proc. R. Soc. **A91**, 410 (1915)
115. C.Y. Wang, Phys. Fluids **27**, 1915–1917 (1984)
116. J. Wang, W. Gao, Z. Zhang, J. Math. Anal. Appl. **233**, 246–256 (1999)
117. J.Y. Wang, Z.X. Zhang, ZAMP **49**, 506–513 (1998)
118. S.K. Wilson, B.R. Duffy, S.H. Davis, Eur. J. Appl. Math. **12**, 233–252 (2001)
119. C. Yang, J. Hou, Bound. Value Probl. **2013**, 142 (2013)
120. G.C. Yang, Appl. Math. Lett. **16**, 827–832 (2003)
121. G.C. Yang, Appl. Math. Lett. **17**, 1261–1265 (2004)
122. L. Zheng, J. He, Dyn. Syst. Appl. **8**, 133–145 (1999)

Correction to: Systems of First-Order Differential Equations

Correction to:
Chapter 5 in: R. P. Agarwal et al.,
500 Examples and Problems of Applied Differential
Equations, **Problem Books in Mathematics,**
https://doi.org/10.1007/978-3-030-26384-3_5

In the original version of the book, the title of chapter 5 has been changed from "Systems of First-Order Differential Systems" to "Systems of First-Order Differential Equations".

The erratum chapter and the book have been updated with the change.

The updated version of this chapter can be found at
https://doi.org/10.1007/978-3-030-26384-3_5

Author Index

A

Abell, M.L., 151, 201
Abo Elrish, M.R., 212
Ackerman, E., 148
Ackroyd, J.A.D., 370
Agarwal, Hans, 293
Agarwal, R.P., 31, 38, 106, 218, 279, 293–
    295, 299, 315, 317, 319, 321, 328,
    331, 333, 334, 341, 343, 346, 351,
    354, 373, 375
Alexander, A.E., 330
Allen, R.G.D., 131
Ames, W.F., 140
Amiot, L.W., 33
Anderson, N., 366, 367
Anderson, R.M., 180, 211
Arthurs, A.M., 366, 367
Ayala, F.J., 205

B

Bailey, N.T.J., 136
Bailey, P., 236, 336
Baker, A.J.M., 213
Banks, W.H.H., 370
Baranyi, J., 158
Barton, C., 74
Batabyal, A., 218
Bateman, H., 307
Bates, P., 375
Batyshchev, V.A., 371
Baxley, J., 317, 322
Belhachmi, Z., 370
Bellman, R., 146
Bender, C.M., 209

Bernis, F., 340
Bickley, W.G., 371
Bisshopp, K.E., 364
Blythe, P.A., 371
Bodmer, W.F., 151, 210
Bogoliubov, N.N., 170
Bougoffa, L., 295
Bradner, H., 14
Bradshaw, A.D., 213
Brady, J.P., 74
Braselton, J.P., 151, 201
Brearley, M.N., 166
Brighi, B., 370
Brokow, C.J., 77
Burley, D.M., 278

C

Caldwell, D.E., 17
Cambi, E., 102
Cameron, A.G.W., 36
Chambré, P.L., 307
Chao, W.W., 16
Chapman, S.J., 151, 210
Charman, S.J., 374
Chattopadhyay, S.K., 218
Chen, L., 219
Chou, S.T., 16
Cohen, H., 152
Constantin, A., 299
Cook, G., 33
Countryman, M., 331
Cunningham, Z., 212

© Springer Nature Switzerland AG 2019
R. P. Agarwal et al., *500 Examples and Problems of Applied
Differential Equations*, Problem Books in Mathematics,
https://doi.org/10.1007/978-3-030-26384-3

D
Daniels, P.G., 371
Danziger, L., 143, 149
Davis, H.T., 310
Davis, S.H., 353
Dobson, A.P., 213
Domar, E.D., 71
Drucker, D.C., 364
Duffy, B.R., 353
Dym, E.D., 77

E
Eckelman, W.R., 13
Edwards, C.M., 151, 210
Elmergreen, G.L., 143, 149
Engel, J.H., 121
Erbe, L.H., 358

F
Fermi, E., 295
Field, R.J., 215
Fife, P., 375
Flesch, U., 366
Foerster, H.V., 33
Fosdick, R.L., 314

G
Gardner, G.A., 212
Gardner, L.R.T., 212
Gardner, R., 375
Gatewood, L.C., 148
Gilpin, M.E., 205
Goh, B.S., 205
Goncerzewicz, J., 350
Gondolfo, G., 76
Gray, B.F., 366
Gregus, M., 327
Griffith, J.S., 217
Grossberg, S., 149
Gurung, D.B., 214

H
Haldane, J.B.S., 35, 218
Hamberg, D., 35
Hanski, I., 218
Hartman, P., 368
Hastings, A., 219
Hearon, J.Z., 38
Herfindahl, O.C., 74
Hethcote, H.W., 167, 211

Ho, D.D., 150
Holling, C.S., 218
Hoppensteadt, F.C., 207
Hou, J., 293
Howard, L.N., 215
Hueter, G.J., 121

I
Ingard, K.U., 179
Ivey, E.S., 77
Iyengar, S.R.K., 218

J
Jackson, H.C., 180
Jacobsen, J., 365
Jacquez, J.A., 149
Jefferys, G.V., 103
Jenson, V.G., 103
Johnson, P., 330
Johnston, M.D., 151, 210
Jones, C.K.R.J., 375
Jones, D.S., 178
Jones, M., 174

K
Kannan, R., 331
Kar, T.K., 218
Kermack, W.O., 136
Kidder, R.E., 336
Klaus, D., 136
Klebanoff, A., 219
Kneese, A.V., 74
Kopal, Z., 73
Krajcinvic, D., 279
Krylov, N.M., 170
Kulp, J.L., 13
Kumar, V., 218

L
Lai, L.S., 16
Lakshmikantham, V., 321
Lanchester, F.W., 121
Lebowitz, J.L., 14
Lee, C.O., 14
Lenbury, Y., 213
Leonard, J.M., 150
Lewis, J.P., 74
Libby, W.F., 5
Linhart, P.B., 14
Love, A.H., 236

M

Mackay, R.S., 14
Maini, P.K., 151, 210
Marcinkowska, H., 350
Markowitz, M., 150
Marmasse, C., 74
May, R.M., 173, 211
May, R.N., 180
McCallion, H., 78
McClelland, M.A., 121
McClure, P., 158
McKendrick, A.G., 136
McLaren, B.E., 202
McLennan, G.A, 77
Miller, F.H., 69
Milne, W., 173
Mirsky, I., 73
Molnal, G.D., 148
Mora, P.M., 33
Morse, P.M., 179
Murray, J.D., 179, 212

N

Na, T.Y., 367
Neumann, A.U., 150
Neupane, S.P., 214
Noyes, J., 215

O

Okrasinski, W., 350
Oomens, W.J., 35
O'Regan, D., 31, 38, 293, 295, 299, 315, 317,
     319, 321, 328, 331, 333, 334, 341,
     343, 346, 351, 354, 367
Orszag, S.A., 209

P

Palamides, P.K., 299
Pandey, L.K., 218
Peletier, L.A., 340
Perelson, A.S., 150
Petersen, E., 72
Peterson, R.O., 202
Phillips, A.W., 82
Pianka, E.R., 205
Planck, B.M., 157
Plank, M.J., 178
Powers, D.L., 279
Precup, R., 351

R

Rai, V., 218
Rajagopal, K.R., 314
Rapoport, A., 38, 149
Rashevsky, N., 17, 74, 143, 148
Rattanakul, C., 213
Raynor, S., 74
Reddick, H.W., 69
Regan, D.O', 106
Reid, A.T., 132
Resner, L.A., 121
Richardson, O.W., 107
Riley, N., 370
Roberts, T.A., 158
Robinson, S.B., 322
Rosevear, J.W., 148

S

Saaty, T.L., 126
Samanta, T., 293
Samson, A.M., 208
Satiracoo, P., 213
Schmitt, K., 365
Schoener, T. W., 205
Scott, A.C., 172
Seda, V., 356
Sen, S.K., 293
Shampine, L., 236, 336
Shin, J.Y., 342
Simpkins, P.G., 371
Sleeman, B.D., 178
Smith, A.M., 180
Smith, H.L., 210
Snell, W.A., 13
Southwick, L., 14
Stark, L., 33
Stewart, S., 53
Stoker, J.J., 31, 175
Strogatz, S.H., 123
Suksamran, J., 213

T

Tabisz, K., 350
Taous, K., 370
Thomas, L.H., 295
Thompson, H.B., 373
Thron, H.L., 366
Thurstone, L.L., 36
Tisdell, C.C., 373
Tocher, K.D., 148
Tryin, L.E., 13

Turchin, P., 218
Twizell, E.H., 212

U
Upadhyay, R.K., 218

V
Volterra, V., 205

W
Waltman, P., 210, 236, 336
Wang, C.Y., 370

Wilson, S.K., 353
Wong, P.J.Y., 293, , 375

Y
Yang, C., 293
Yule, G.U., 157

Z
Zevallos, M.G., 121
Zhang, S., 219
Zill, D.G., 44
Zimmerman, S.O., 158
Zionts, S., 14

Subject Index

A

Absolutely convergence, 86, 227, 266, 268, 269, 271, 272, 289, 290
Absolutely integrable, 227, 266, 269, 271, 272, 289, 290
Acoustic resonator, 77
Air resistance, 30, 159
Airy functions, 89
Airy, George Biddle, 88
Airy's differential equation, 89
Airy's stress function, 288
Allee effect, 205
Allometric equation Ambient temperature, 33
Amplitude, 50, 52, 55, 56, 59, 75, 79, 105, 207, 231, 234, 365
Analytic, 85, 87, 92, 93, 96–98, 183
Angel, of reflection, 29
Aquaculture, 33
Archimedes, 72
Associated Legendre equation, 95
Astroid, 43
Auxiliary equation, 253

B

Beams, 81, 112, 228–231, 274–276, 279–281, 285, 364
Beats, 55
Bending, 110
Bernoulli, 21, 22, 24, 25, 31, 97, 109, 252, 284, 309
Bessel, 97–101, 108, 109, 236, 282
Beta function, 86, 87

Bifurcation, 216–219
Biharmonic equation, 288
Boundary conditions, 221, 222, 229, 230, 232, 235, 241, 243–245, 249, 251, 253, 254, 256, 258, 261–263, 271–274, 277, 279–287, 295, 307, 314, 322, 336, 342, 353, 360–362, 366, 375
Boundary value problem, 221–223, 228, 229, 235, 236, 241, 244, 249, 251, 252, 254, 258, 261, 274–278, 283–286, 290, 291, 293–296, 298, 299, 303, 304, 309, 310, 312, 314, 315, 317, 322, 327, 328, 331, 332, 334, 337, 338, 341, 346, 350, 353–357, 361, 364, 366–368, 370–374
Brachistochrone problem Buckling, 39
Burnout velocity, 11

C

Cable, 11, 12, 43, 81, 107, 273, 277
Cantilever, 274, 275, 364
Capacitance, 58, 250
Carrying capacity, 177, 180, 202, 214
Catenary, 43, 277
Cauchy, Augustin-Louis, 313
Central force, 66–68
Chaos Theory, 215
Chaotic behavior, 105
Characteristic equation, 48, 49, 51, 71, 73, 156
Chebyshev equation, 91–93
Chebyshev, Pafnuty, 91

© Springer Nature Switzerland AG 2019
R. P. Agarwal et al., *500 Examples and Problems of Applied Differential Equations*, Problem Books in Mathematics,
https://doi.org/10.1007/978-3-030-26384-3

Chebyshev polynomial, 92, 93
Chemical, 27, 34, 35, 178, 201, 202, 204, 210, 215, 216, 241, 306, 373
Circuits, electrical, 55
Circular frequency, 56
Circular orbits, 68, 181
Competition, 31, 134
Concentration, 9, 14, 16, 17, 24, 34, 35, 72, 132, 140–142, 144, 147–151, 159, 175, 201, 217, 276–278, 306, 330
Constant, damping, of recombination, 31, 51, 52, 65
Continuous, 1, 18, 47, 77, 108, 115, 120, 121, 124, 137, 185, 186, 194, 201, 203, 221, 223–227, 248, 267, 294, 295, 322, 354, 358, 361, 362, 368
Convergence, 86
Convolution, 227
Cooling fin, 111, 278
Cooling, Newton's Law, 7
Coulomb wave equation, 101
Couple spring system, 130
Critically damped, 52, 195
Critical points, 133, 166, 184–187, 189–193, 195–200, 202, 203, 205–209, 212–214, 216–219
Curvature, 77, 229
Curve, pursuit, 43
Cycloid, 39, 154
Cylindrical functions, 97

D
D'Alembert's solution, 252, 284
Damped, 51, 52, 72, 105, 195, 198
Damping constant, 51, 52, 65
Decay, 3, 5, 6, 128, 129, 134, 152, 176, 204
Deflection, 70, 78, 81, 109, 112, 156, 203, 228, 230, 231, 234, 235, 275, 276, 280, 281, 317
Density, 3, 12, 17, 37, 43, 69, 72, 75, 77, 80, 103, 105–107, 166, 177, 205, 208, 234, 240, 257, 284, 310, 330, 346, 352
Derivative, 86, 115, 121, 133, 185, 186, 194, 203, 208, 225–227, 266, 283, 286, 314
Differential equations, 1–4, 6, 7, 9, 11–16, 18, 19, 21–28, 31–33, 35–44, 47, 48, 50, 53, 56–58, 61, 64, 65, 69–74, 76, 77, 80, 81, 85, 87, 88, 91–94, 98–104, 106, 108–112, 115, 119–121, 123, 130, 132, 135, 136, 143–145,

147, 152, 156, 158, 159, 163, 164, 172–174, 176, 178–181, 183, 185, 194, 200–207, 216, 221–223, 228–231, 234, 235, 238, 239, 241–244, 249, 253, 259, 260, 267, 269, 271, 275–277, 279, 281–285, 288, 295, 308, 310–312, 317, 321, 336, 340, 351, 354, 362, 364, 365, 368, 369, 375
Diffusivity, 241, 258, 370
Dirichlet, conditions, problem, 222, 241, 258, 261, 262, 287
Distribution, 13, 93, 105, 158, 229, 244, 258, 287, 307, 312, 330, 365–367, 371
Domain, 120, 185, 193, 221
Double pendulum, 144
Draining, 340, 367
Duffing's equation, 105, 113, 181, 206, 215

E
Eigenfunctions, 231–234, 236, 242, 243, 247, 251, 255, 256, 259, 263, 280–282
Eigenvalues, 118, 119, 122, 125–127, 129, 152, 157, 184, 187, 190, 192, 195–200, 203, 231–236, 241–243, 247, 251, 254–256, 259, 263, 280–282
Eigenvectors, 118, 187, 190
Elastic, 13, 34, 51, 65, 73, 78, 178, 228, 231, 232, 256, 274, 275
Electrical network, 126, 152, 153
Elementary functions, 85, 279
Elliptic integral of the first kind, 80
Embedded, 229, 230, 275, 370
Emden differential equation, 80
Emigration, 6, 7, 164
Envelope, 55
Equation of continuity, 8, 36
Equidimensional equation, 73
Equilibrium, 8, 17, 49, 50, 54, 65, 72, 74, 78–82, 105–107, 130, 155, 174, 178, 184, 193, 195, 201, 202, 204, 210–213, 228, 233, 234, 248, 273, 275, 281, 284, 310, 311
Equipotential lines, 40
Escape velocity, 29, 30
Euler, 73, 87, 97, 100, 231, 232, 284
Evolution, 35, 117, 178, 215
Existence, 121, 145, 193, 222, 286, 293–296, 298–301, 303, 309, 312, 317, 321, 322, 328, 331, 332, 334, 336–338, 340, 341, 343, 345, 346, 348, 350, 354, 357, 362, 366, 367, 370, 371

Explicit, 92, 141, 163, 272, 295, 307, 309
Exponential growth, 126
Exponents, 88, 96, 97, 99, 100, 103, 353
Exterior problem, 266

F
Falling droplets, 42
Family of curves, 39, 40
Fick's law, 16
First-order, differential equations, system, 4,
      82, 129, 144, 145, 155, 158, 166, 177,
      183, 184
FitzHugh-Nagumo equation, 142, 208
Flexural rigidity, 81, 236, 364
Forced response, 56, 76
Fourier
   law, 240
   method, 221
   series, 224
   transform, 221, 227, 266, 269, 289
Fourth-order differential equation, 144
Free end, 275, 284, 285, 364
Free frequency, 207
Free oscillations, 173
Frequency
   forced mode, 54
   natural mode, 54
   response curve, 57
Frobenius, Ferdinand Georg, 87
Fundamental
   matrix, 117
   solution, 267
   system of solutions, 116

G
Gamma function, 86
Gauss, Frederick Carl, 100, 267
General solution, 1, 12, 14–16, 21, 32, 34, 42,
      43, 48, 49, 53, 56, 58, 71–75, 77, 78,
      80–82, 96–99, 101, 109–113, 117–
      119, 147–150, 155, 187–189, 223,
      230, 232, 242, 251, 260, 271, 279,
      366
Global existence theorem, 121
Gompertz, Benjamin, 32
Green's theorem, 286
Growth, 3, 6, 16, 17, 22, 31–33, 35, 74, 126,
      132–134, 151, 159, 172, 176, 180,
      202, 211, 213, 214

H
Half-life, 4, 5, 130
Hankel functions, 99
Harmonic oscillator, 49, 102
Harvesting, 200, 205
Heart, 9, 178
Heat transfer, 103, 306, 334, 367, 374
Hermite, 90, 91, 94, 102, 282
Higher order, 221
Homeostasis, 211
Homogeneous, 3, 8, 25, 47, 56, 71, 116, 117,
      148, 221–223, 240, 241, 313, 348
Hooke's law, 49, 105
Hopf bifurcation, 218
Hurwitz matrix, 184
Hypergeometric differential equation, 100

I
Immigration, 6, 7, 13, 164, 172
Implicit, boundary condition, 263
Indicial equation, 88, 96, 97, 100
Inductance, 19, 58
Inflection points, 32
Initial condition, 1, 13, 15, 36, 38, 39, 44, 48,
      49, 54, 57, 76, 77, 80, 83, 87, 105–
      107, 113, 117, 122, 125, 129, 137,
      147, 154, 155, 158, 165, 166, 173–
      182, 200, 203, 215, 218, 241, 244,
      249, 256, 258, 267, 271, 283–285
Initial phase, 50
Initial-value problem, 168, 266
Input, 8, 35, 149, 205
Integral equations, 120, 343, 344
Integrating factor, 1
Interior problem, 263
Inverse square law, 29, 67
Irregular singular point, 87
Isogonal trajectories, 40
Isolated critical point, 186
Isotherms, 40

J
Jacobi's equation, 112
Jacobian matrix, 192, 203
Jacopo Francesco, 37

K
Kinetic, 16, 50, 75, 215, 306, 367
Kirchhoff, Gustav Robert, 19
Kirchhoff's law, 19, 126

L

Laguerre polynomial, 97, 282
Lane–Emden equation, 107, 175
Langmuir equation, 107
Laplace equation, 265, 271, 272, 286, 287, 290
Lascaux, 5
Law
    Newton's, 36, 145, 244
    Stefan–Boltzmann, 74
    Torricelli's Evangelista, 27
Legendre, Adrien Marie, 93
Legendre equation, polynomial, 95
Libby, Willard Frank, 5
Limit cycle, 172, 178, 193, 203, 215, 218
Linear differential equations, functions, 21, 48
Linearity property, 227
Linearly independent, 47–49, 71, 88, 90–92, 94, 98, 100, 116, 118, 191, 223, 263, 264
Logarithmic decrement, 52
Logistic equation, function, growth, 32
Lorenz equations, 215
Lotka–Volterra equation, 134, 177, 199, 208, 218
LRC- circuit systems, 58

M

Mass action, 27, 34
Mathematical models, 15, 33, 36, 115, 142
Mathieu, Emile Leonard, 102
Matrix
    fundamental, 116, 117
    transition, 117
Memorization
    mathematical model, 15
Mesh spacing, 163
Method
    of images, 268
    Runge–Kutta, 163, 165, 166, 168–170, 172–182
    separation of variables, 221, 223, 241, 246, 251, 258, 284
    undetermined coefficients, 57
Mixtures, 8, 14, 72, 202, 204, 216
Modified
    Bessel functions, 100
    logisitic population, 172
    differential equation, 32, 364, 365
    Samuelson's investment model, 52

N

Natural frequency, 50, 54–56, 77, 231, 280
Network, electrical, 126, 152, 153
Neumann
    conditions, 243, 254, 286, 287
    function, 243
Newton's second law, 10, 15, 49, 59, 61, 72, 130, 155, 216, 235, 248
Node, stable proper (star-shaped), 189
Nonhomogeneous, 53, 56, 112, 117, 119, 221, 223, 229, 289, 290
Nonlinear differential equation, 27, 36, 64, 135, 163, 174, 176, 180, 207
Nonsingular, 185, 192, 337
Numerical
    methods, 163
    solutions, 165, 168–170, 173–183, 215, 216
Nutrient, 210

O

One-dimensional, 72, 249, 253, 336
Order
    of differential equations, 85
Ordinary point, 87, 92, 93, 100
Orthogonal trajectories, 39, 40
Overdamped, 52, 196

P

Painleve's equations, 104, 374
Parametric
    equation, 39
    form, 39
Partial
    derivatives, 115, 121, 185, 186, 283
    differential equations, 91, 115, 221, 223, 242, 244, 249, 253, 283–285, 288, 310
Particular solution, 57, 71, 126, 223, 297
Pendulumm
    physical, 61
    simple, 61, 173, 185, 197, 198, 361
Period
    of oscillatory motion, 52, 55
Periodic functions, 54, 56, 76, 169, 200, 224
Permeability, 142, 147, 348, 367
Phase
    angle, 50
    initial, 50
    plane, 185, 207, 218
    portrait, 187, 189, 191, 192, 217
Pitchfork bifurcation Physical pendulum, 216

Point of expansion, 85
Poisson's integral formula, 265, 272
Population
    growth, 172, 180, 213, 214
    of colony, 2
Potassium ions, 142
Potential functions, 258, 287
Power series, 85–87, 89, 90, 102, 104, 105,
    107, 109
Predation, 134
Predator–prey, 147, 218
Principal fundamental matrix, 117
Projectile, 154, 159
Pursuit
    curve, 43

Q
Quasi
    - steady, 151
    -periodic, 76

R
Radioactive, 4, 5, 128, 129
Radiocarbon
    dating, 5
Radium, 4
Radius of convergence, 86
Radon, 4
Rayleigh's equation, 105, 206
Reactions
    chemical, 82, 174, 177
Recurrence formula, 86
Reflection
    angle, 29
Regular singular point, 87, 96, 97, 100, 101,
    103
Resistance, 19, 30, 36, 51, 52, 58, 159, 198,
    250, 293
Resonance, 56–58
Resonant circular frequency, 56
Response
    frequency curve, 57
Riccati's differential equation
    -Bessel equation, 101
Richardson's equation, 107, 124, 310, 312
Robin's conditions, 244, 256
Runge–Kutta method, 163, 165, 166, 168–
    170, 172–176, 178–182
Rutherford, Lord Ernest, 4

S
Satellite equation, 83
Second order, 47, 69, 85, 87, 132, 145, 163,
    221, 295, 299, 314
Self-excited oscillation, 193
Self-orthogonal, 40
Semi-differentiated cells, 151, 210, 211
Separable
    equations, 21
    variables, 26, 30
Series, 85, 86, 88, 92–94, 100, 102–104, 107,
    109, 174, 224–226, 242–244, 248,
    252, 265, 284
Shroud of Turin, 5
Simple
    beam, 274
    harmonic, 90, 102, 105
    pendulum, 61, 80, 111, 173, 185, 197,
    198, 361
Singular point, 87, 101, 184
Sinusoidal force, 54, 57
SIR model, 139, 140, 211, 212
Solution
    general, 1, 12, 14–16, 21, 32, 34, 42, 43,
    48, 49, 53, 56, 58, 71–75, 77, 78, 81, 82,
    96–99, 101, 109–113, 117, 119, 147–
    150, 155, 187–189, 230, 232, 242, 251,
    260, 271, 366
    transient, 57
    trivial, 49, 222, 223, 233, 236, 375
    unique, 2, 48, 87, 121, 185, 222, 223, 245,
    279, 293, 294, 312, 362–365, 367, 372
    unstable, 72, 186
Solution curve, 143, 178, 215
Spring constant, 52, 73, 130, 155, 206, 231
Square
    inverse law, 29, 67
Stable
    initial value problem, 183, 185
    proper (star-shaped) node, 189
Stationary points, 184, 194, 195, 210
Steady-state solution, 57–59, 75, 76, 244
Stefan
    −Boltzmann law, 74
    Joseph, 36
Stiffener ring, 103
Streamlines, 40
Sturm–Liouville conditions, 222
Superposition, 54, 267
Symbiosis, 134
Systems of first-order linear differential
    equations
    LRC circuit, 58

T

Tacoma Narrows bridge, 56
Taylor
 expansion, 85
 formula, 186
Telegraph equations, 250
Telephone cable, 273, 277
Terminal velocity, 42, 207
Thermal conductivity, 103, 111, 240, 257,
 306, 366, 367
Threshold
 models, 161
 value, 205
Torricelli's Evangelista law, 27
Tractrix, 39
Trajectories
 orthogonal, 39, 40
Transfer
 heat, 103, 306, 334, 367, 374
Transient solution
 state, 375
Transition matrix, 117
Trivial solution, 49, 222, 223, 233, 236, 375
Tsunami model, 44
Two-dimensional
 motion, 175
 system, 185, 194

U

Undamped, 53, 72, 77
Underdamped, 52
Undetermined
 coefficients, 57
Uniqueness theorem, 145
Unique solution, 2, 48, 87, 121, 185, 222,
 223, 245, 279, 293, 294, 312, 362–
 365, 367, 372

Unstable

Unstable
 critical point, 189
 solution, 186

V

Van der Pol's equation, 113, 173, 181, 196,
 203
Variable mass, 10
Variables
 separable, 26, 30, 70
Velocity, 10–12, 15, 19, 27, 30, 37, 41, 42,
 51, 59–62, 73, 78, 81, 104, 154, 155,
 159, 166, 208, 248, 249, 253, 277,
 287, 306, 314, 346, 353
Verhulst
 model, 22, 136, 194
Vibrations, 52, 56, 248, 249, 285
Viscous damping, 15, 80
Volterra
 Lotka system, 134

W

Weber's Bessel function of the second kind,
 99
Weight function, 247
Whirling shaft, 281
White-dwarf equation, 107
Wronskian, 47, 48, 116, 117

Y

Young's modulus, 37, 81, 110, 112, 232, 281,
 284, 322

Printed in the United States
By Bookmasters